Mobile Telecommunications Technology

Mobile Telecommunications Technology

Edited by **Adam Houle**

CWILLFORD PRESS

New York

Published by Willford Press,
118-35 Queens Blvd., Suite 400,
Forest Hills, NY 11375, USA
www.willfordpress.com

Mobile Telecommunications Technology
Edited by Adam Houle

International Standard Book Number: 978-1-68285-162-3 (Hardback)

The publisher's policy is to use permanent paper from mills that operate a sustainable forestry policy. Furthermore, the publisher ensures that the text paper and cover boards used have met acceptable environmental accreditation standards.

Trademark Notice: Registered trademark of products or corporate names are used only for explanation and identification without intent to infringe.

Printed in the United States of America.

Contents

Preface

Over the recent decade, advancements and applications have progressed exponentially. This has led to the increased interest in this field and projects are being conducted to enhance knowledge. The main objective of this book is to present some of the critical challenges and provide insights into possible solutions. This book will answer the varied questions that arise in the field and also provide an increased scope for furthering studies.

Mobile telecommunications technology has gained immense popularity in the last three decades. There has been a lot of research and investigation in the field of mobile telecommunications technology and its applications for diverse group of users. The book encompasses various principles, concepts, techniques and advancements in this field. Mobile broadband and internet services in particular have been discussed at length. It aims to provide an overview of the present status of mobile telecommunications technology while bringing forth future prospects and avenues for further research in this discipline. This book, with its detailed analyses and data, will prove beneficial to professionals and students involved in this area at various levels.

I hope that this book, with its visionary approach, will be a valuable addition and will promote interest among readers. Each of the authors has provided their extraordinary competence in their specific fields by providing different perspectives as they come from diverse nations and regions. I thank them for their contributions.

Editor

Extending mobility to publish/subscribe systems using a pro-active caching approach

Abdulbaset Gaddah and Thomas Kunz*

Department of Systems and Computer Engineering, Carleton University, Ottawa, Canada

Abstract. The publish/subscribe communication paradigm has many characteristics that lend themselves well to mobile wireless networks. Our research investigates the extension of current publish/subscribe systems to support subscriber mobility in such networks. We present a novel mobility management scheme based on a *pro-active* caching approach to overcome the challenges and the performance concerns of disconnected operations in publish/subscribe systems. We discuss the mechanism of our proposed scheme and present a comprehensive experimental evaluation of our approach and alternative state-of-the-art solutions based on *reactive* approaches and *durable subscriptions*. The obtained results illustrate significant performance benefits of our proposed scheme across a range of scenarios. We conclude our work by discussing a modeling approach that can be used to extrapolate the performance of our approach in a near-size environment (in terms of broker network and/or subscriber population) to our experimental testbed.

Keywords: Message-oriented middleware, publish/subscribe paradigm, mobility management, mobile computing, wireless networks

1. Introduction

A publish/subscribe (pub/sub) system is a push-based information dissemination model that inherently decouples communication between publishers and subscribers in *time*, *space*, and *flow* [2,13]. In such a system, *publishers* are the information producers that deliver information to a distributed set of brokers in the form of *messages* (or *events*), *subscribers* are the information consumers that subscribe to receive a selective set of messages within the system, and *brokers* are the routers that ensure the reliable and timely delivery of published messages to all interested subscribers. The pub/sub-based architecture is recently considered as a promising communication paradigm for future mobile information dissemination applications [22,27]. This is due to the advantages of this paradigm, including *decoupling*, *anonymous*, and *asynchronous* many-to-many information dissemination.

Most existing pub/sub systems [8,9,39,54] are designed for fixed wired networks, where both publisher and subscriber clients are usually stationary and have reliable low-latency high-bandwidth connections. Support and optimizations for client mobility are not built-in features of their formal semantics. Instead, it is left to the applications to adapt to the conditions of dynamic environments. This can significantly complicate the development of information dissemination applications. Recently, some literature [11, 42,46] has taken a first step towards supporting mobility in pub/sub systems. There is hence a pressing need for add-on protocols to extend these systems to operate in mobile wireless environments that

---

*Corresponding author: Thomas Kunz, Systems and Computer Engineering, Carleton University, 1125 Colonel By Drive, Ottawa, ON, Canada K1S 5B6. E-mail: tkunz@sce.carleton.ca

are characterized by frequent and unpredictable disconnections of participants due to wireless channel impairments or client mobility.

Although mobility management is widely studied by mobile computing researchers, the indirect communication paradigm of pub/sub systems introduces new challenges in designing handoff management solutions [11,41]. In a pub/sub system, published messages do not rely on an explicit destination address set by the publishers. Instead, they are routed to the end points (*subscribers*) based on their content and the subscriptions in the system. In other words, publishers do not know the explicit addresses of subscribers and therefore acknowledgement mechanism cannot be used to identify message loss. Subscribers cannot also depend on the sequence numbers of the received messages to detect message loss as they receive a selective set of the published messages. As a result, without any coordination between network brokers, subscribers may miss some or all the messages that were published during their movements from one broker to another. This can be a serious issue for some applications that do not tolerate message loss.

Since the migration of the subscribers is transparent to the system, the network brokers end up managing a large number of inactive subscriptions and tracking their corresponding messages. As perpetually caching messages for migrated subscribers imposes a substantial overhead on the brokers, the overall system performance may gradually degrade to the point of failure. Moreover, the subscribers may receive duplicated messages when they reconnect to the previously visited brokers. Such duplication may result in flooding the wireless channel and wasting a considerable amount of the bandwidth. Reported studies [14,15,32] discuss the above issues in details. Thus, the handoff management solutions for pub/sub systems should take into account these factors in addition to the conventional objectives such as low handoff latency and message overhead to guarantee reliable message delivery semantic and to hide the interruption of message dissemination.

In recent years, several mobility management solutions [7,11,14,29,36,52] have been proposed for pub/sub systems deployed on various wireless environments. One most commonly-used solution is based on a *reactive* scheme [7,11,53]. The reactive scheme works as follows. Once a mobile subscriber disconnects from source broker B_i, the broker starts to locally store published messages that match the subscriber's subscriptions. When the subscriber reconnects to target broker B_j, it first informs B_j that it was previously connected to B_i. Then B_j contacts B_i to fetch the subscriptions associated with the mobile subscriber. After B_j obtains all the subscriptions, it subscribes these subscriptions and informs B_i to remove them. Then B_j begins to store in a temporary queue all the new messages it receives for the moving subscriber. Meanwhile, B_i sends all the subscriber messages to B_j. After all the messages are forwarded, B_j simply replays the set of locally stored messages and received messages from B_i to the subscriber, potentially after removing duplicates from both set of messages. This scheme may result in a drastic increase in the network traffic load since the subscriptions and actual messages need to be transferred between the brokers [5]. It also imposes high handoff latency that may not be acceptable by applications requiring fast handoffs between brokers to maintain high communication quality.

Another recently-used solution, based on a *durable subscription-based* approach [14,20,31,32,36], is proposed to cope with the connection/disconnection operations. This approach is believed to be highly reliable and is typically used for applications that cannot tolerate message loss. In the absence of any mobility management mechanism, the durable subscription-based approach suffers from the issues described early when it is deployed on a mobile wireless domain in addition to the fact that frequent mobility of subscribers significantly degrades the system performance. In such a scheme, as broker B_i has no knowledge about the state of the mobile subscriber, which has already reconnected to broker B_j, it will keep buffering messages for that subscriber, causing B_i a significant performance overhead.

Also, the durable subscription-based scheme does not support a mechanism that removes the subscriber subscriptions from the previously visited brokers. In this case, each broker will end up managing a large number of inactive subscribers that may not reconnect again to that broker. Therefore, the durable subscription-based scheme gains by not propagating subscriptions and messages between the brokers, at the expense of perpetually caching messages for inactive subscribers. Results reported in [14,15,20, 32] show that the system's performance gets increasingly worse as the population of inactive subscribers increases in the system.

In this paper, we propose a novel and efficient mobility management scheme for current pub/sub systems to support subscriber mobility and to provide fast handoffs. The core idea of this scheme is to intelligently transfer and cache subscriber contexts (its actual subscriptions) one broker-hop ahead of its current broker in a *pro-active* manner (i.e., context transfer/caching occurs before the subscriber movement). Since it is difficult to predict the subscriber's movement, we need to identify the *set* of potential next brokers without examining the brokers' topology and manually creating the set. We exploit a data structure, called *neighbor graph*, which forms the basis for our proposed pro-active scheme as it dynamically captures the potential mobility graph of mobile subscribers. Each broker over time learns about its immediate neighbors; thus, only these neighbors will receive/cache the subscriber context prior to the occurrence of handoffs.

We have comprehensively evaluated the performance of our proposed pro-active scheme through testbed experiments, comparing it to alternative solutions: reactive and/or durable subscription-based. The obtained results show that our pro-active scheme achieves superior performance across a range of scenarios over the other solutions in terms of message loss, message duplication, and handoff latency. Using two different mobility models, we demonstrate that the pro-active approach can outperform the other solutions even when the neighbor graph is a relatively weak predictor of mobility (i.e., each broker has many neighbors). As the neighbor graph narrows the choice of potential next-hop brokers, the performance improvements become even more noticeable. We conclude our work by discussing a modeling approach that can be used to extrapolate the performance of our proposed mobility management scheme in a near-size environment (in terms of broker network and/or subscriber population) to our experimental testbed.

The rest of the paper is organized as follows. Section 2 discusses the related work. Section 3 describes the system model and assumptions. Section 4 presents the proposed pro-active mobility management scheme. Section 5 describes the experimental setup and discuses the evaluation results. Section 6 applies a modeling approach to extrapolate the performance of our proposed pro-active scheme. Section 7 concludes the paper.

2. Related work

In the recent years, several mobility management solutions have been proposed for well-known distributed pub/sub systems, like JEDI, SIENA, REBECA, and ELVIN. JEDI [11] is one of the first pub/sub systems to add support for subscriber mobility that is based on explicit *moveIn* and *moveOut* operations. The mobile subscribers explicitly invoke these operations during the handoff process, which can be problematic if a wireless link breaks down suddenly due to physical mobility or interference. Also, JEDI adapts a hierarchical topology of event brokers, which has a potential performance bottleneck at the root node of the hierarchical tree [5].

SIENA [7] is a scalable messaging system that has been extended to support subscriber mobility. The extension is typically presented in the reactive fashion and evaluated in wired and GPRS-based networks.

Although their reported results have demonstrated the applicability of their mobility extension, they are limited to narrow evaluations of a single mobile subscriber roaming across the network. This limits the value of their results since their proposed extension never needs to transfer a large volume of messages between brokers.

REBECA [53] incorporates a reactive-based solution to support physical mobility in an acyclic event topology. The mobility support scheme uses an intermediate node between the source and target broker, called *Junction*, for synchronizing the brokers. The source broker routes subscriber messages through the Junction to reach the target broker, and then the subscriber. The proposed extension relies on tracking the Junction broker that significantly increases the handoff delay, particularly in a large-scale network, and the overhead on the Junction [35]. It has not been justified why subscribers cannot maintain the information about the source broker, necessitating an intermediary to manage mobility.

ELVIN [46] is an event-based system that supports disconnected operation using a central caching proxy server but does not support mobility between proxies. In a wide-area system, mobility support between proxies is needed and may also be useful for load balancing purpose. The central proxy server tends to become a performance bottleneck and the system is not scalable.

Farooq et al. [14] presented their experience in evaluating the performance of a commercial JMS-based pub/sub system in wired and mobile cellular networks. The nature of their work differs from ours since it mainly focuses on studying the impact of certain mobility factors on the performance of reactive and durable subscription-based schemes without proposing a new mobility management scheme.

Chelliah et al. [10] proposed a distance-based cache relocation scheme to support continuity of service in a cellular network. To some extent, their approach is similar to the one proposed in [25]. The cache is relocated to the next base station once the mobile user reaches the relocated point in the cell. This mechanism relies mainly on predicting the path of the mobile user accurately. A combined approach of LA [45] and PM [28] techniques is used for path prediction. A distance-based relocation scheme is used to identify the time at which the relocation has to be performed. Distance between the mobile unit and base station is regularly monitored to make such a decision. The cost of prediction methods is high as they are repeatedly applied to every individual mobile user that enters to a cell. They also show low accuracy, which is not justified, even for a small number of visited cells. Moving close to the boundary of a cell initiates unnecessary cache relocation. Due to the fact that relocation takes place just after the handoff, the mobile user may experience service disruption after entering the new cell.

Katsaros et al. [2] introduced a prototype based on SCRIBE [42], an overly multicast routing system, to support mobility in pub/sub systems. In their prototype architecture, a mobile client is connected to an Overly Access Router (OAR) through the currently associated Access Point (AP). Since the mobile clients move from one AP to another, it is possible that they will be served by a different OAR. In such a case, the mobile client should inform the new OAR about the publications of its interest in order for the OAR to join the appropriate trees. The cost of finding and joining the appropriate multicast trees should be investigated as it has a direct impact on the handoff latency. Also, mobility prediction is absent in their proposed solution and thus proactive multicast group joins cannot be achieved to reduce the handoff latency. Caching mechanisms should be utilized to improve the efficiency of their approach.

Podnar et al. [41] discussed a persistent notification scheme to support subscriber mobility. In this scheme, each broker maintains a list of the IDs for the events routed to its neighbor brokers and the interested subscribers. It also stores the published events in a persistent buffer according to their lifetime. When a mobile subscriber reconnects to a new broker, the subscriber will provide the IDs of the latest received events to the new broker in order to avoid duplicated events. This scheme constantly burdens neighbor brokers with event transfer/caching that leads to a significant degradation in the system performance. This is due to the lack of coordination between subscriber mobility and the caching process. The scheme also adds extra load on the brokers as they are required to maintain a large volume of IDs and validate the lifetime of a tremendous number of events.

Burcea et al. [5] deployed a simple handoff management scheme that is based on the successful prediction of the subscriber destination. In the proposed scheme, the subscriber context is transferred to the destination brokers once the mobile subscriber disconnects from the network. The entire context will then be removed from the source broker. This limits the proposed scheme to only support mobility but not disconnect/reconnect operations that may occur frequently due to the loss of connectivity. Our proposed approach takes into consideration such behavior and can handle it well. The proposed scheme may also not be adequate for supporting fast handoffs particularly in large-scale networks as the context pre-fetching takes place only after the mobile subscriber disconnects from the network. In our scheme, the subscriber context is always transferred prior to the subscriber movement to support fast handoffs. The authors have not evaluated their approach in the presence of multiple brokers, which severely limits the applicability of their approach.

Wang et al. [49] provided a handoff management scheme, which is called *multi-hop handoff* (MHH), to offer reliable and ordered delivery of messages to mobile subscribers with minimized cost (in terms of message loss and duplication). In MHH, when a mobile subscriber disconnects from the system, the new incoming messages will be buffered at the subscriber's last visited broker. Once the subscriber reconnects to a new broker, the subscriber context (subscriptions and messages) is migrated in parallel. The context is typically moved hop-by-hop along the path from the last visited broker to the new broker in a reactive manner. In general, the proposed scheme will introduce high handoff latency as it may take a long time for the new broker to receive the subscriber context upon its reconnection, particularly when the network is congested or is large.

Hu et al. [21] addressed subscriber mobility in a distributed content-based pub/sub system. They mainly focus on the transactional semantics required by a mobile subscriber (i.e., a subscriber who wishes to disconnect from a source broker and reconnect to a new broker in the overlay as part of a transaction). They identified the transactional semantics for a mobile subscriber and outlined the transactional concerns at various layers, focusing on the subscriber movement and routing protocol layers. The proposed solution requires the system to reconfigure and update the routing tables of all the brokers on the path from the source to the destination broker. Such a behavior in the high mobility scenarios can be very costly from a performance perspective.

Tarkoma et al. [47] discussed a formal discrete model to study the safety and cost of handoff protocols for both publisher and subscriber mobility in content-based routing networks. Three new properties are defined to improve mobility support in the pub/sub topologies, namely *overlay-based routing*, *rendezvous points*, and *completeness checking*. Overlay-based routing prevents the content-based flooding problem. It abstracts the communication used by the pub/sub system from the underlying network-level routing and enables the system to cope with network-level routing errors and node failures. Rendezvous points simplify mobility by allowing better coordination of topology updates. Completeness checking ensures that subscriptions and advertisements are fully established (propagated) in the topology. The reported results are limited to a single performance metric (i.e., message cost), which is not sufficient to completely evaluate the impact of different protocols and topologies in managing mobile pub/sub clients.

3. System model

This section gives an overview of the system model and outlines some of the properties and assumptions of the used pub/sub system. In our system model, the broker network is modeled as an undirected *general peer-to-peer* graph $G = (V, E)$, where V is the vertex set of all brokers, $V = \{B_1, \cdots, B_k\}$, and E is the set of unordered distinct pairs of edges (or *links*), $e = \{B_i, B_j\}$ where $B_i \neq B_j$. Two brokers B_i and $B_j \in V$ may communicate directly only if they are linked by an edge e. We formally say that B_i and B_j have a neighboring relationship if $\{B_i, B_j\} \in E$. We thus define the set of all B_i neighbors in G as follows: *Neighbor* $(B_i) = \{B_j: B_j \in V, (B_i, B_j) \in E\}$. The set of brokers in G are assumed to be distributed in the same geographical neighborhood region and the mobile subscribers may reconnect to the physically closest brokers. This is an essential assumption in order to support emerging location dependent services (i.e., messages routed to the subscribers based on their current locations) and reduce the network overheads. It is assumed that G is a static routing topology [7] that routes publications to all interested subscribers.

The traffic on G travels via reliable and authenticated communication mechanism on each edge e and is classified into two different categories: *control* and *content* messages. The control traffic consists of the subscription and mobility update messages that are mainly exchanged by peer brokers whereas the content traffic consists of the actual data that are forwarded from publishers to interested subscribers. Each broker maintains two tables: a Subscription Table (*ST*) and a Neighborhood Table (*NT*). The ST consists of a set of $\{(E, Sub)\}$ pairs that record the subscriptions (*Sub*) received from or delivered to the immediate neighbor brokers. The edges of the neighbor brokers are recorded into the NT as a $\{(E,T)\}$ pairs that are used for the control traffic routing. T is a timestamp used to ensure the correctness and freshness of the NT as discussed later.

In our system model, a subscriber handoff is performed between two brokers in the network: B_i and $B_j \in V$ B_i is the source broker and B_j is the destination broker that is usually unknown to the mobile subscriber. After the subscriber disconnects from B_i, the handoff is negotiated and a new connection is established with B_j. Some systems limit subscriber mobility between only border brokers, the leaves of a routing tree. In contrast, we allow mobile subscribers to commute between any two brokers. The edge e between B_i and B_j is unique since G is a peer-to-peer graph. We refer to an edge e as *active* if a subscriber logically moves across that edge. It is necessary to keep track of *inactive* edges as they drastically impact the system's performance due to the overhead of propagating unnecessary subscriptions and messages in the system. Hence, we regularly update NT to capture the active edges and eliminate inactive ones. Broker B_j uses the propagated subscriptions, received from broker B_i, to store messages for offline subscribers in a dedicated buffer. All stored messages will then be routed directly to the mobile subscriber once it hands off to B_j. Subscription covering [26] is a key technique to quench the subscription propagation and hence reduces the propagation overhead. We give no consideration to such technique due to its high cost in an environment characterized by frequent movement [21]. In the described model, the mobile subscribers do not change their subscriptions during the course of disconnect/reconnect operations, but they are free to change them when connected to a broker.

The work described in this paper is based on the content-based subscription system that enables subscribers to express their interests with a finer level of granularity. It is also based on the use of the flooding strategy to steer message delivery to the distributed brokers. This strategy is strongly recommended by [3,39] in highly mobile environments since mapping content-based pub/sub systems on top of IP Multicast (*topic-based*) results in an explosion of multicast groups. Like most previously discussed work on supporting subscriber mobility in pub/sub systems, our work assumes that there are no failures at the pub/sub routing layer (broker nodes and their links) as the development of fault-tolerant pub/sub protocols [23] is out of the scope of this paper. Security has also not been addressed in conjunction with our proposed mobility management scheme. We assume that previously proposed techniques [44,48] can be also applied to our approach. In this work, we are only interested in managing subscriber mobility and not concerned about publisher mobility that has been addressed by others [37]. Unlike subscribers, there is no specific information that the publishers would miss during the period of their handoffs.

4. A pro-active mobility management scheme

Although the lower layers (such as *link-layer* and *IP-layer*) are conceptually the *"right layers"* to express the context of mobility support, the lower-layer mobility protocols have not been widely accepted and deployed for several reasons [4,47]. The application-layer mobility solutions on the other hand can easily remove the major drawbacks of the lower-layer mobility protocols and offer better mobility solutions for the next-generation heterogeneous networks. This motivates our choice to solve the mobility problem at the application-layer.

The core idea of our pro-active mobility management scheme is to intelligently transfer/cache a subscriber's context (subscriptions/messages) one broker-hop ahead of its current broker in a *pro-active* fashion (i.e., context transfer/caching occurs prior to the handoff of the mobile subscriber). A *neighbor graph*, which forms the basis of our proposed scheme, is used to dynamically capture the candidate set of brokers to which subscriber context should be pro-actively transferred to and cached at.

We model subscriptions to be in either *active* or *passive* mode. When a subscription is in active mode, it is used for filtering incoming messages and only those messages that match its filter are either routed to the subscriber or cached locally for future use. A passive subscription, on the other hand, is simply ignored in the filtering process. Initially, a subscriber submits an active subscription to a broker to receive the messages of its interest.

4.1. The pro-active context distribution algorithm

The pro-active context distribution algorithm is at the core of our pro-active mobility management scheme and described here at a high level as separate cases that correspond to various *connection*, *disconnection*, and *handoff* scenarios. Figure 1 depicts a simplified finite state machine (FSM) diagram that more formally describes the protocol. The following notation is used throughout the description:

S: a subscriber who is potentially mobile.
B_j: a broker who is initially serving S.
B_i: a neighbor broker of B_j.
$Sub(S)$: S' subscriptions.
$Msgs(S)$: S' messages.

$Neighbor(B_j)$: set of neighbor brokers of B_j.
$Context(S)$: $\{Sub(S)+Msgs(S)\}$.
$Timeout(S)$: a chosen timeout for managing $Context(S)$. When it expires, $Context(S)$ is garbage collected.

The following pseudo-code summarizes the algorithm, as executed on each broker:

Case 1: When S initially connects to broker B_j, B_j sends a passive copy of $Sub(S)$ to each broker $B_i \in Neighbor(B_j)$, which locally stores $Sub(S)$. In the meantime, broker B_j routes the published messages to S.

Case 2: When S disconnects from the network due to poor network connectivity or a handoff, broker B_j detects this (through receiving generic ping replies from S periodically) and consequently sends an activate request to each broker $B_i \in Neighbor(B_j)$. Following this request, broker B_j forwards $Msgs(S)$ to its neighbors B_i until the activation request is acknowledged, to avoid message loss that may occur due to the activation latency. As $Sub(S)$ is activated, $Neighbor(B_j)$ will locally store all the incoming messages that match $Sub(S)$. The ID of the latest message consumed by S (for each subscription) is

Algorithm 1: Pro-active Context Distribution – *executed on broker (B_j)*

Case 1: **Initial connection**
```
IF subscriber S connects to Bj THEN
    FOR all Bi ∈ Neighbor(Bj) DO
        Forward Sub(S) to Bi
    ENDFOR
ENDIF
```
Case 2: **Subscriber disconnects**
```
IF subscriber S disconnects from Bj THEN
    FOR all Bi ∈ Neighbor(Bj) DO
      Activate Sub(S) stored at Bi
      Forward Msgs(S) to Bi stored during Sub(S) activation
    ENDFOR
ENDIF
```
Case 3: **Subscriber reconnects**
```
IF subscriber S reconnects to Bj THEN
    FOR all Bi ∈ Neighbor(Bj) DO
      Deactivate Sub(S) stored at Bi
    ENDFOR
ENDIF
```
Case 4: **Subscriber moves out to a peer broker**
```
IF subscriber S hands off to Bk from Bj THEN
    FOR all Bi ∈ Neighbor(Bj) and Bi ≠ Bk DO
      IF Bi ∉ Neighbor(Bk) THEN
          Delete Context(S) stored at Bi
      ELSE
          Deactivate Sub(S) stored at Bi
      ENDIF
    ENDFOR
ENDIF
```
Case 5: **Subscriber moves in from a peer broker**
```
IF subscriber S hands of to Bj from Bk THEN
    IF Sub(S) is not in Bj buffer THEN
      Obtain Context(S) stored at Bi
    ENDIF
    FOR all Bi ∈ Neighbor(Bj) DO
      IF Bi ∉ Neighbor(Bk) THEN
          Forward Sub(S) to Bi
      ENDIF
    ENDFOR
ENDIF
```
Case 6: **Subscriber unsubscribes**
```
IF subscriber S unsubscribes from Bj THEN
    FOR all Bi ∈ Neighbor(Bj) DO
      Delete Context(S) stored at Bi
    ENDFOR
ENDIF
```
Case 7: **Subscriber context obtained**
```
IF Bj obtains Context(S) from neighbors THEN
    Buffer Context(S) at Bj
ENDIF
```
Case 8: **Subscriber times out**
```
IF Bj triggers Timeout(S) THEN
    FOR all Bi ∈ Neighbor(Bj) DO
      Delete Context(S) stored at Bi
    ENDFOR
ENDIF
```

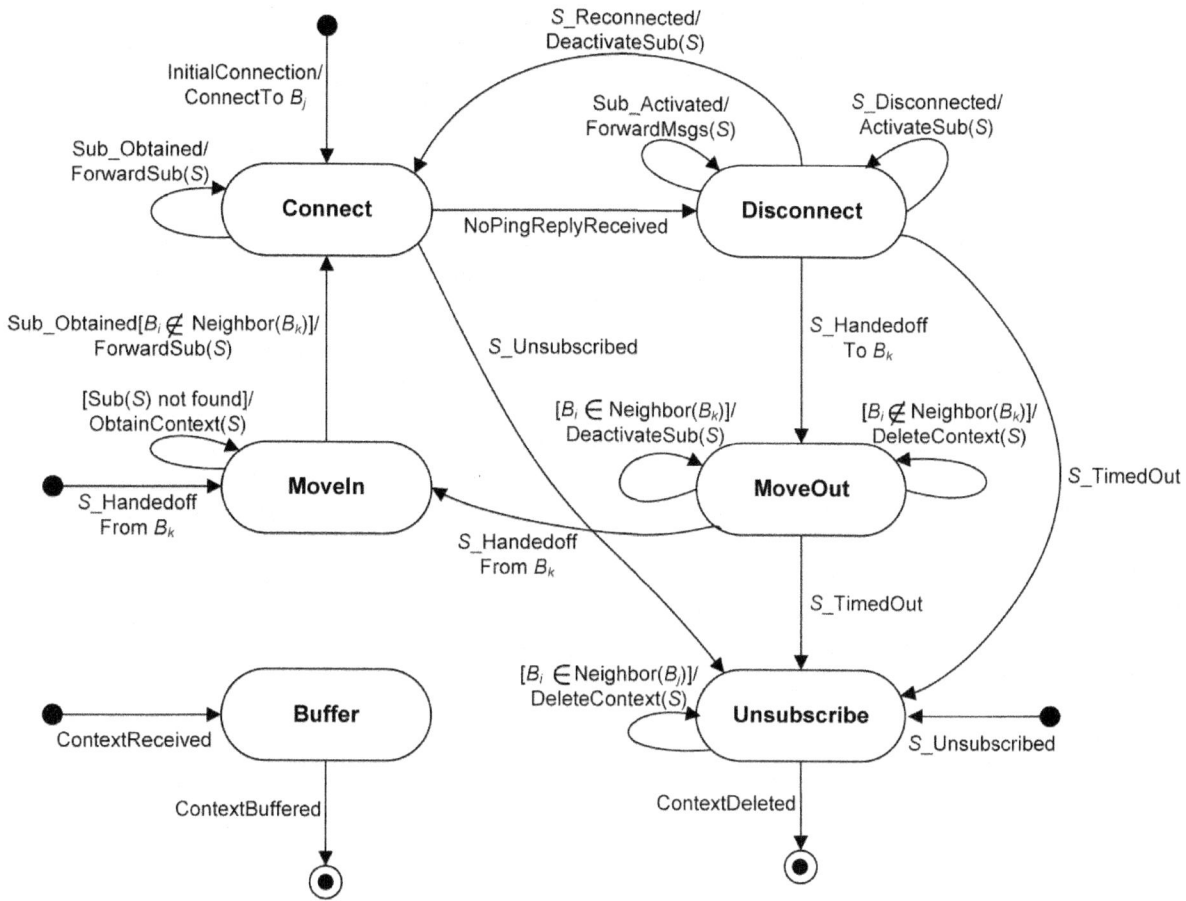

Fig. 1. FSM diagram for the pro-active context distribution algorithm.

enclosed with the activation request and thus only the messages with higher IDs are stored. Broker B_j similarly keeps buffering $Msgs(S)$ as it may reconnect to B_j again.

Case 3: When S reconnects to the same broker B_j, rather than moving to a neighbor broker B_i, B_j sends a deactivate request to its neighbors B_i, informing them to deactivate $Sub(S)$, terminate the caching process, and clean up their local buffers. In the meantime, broker B_j routes all buffered messages to S.

Case 4: When S moves from broker B_j to broker B_k, B_k informs B_j that is currently serving S. Accordingly, broker B_j requests its neighbors B_i either to delete *Context(S)* or deactivate *Sub(S)*, excluding broker B_k. To reduce the overhead of context transfer, B_k and B_j exchange *Neighbor(B_k)* and *Neighbor(B_j)*. Throughout these lists, broker B_j can decide which *Sub(S)* should be deleted and which *Sub(S)* should be deactivated for future use by broker B_k. Similarly, broker B_k can identify which *Sub(S)* should be routed to its neighbors B_i. Neighbor brokers typically exchange neighbor information whenever an edge is added to or deleted from their lists.

Case 5: When S moves from broker B_k to broker B_j, B_j first checks if *Context(S)* is available in its buffer. If it is not found in the buffer, B_j requests *Context(S)* from B_k. If *Context(S)* is found in the local buffer of B_j, similar actions to *case 1* will be performed.

Case 6: When S unsubscribes from broker B_j, B_j instructs its neighbors B_i to delete *Context(S)* from their buffers.

Case 7: When broker B_j receives *Context(S)* from its neighbors, it locally stores it in a persistent buffer.

Case 8: When S disconnects from broker B_j and *Timeout(S)* is reached, B_j informs its neighbors B_i to delete *Context(S)* from their buffers. This is a necessary task to manage mobile subscriber failures/crashes.

As this algorithm executes in parallel on all brokers, concurrency issues need to be dealt with, in particular potential race conditions due to network latency or/and delayed response of overloaded brokers. Race conditions appear when two concurrent operations, initiated by two different brokers, are intended to change the state of the same subscription(s) (e.g., activate, deactivate, or delete states). This could then lead to an inconsistent state among the same subscription(s) copies stored at neighbor brokers, and hence impact the performance of the pro-active approach. For example, when S enters the *MoveOut* state, broker B_j sends a deactivate request to the neighbor brokers B_i that are also neighbors of broker B_k. It may happen that S disconnects from B_k at the same time that B_j issued the deactivate request. In this case, B_k also needs to send an activate request to its neighbors B_i. Due to the previously mentioned delays, it is possible that the *Sub(S)* are activated to support the movement of S from B_k and then deactivated based on the request issued by B_j. To address this race condition, we have integrated the broker ID with the propagated subscription(s) to indicate which broker has the control to deactivate or delete the activated subscription(s). The integrated ID is frequently updated using the activate request received from the most recent broker that served the moving subscriber. Looking at the previous scenario, B_j would not be allowed to deactivate the *Sub(S)* since B_k gained control of the *Sub(S)* just after performing the activate request. This keeps the *Sub(S)* in a consistent/appropriate state and eliminates the overhead of the reactive method.

Similarly, race conditions may occur when S hands off to broker B_k. In this case, broker B_j needs to issue a delete request to remove the *Sub(S)* from its neighbor brokers B_i (that are currently not broker's B_k neighbors) once it is informed by B_k about the subscriber's handoff. It may happen that one or more of those brokers become broker's B_k neighbors just after notifying B_j. B_k does not need to propagate a copy of the *Sub(S)* to the new added neighbors as it has previous knowledge (through the exchanged neighbor information) that there are previously propagated copies of the same *Sub(S)* at those neighbors. Thus, if S disconnects from its current broker B_k and an activate request is sent by B_k to its neighbors, we may experience concurrent requests sent by B_k and B_j (activate and delete requests). The broker ID can again help to control such a situation. When the activate request is performed, broker B_k gains control over the *Sub(S)* and thus the delete request is ignored. Note that in case the activate request could not find the *Sub(S)*, B_k will be notified and asked to deliver *Context(S)* prior to S arrival.

Finally, race conditions may occur in the *unsubscribe* state. If S hands off to broker B_k and just after the old broker B_j has been informed, S may decide to unsubscribe from the system. Thus, the neighbor brokers of B_j and B_k may receive concurrent requests (deactivate and delete). In this scenario, the delete request will fail if it gets executed before the deactivate request as broker B_j still has control over the active subscription(s). This leads to having a number of unclaimed subscriptions in the system. Note that the deactivate request disables the control attribute (broker ID) and therefore subscription(s) can be removed. To manage this race condition, the delete operation here has a special privilege to remove subscriptions without examining the control attribute of the subscriptions since the subscriber is leaving for good. A subscriber timing out is an alternative situation where a race condition may occur in the *unsubscribe* state. Consider the scenario when S disconnects from broker B_j and then hands off to broker B_k. If B_j for some reasons has not been informed about the subscriber movement, it will send a timeout request to the neighbors once the timeout interval is reached, deleting *Sub(S)*. If S disconnects from B_k and at the same time B_j reaches its timeout interval, we again have conflicting concurrent requests (activate and delete). This race condition can be handled by using the control attribute (broker ID) as discussed earlier. Note that the delete request here does not have special privilege to ignore the control attribute as with the previous scenario.

4.2. Neighbor graph

The effectiveness of our proposed pro-active mobility management scheme largely relies on the successful approximation of the subscriber's movement. For a better chance of success, broker B_j can approximate a set of potential brokers that are most likely to be the next-hop destination of S. This approximation can be achieved, e.g., through observations of the mobility patterns of subscribers. We hence make full use of a data structure, called *neighbor graph*, which provides the abstractions to achieve this goal.

A neighbor graph is basically a geometrical representation of a network (a collection of vertices V connected in neighborhood fashion). The graph contains a set of edges E (or *mobility paths*) that directly connect every vertex v (broker) to each of its neighbors. As a result, the neighbors of a given broker B_i in the graph correspond to the set of potential next-hop brokers. The neighboring relationship among the brokers can be represented by *undirected* edges if it is reflective. In other words, if S travels from broker B_j to B_i or vice versa, we then connect B_j and B_i with a single undirected edge. As the mobile subscribers in our experimental testbed are allowed to travel in a bidirectional way, we choose to use an undirected graph to represent the mobility paths between the brokers.

One way of building the neighbor graph is to allow individual subscribers to capture their own mobility graphs and offer them to the brokers upon their connections. Building the graph in such a way has several drawbacks. Mobile subscribers presumably use portable devices with limited capability (in terms of CPU and memory) to interact with the distributed brokers in the backbone network. Hence, the task of capturing the mobility graph by subscribers adds additional load on these devices; potentially degrading their performance, especially in a large-scale network where the size of the global neighbor graph is large. Each mobile subscriber needs to repeatedly submit its global graph upon its connection to the target broker. This may result in consuming a considerable amount of bandwidth and potentially congesting the wireless channel, particularly with a large subscriber population. To approximate mobility in a neighborhood fashion, brokers need to acquire knowledge only about the local view of the complete graph (i.e., its subset of neighbor brokers). As a result, forwarding the global view to the target brokers is indeed inadequate as it may include a number of non-neighbor brokers, hence adding irrelevant load on the system. Every broker also needs to separately deal with (e.g., store, search, and update) the graphs of individual subscribers, which may complicate and increase the overhead of processing the pro-active approach. In addition, building subscriber-specific neighbor graphs will prevent subscribers to benefit from common movement patterns, which get reinforced as different subscribers migrate between specific brokers. Thus, we choose to allow individual brokers (typically running on machines with high capabilities) to automatically build the neighbor graph that captures the local view of their neighbors.

The neighbor graph can be constructed either in a static manner (i.e., manually created once and never changes over time) or in a dynamic manner (i.e., automatically generated and adaptively changes according to the mobility pattern). A static neighbor graph is problematic as it fails to adapt itself to the dynamic changes in the mobility pattern and/or broker topology. The neighbor graph also can be maintained either in a centralized manner (i.e., a single server stores the entire graph) or in a distributed manner (i.e., each broker stores a local view of the entire graph). A centralized neighbor graph raises

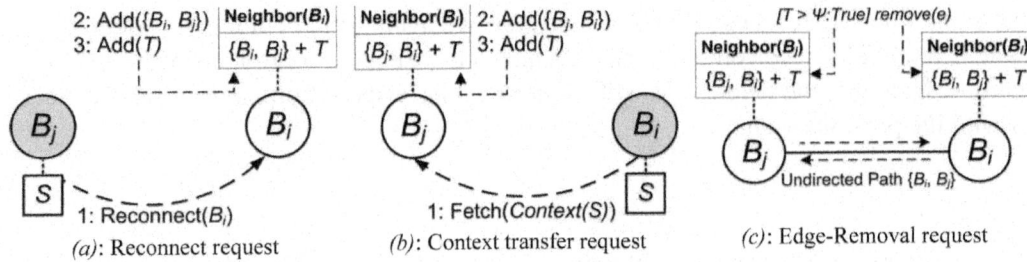

Fig. 2. The construction of the neighboring relation.

scalability and performance concerns as the network grows. Thus, we opted for a dynamic and distributed manner when generating the neighbor graph.

Several techniques [30,50] were proposed in the literature to build the neighbor graph, which are largely based on neighborhood search in the proximity space. In this paper, we adopt a different methodology for constructing the neighbor graph, which depends mainly on capturing the movement of mobile subscribers. Figure 2 illustrates how the adopted methodology works. Algorithm 2 summarizes the scenarios of the adopted methodology for constructing the neighboring relationship at each broker.

Algorithm 2: Construction of the Neighboring Relation – *executed at each broker*

Case a: Receive a reconnection request
```
1: subscriber S reconnects to Bᵢ AND
   submits the address of Bⱼ
2: IF an edge e={Bᵢ, Bⱼ} is not in Neighbor(Bᵢ) THEN
       Bᵢ adds e in Neighbor(Bᵢ)
   ENDIF
3: Bᵢ attaches a timestamp T to e OR
   updates an existing T
```

Case b: Receive a context transfer request
```
1: Bᵢ requests Bⱼ to transfer the Context(S) AND
   encloses its address with the request
2: IF an edge e={Bⱼ, Bᵢ} is not in Neighbor(Bⱼ) THEN
       Bⱼ adds e in Neighbor(Bⱼ)
   ENDIF
3: Bⱼ attaches a timestamp T to e OR
   updates an existing T
```

Case c: Edge-removal request
```
IF a timestamp T > a given time Ψ THEN
    remove an edge e from the neighborhood table
ENDIF

* Ψ ≥ the average frequency interval of edge-addition operations (cases
  a and b) at all broker nodes.
```

Two complementary methods can be used by each broker to effectively learn the edges in the neighbor graph. The *first* method is to attach the address of the old broker B_j with the reconnection request sent by the mobile subscriber S to the new broker B_i, hence building the neighboring relationship between the two brokers. The *second* method is to use the request for context transfer received from the new broker B_i to establish the relationship. This request is usually triggered whenever the Context(S) is not found at that broker. It is worth mentioning that some *outlier* edges (i.e., the ones that do not correctly model the neighboring relation) may be added to the neighborhood table. The table may also hold some *unused* edges that are created through rarely used paths. The impact of the outlier and unused edges on the performance of our pro-active approach can be significant due to the additional overhead required to cache *Context(S)* over time. It is thus essential to remove such edges from the graph table over time. To this end, we use a timestamp-based Least Recently Used (LRU) method to ensure the correctness and freshness of the graph. It should be apparent from that description that the autonomous creation of the graph makes it self-adaptive to dynamism in the neighboring relation (e.g., adding and deleting brokers, changing network topology, changing user behavior, etc.). Each broker independently builds and locally stores a subgraph of the complete graph of all brokers.

As the neighbor graph is initially an empty graph, the majority of handoffs, based on our creation algorithm, cause edge-addition during the early age of the graph, thereby reducing its benefit in our pro-active approach. Also, a mobile subscriber performing the first handoff along a path not in the graph may miss its messages, as the graph fails to provide information about the potential next brokers. To avoid this, the first mobile subscriber to cross over a path will receive its context in *reactive* manner. This will be gradually changed to the pro-active manner as the edges are added to the graph.

5. Performance evaluation

In this section, we describe our experimental environment and performance results for our pro-active scheme with various workload conditions. We compare our proposed scheme with two major existing schemes: *reactive* and *durable subscription-based*.

5.1. Experimental environment

The selected pub/sub system for our experimental study is based on a recent middleware technology called Java Message Service (JMS) [31]. JMS is a service-oriented API specification introduced by Sun to provide a standard platform for Java applications to create, send, and receive messages. Detailed descriptions of the JMS features can be found in [31,36]. We extended the selected JMS-based pub/sub system with the core functionalities of the pro-active and reactive schemes, the implementation details are more fully described in [17]. The durable subscription-based scheme is a built-in feature of all JMS-based pub/sub systems.

5.1.1. Testbed

Our testbed consisted of a dedicated network of *ten* Intel based Pentium 4 machines running RedHat 9, inter-connected by a 100 Mbps switch. Six machines were used for running six instances of the JMS broker with default configuration values. This work considers the distribution of these brokers in a simple in-building scenario as shown in Fig. 3 (left-side). The dotted lines represent a potential path of motion and the square boxes show the placement of broker nodes. Figure 3 (right-side) shows the complete experimental network. A router machine was used for running a wireless network emulator NistNet [38], with all the communication between the subscribers and the brokers tunneled through this router. All the configuration parameters for NistNet, like packet delay, packet loss, packet duplication, and network bandwidth, were set to the most commonly used values reported for IEEE 802.11 wireless LAN networks [19,51]. One machine was used for running a single message publisher $P1$ to inject messages in the broker network. The remaining two machines ($S1$ and $S2$) were used for running

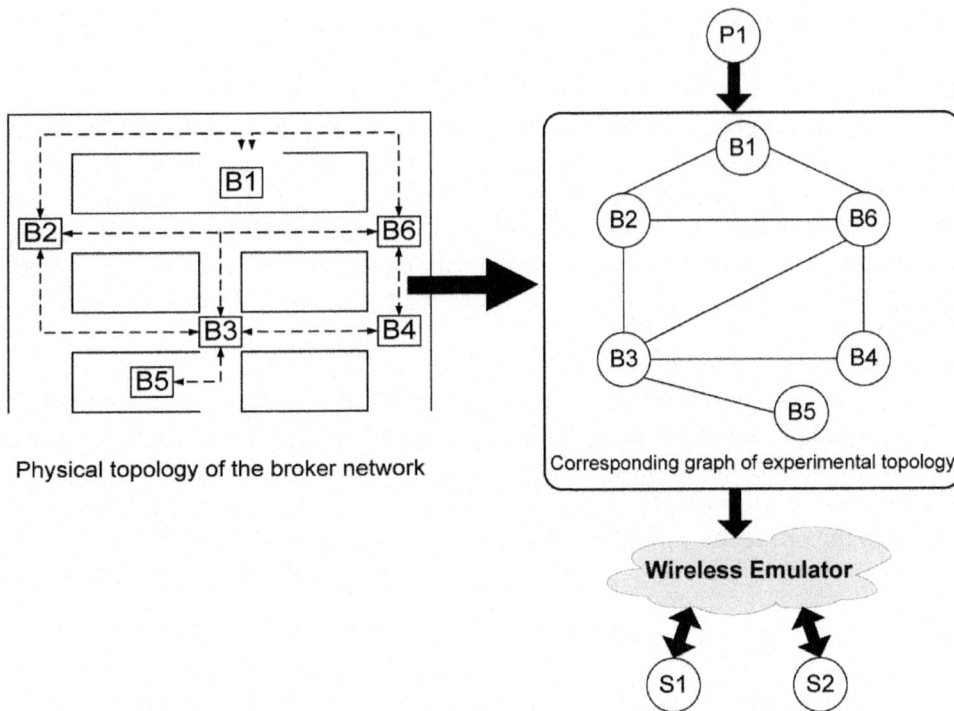

Fig. 3. A general view of experimental setup constituting the network under consideration.

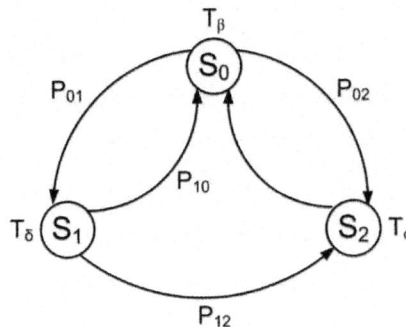

Fig. 4. Subscriber mobility model.

multiple subscribers. Subscribers that share the same machine run in separate threads and establish independent connections. Each subscriber subscribes to a single broker to receive approximately 20% of the published messages.

5.1.2. Mobility model

We developed a Java program to emulate subscriber mobility. Each subscriber goes through three mobility states: *connect* (S_0), *disconnect* (S_1), and *handoff* (S_2), as shown in Fig. 4. Initially, a subscriber enters to state S_0 to receive its messages from a uniformly chosen broker. The subscriber remains in S_0 for a randomly generated, exponentially distributed time with a mean of T_β seconds. With an equal probability $(P_{01} = P_{02})$, the subscriber either moves to state S_1 or S_2. S_1 reflects the case of signal

Table 1
Workload parameters

Workload parameters	Input values	Default values
Number of subscribers	10, 50, 100, 150, and 200	200
Sleep time	0, 0.5, 1, 1.5, and 2 seconds	0 seconds
Network bandwidth	1Mbps and 11Mbps	11 Mbps
Queue size	1, 2, 3, and 4 Mbytes	1 Mbytes
Mean disconnect interval (T_δ)	12 and 24 seconds	12 seconds
Mean connect interval (T_β)	60 and 150 seconds	60 seconds
Mean handoff interval (T_α)	1.5 and 3 seconds	3 seconds

breakdowns due to poor network connectivity. The subscriber remains in S_1 for a randomly generated, exponentially distributed time with a mean of T_δ seconds. With an equal probability $(P_{10} = P_{12})$, the subscriber either moves back to S_0 and reconnects to the same broker or goes to S_2. S_2 represents the case when a subscriber moves out of the covered area of its previous broker. After staying in S_2 for a randomly generated, exponentially distributed time with a mean of T_α seconds, the subscriber moves to S_0 and reconnects to a different broker.

We applied two mobility models, *random-based* and *neighboring-based* [1,12,18,24,33,43] to gauge how the three mobility management schemes react to different forms of subscriber mobility. In the random-based model, there are no dependencies or any other restriction modeled. The subscribers randomly move to any new target brokers deployed in an open geographical area. The target brokers are selected independently and uniformly, for every handoff, over the set of all six brokers. The subscribers pause at the selected brokers for randomly chosen time and then resume their movements. In the neighboring-based model, the area in which the subscribers are allowed to move is restricted due to obstructions. In other words, the subscribers can only move to some specific set of other locations from any given location, i.e., there are only a certain number of paths leaving each location. Thus, the subscribers in this model move to new neighbor brokers every handoff independently and uniformly over a specific set of neighbor brokers. Similarly, the subscribers pause at the selected neighbors for their randomly chosen time and then resume their movements.

5.1.3. Test conditions

Each experiment was run for a duration that was long enough to reach a steady state and repeated several times for verification purposes. Each broker machine was fully dedicated to run a single instance of the JMS broker. The CPU and memory usage of the broker machines were kept below 75% in any saturated modes (maximum publication rate and/or large subscriber population) to prevent performance bottlenecks. The publisher and subscriber machines must also not be bottlenecks, either, we ensured that their CPU and memory utilizations similarly stayed below 75%. We used the Linux tool "*sar*" to monitor the CPU and memory utilizations for each measurement run. Topic destinations and message stores are purged and reinitiated to start each test with a clean slate. Clock synchronization of the publisher and subscriber machines is required to calculate the end-to-end latency of message delivery and was done using NTP.

The reported results were captured from the measurement data obtained under different workloads, as summarized in Table 1. Unless otherwise stated, experiments were conducted using the default values listed in Table 1. The following metrics were used to evaluate the performance of the three mobility management schemes:

- *Subscriber throughput (Ts)*: total number of messages received per second.
- *Broker throughput (Tb)*: total number of messages routed per second to subscribers.

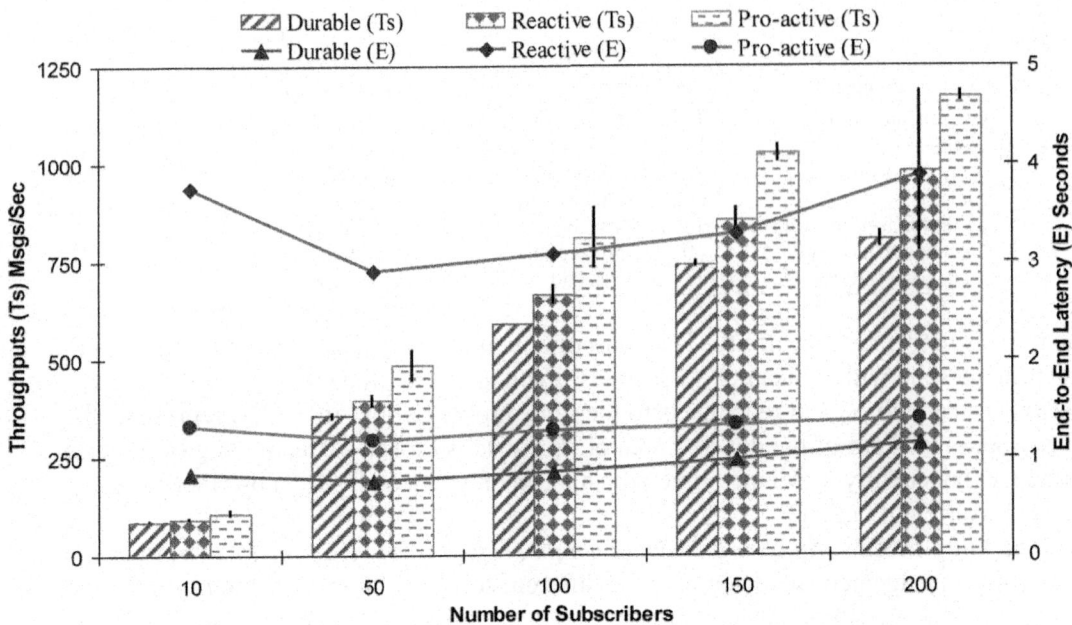

Fig. 5. Overhead of mobility support schemes.

- *Message loss* (*L*): percentage of missed messages by subscribers.
- *Message duplication* (*D*): percentage of duplicated messages received by subscribers.
- *End-to-end latency* (*E*): average time (in seconds) that it takes a message to travel from the publisher to the subscriber.
- *Handoff latency* (*H*): time (in seconds) between sending the reconnect request and receiving the first message of the subscriber at its new broker.
- *Message processing time* (*M*): average time (in milliseconds) that it takes a broker to filter messages and to route them to interested subscribers.

5.2. Performance results for random mobility model

We first present the performance results and comparisons of the pro-active, reactive, and durable subscription-based approaches in terms of overall subscriber throughput, end-to-end latency, handoff latency, message loss, and duplication. All the results presented are averages over 5 runs. For validity purposes, we plot the 95% confidence interval on top of each data point. The results were obtained using the random-based mobility model and default input values listed in Table 1. The random-based model presents the worse-case scenario for our pro-active approach: hard to predict mobility can result in the maximal protocol overhead. Due to space limitation only a representative set of the results are presented here. Interested readers are referred to [17] for more details.

5.2.1. Overhead of mobility support

Figure 5 shows how the three schemes compare in terms of end-to-end latency (*E*) and subscriber throughput (*Ts*) with the increase of subscriber population from 10 to 200. From the graph, we note that the *reactive* latency is by far the highest. A large portion of this latency is due to the overhead imposed by the state transfer semantic adopted by this scheme. During the handoffs, several messages may travel

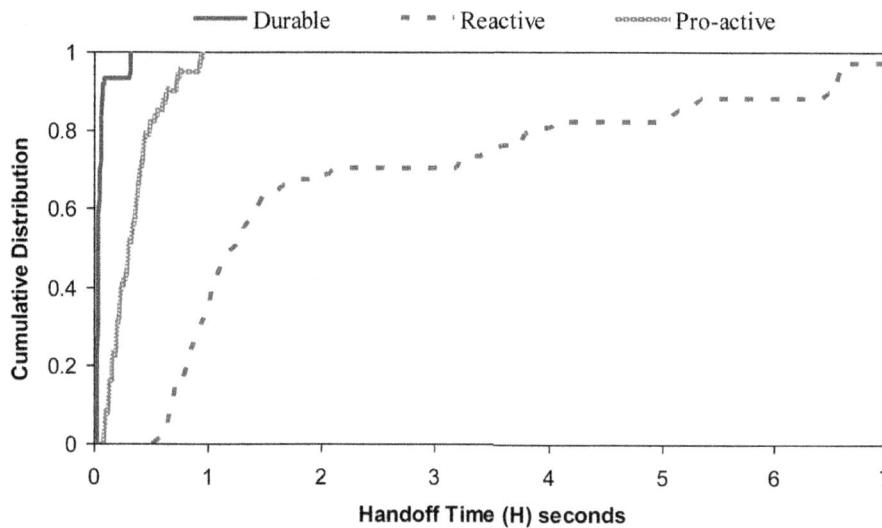

Fig. 6. Cumulative distribution of handoff times.

along multiple brokers as the state transfer protocol cannot catch up with the moving subscribers. We observed that the message overhead accounts for 37%–42% of the total consumed messages (i.e., the percentage of messages consumed via state transfer protocol) with 200 subscribers. The *pro-active* latency is much lower than the *reactive* as the *pro-active* scheme has almost no message overhead. In the *pro-active* scheme, messages are always available at the neighbor brokers and can be routed directly to the subscribers upon their reconnections. A small message overhead is imposed by this scheme to prevent message loss that may occur due to the latency of subscription activation request. The *durable-based* scheme shows the lowest latency as it has no state transfer overhead. However, it shows the lowest throughput results due to constantly caching messages at multiple brokers. The *pro-active* scheme also shows the highest throughput, as it caches messages on-demand and almost has no message overhead.

5.2.2. Handoff latency

Figure 6 shows the cumulative distribution of the handoff time. From the figure, we can note that the *reactive* scheme's latency is by far the highest among the three, with almost 60% of the handoffs taking more than 1.2 seconds. In contrast, almost 60% of the handoffs take less than 0.35 and 0.05 seconds with the *pro-active* and *durable-based* schemes, respectively. The observed results confirm that these two schemes can provide fast handoffs since the subscriber context is (almost) always ready at its new target broker prior to a subscriber's movement.

5.2.3. Frequency of handoffs

Figure 7 shows the performance results of the three schemes with low and high frequency of handoffs. We used a mean connect time interval (T_β) of 60 and 150 seconds in the low and high frequency of handoffs, respectively. From the figures, we observe that the overall throughputs of the *pro-active* and *durable-based* schemes slightly increase with the low handoff frequency. This is due to a reduced caching overhead in the two schemes. This can be clearly observed in the significant reduction of their message loss and duplication. Surprisingly, the overall throughput of the *reactive* scheme has not improved in the low frequency handoff scenario shown in Fig. 7(a), nor has its message loss ratio. In the high frequency handoff scenario, mobile subscribers connect with each broker for a short period. They may reconnect

(a) Low frequency of handoff operations

(b)High frequency of handoff operations

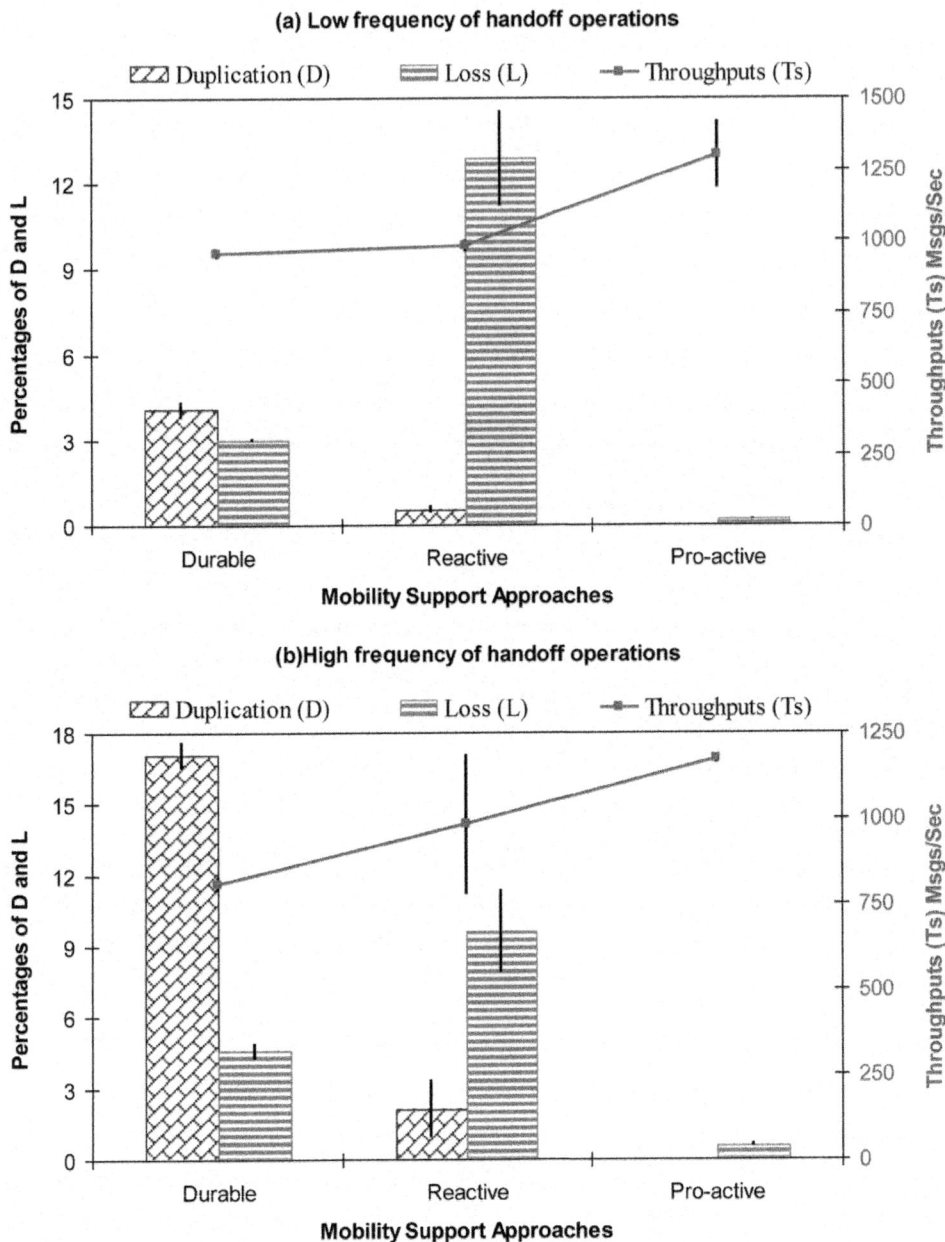

Fig. 7. Overall performance at a given handoff frequency.

back to their old brokers before the completion of their reactive state transfer. This leads to transferring a small number of messages between the old and new broker, but occurs quite often due to the high frequency of handoffs. In the scenarios with low handoff frequency, most mobile subscribers complete their state transfer before migrating to new locations. As a result, the full cost of the state transfer is incurred in such scenarios. For this reason, the throughput remains almost similar in the high and low frequency handoff scenarios. However, the low frequency handoff scenario shows higher message loss as the overhead of message transfer is high, with more messages discarded by the brokers.

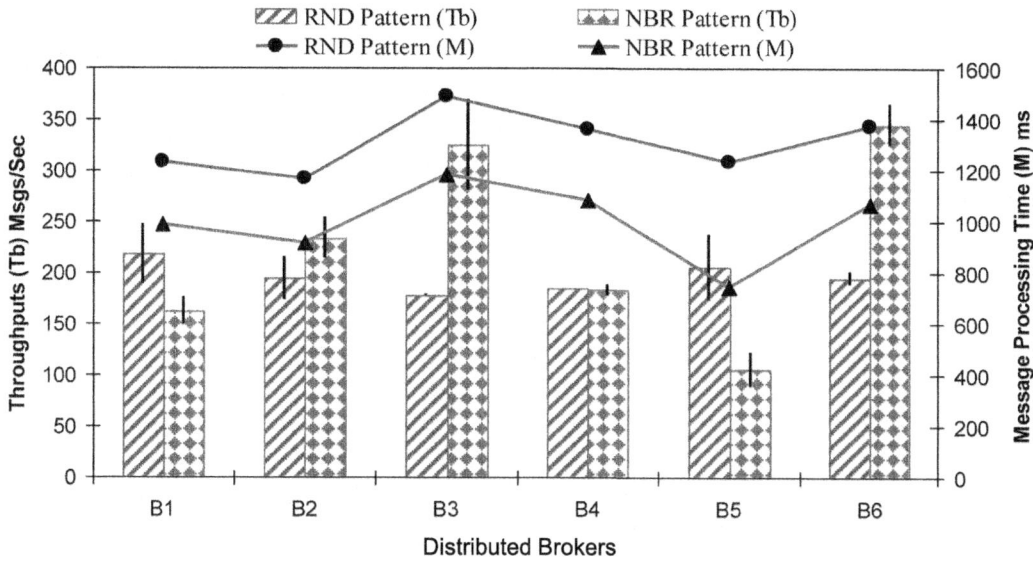

Fig. 8. Performance of pro-active approach in RND and NBR mobility patterns.

5.3. Comparing random and neighboring mobility patterns

Next we evaluate and compare the performance of the *pro-active* approach in the random (RND) and neighboring (NBR) mobility patterns in terms of message processing time (M) and individual broker throughput (Tb). Figure 8 shows the results of the selected metrics in the random (RND) and neighboring (NBR) mobility patterns: the left y-axis presents broker throughput (Tb) and the right y-axis shows message processing time (M). The x-axis category corresponds to the set of brokers presented in Fig. 3, describing the physical layout.

We observe that the *pro-active* approach shows lower message processing times under the NBR mobility pattern with all brokers. This is a good indicator of the overhead reduction due to limiting the mobility prediction to the (true) neighbor brokers. This results in reducing the number of immediate neighbors of each broker and therefore improves the performance of the *pro-active* approach. For example, broker $B5$ shows the lowest message processing time among all the brokers because it has only one neighbor broker $B3$. In the RND pattern, broker $B5$ shows higher processing time, as it has 5 neighbor brokers and needs to buffer messages for each disconnected subscriber at one of these neighbors. We can also observe that broker $B5$ experiences much lower throughput results in the NBR pattern. Since broker $B5$ has only one neighbor broker, the number of subscribers visiting broker $B5$ is much less than in the case of the random pattern. Therefore, broker $B5$ routes fewer messages to the overall subscribers. In contrast, brokers $B3$ and $B6$ have the largest number of neighbors (4 neighbors each) among all the brokers and show the highest throughput results. These brokers are visited by a large number of subscribers due to the NBR mobility pattern, thus many messages are routed through these brokers (traffic hotspots).

The remaining section presents and compares the performance of the three mobility management schemes using the RND and NBR mobility patterns and the default input values listed in Table 1. Three different performance metrics are used to evaluate the behavior of the three approaches: the overall subscriber throughput (Ts), message loss (L), and message duplication (D). We used different subscriber populations (10, 100, and 200) to investigate the overall performance of the three approaches under different workload conditions. Tables 2 through 4 summarize the respective results.

Table 2
Subscriber throughput (*Msgs/Sec*) in the RND and NBR mobility patterns

Subscribers	Durable-based		Reactive		Pro-active	
	RND	NBR	RND	NBR	RND	NBR
10	86.99	86.28	92.39	90.53	108.90	124.72
100	589.33	574.32	664.39	663.63	810.26	957.32
200	806.81	791.74	980.77	953.17	1172.77	1420.26

Table 3
Message loss (%) in the RND and NBR mobility patterns

Subscribers	Durable-based		Reactive		Pro-active	
	RND	NBR	RND	NBR	RND	NBR
10	4.78	3.13	3.65	5.14	2.25	1.26
100	5.14	4.06	8.92	7.03	0.86	0.72
200	4.56	3.51	9.62	9.44	0.58	0.51

As shown in Table 2, the *pro-active* approach shows superior throughput results under the RND and NBR mobility patterns, compared to the other approaches. As we expected, the *pro-active* approach under the NBR mobility pattern shows a noticeable improvement in the throughput results compared to the results achieved under the RND mobility pattern. This is because each broker in this pattern has fewer neighbors compared to the scenario in the RND pattern (brokers have probabilistically the same, maximum number of neighbors). As a result, the overhead (subscription propagation and message caching) of the *pro-active* approach significantly decreases and hence subscriber throughput improves. In contrast, the throughput results of the remaining approaches show a slight decrease with the various subscriber populations as shown in Table 2. We suspect that such behavior is due to the increased load on the central brokers (brokers that have a large number of immediate neighbors) that may become performance bottlenecks. In the *pro-active* approach, we found the overhead on the central brokers has not increased and is always less than the overhead under the RND pattern, as shown in Fig. 8.

Table 3 shows that the *pro-active*, *durable-based*, and *reactive* approaches demonstrate a slightly lower percentage of message loss under the NBR mobility pattern. This is due to the fact that some brokers' buffers (central brokers) are heavily utilized, but not others. Under the RND pattern, all the brokers' buffers experience almost similar (and high) buffer utilization. An interesting observation is that the percentage of message loss of the *pro-active* approach decreases gradually with the increase of subscriber population under both mobility patterns. With larger populations, the probability of having similar interest among the subscribers increases. This leads to a significant reduction in the caching overhead of the *pro-active* approach as one copy of each message can be stored for many subscribers. Therefore, message loss decreases with the increase of subscriber population. The *durable-based* approach shows an approximately similar percentage of message loss with the increase in the subscriber population. Although the *durable-based* approach benefits from the similarity of interest, its caching overhead is almost constant with different population sizes. This can be attributed to the continuous caching process adopted by this approach. In contrast, the *reactive* approach does not benefit from the similarity of interest as it does not reduce the overhead of state transfer. During the handoffs, state transfer has to be performed individually for every moving subscriber and hence its overhead increases proportionally with the subscriber population. Accordingly, message loss increases with the increase of the overhead imposed by subscriber population.

Table 4 shows that the message duplication slightly decreases under the NBR mobility pattern for the *durable-based* and *reactive* approaches. The pro-active approach shows zero message duplication under

Table 4
Message duplication (%) in the RND and NBR mobility patterns

Subscribers	Durable-based		Reactive		Pro-active	
	RND	NBR	RND	NBR	RND	NBR
10	11.35	12.10	0.80	1.52	0.0	0.0
100	12.01	11.68	1.70	0.85	0.0	0.0
200	17.05	16.97	2.13	1.90	0.0	0.0

both mobility patterns as it keeps track of the last consumed message by each subscriber and only buffers messages with higher ID than the last consumed message. Summarizing the results presented in these three tables, we can conclude that the NBR mobility model improves the performance of the *pro-active* approach in terms of overall throughput, message loss, and duplication. The two remaining approaches have not shown a significant difference in their performance results when using either the RND or the NBR mobility pattern.

6. Analytical approach for performance extrapolation

In this section, we present a modeling approach that can be used to extrapolate the performance of our proposed pro-active scheme in a near-size environment (in terms of broker network and/or subscriber population) to our experimental testbed. The general approach for performance extrapolation is as follows: most performance metrics are a function of the number of active subscribers at each broker. Thus, we first describe how to analytically derive the expected number of subscribers for a given broker topology, overall subscriber population, and mobility model. We then use a curve-fitting approach to relate the expected number of subscribers with a performance metric of interest: per-broker throughput. The approach can be generalized to other performance metrics. To validate our approach, the fitted curve derived from our experiments (using the random mobility model) is used to approximate the results in the neighboring mobility model.

Here we use continuous-time Markov chains (CTMC) to model subscriber mobility in the broker network topology presented in Fig. 3. This is due to the fact that we have a discrete state space of brokers (i.e., countable state space S = $\{1, \ldots, H\}$) and the *sojourn* times (holding time in one state before moving to another) are exponentially distributed. In the following two subsections, we describe in detail how CTMC is used for modeling the random and neighboring mobility of a mobile subscriber.

6.1. Modeling random mobility

In the random mobility model, a mobile subscriber has the freedom to randomly and uniformly move to one of N brokers available in the system. Thus, the state space S of this model can be defined by N *connect* states (brokers), a single *handoff*, and a single *disconnect* state. Hence, $S = \{1, \cdots, H\}$, where $H = N + 2$. The transition states of the random model can then be presented by an H-state Markov chain, as shown in Fig. 9.

In Fig. 9, states $\{1\}$ and $\{2\}$ represent the handoff and disconnect states, respectively. The rest of the states $\{3, \cdots, H\}$ correspond to the connect states. The arrows in the figure depict the subscriber mobility between different states while the parameters α, δ, and β represent the transition (departure) rates from state$\{1\}$, $\{2\}$, and $\{3, \cdots, H\}$, respectively. Based on the state diagram in Fig. 9, the values of these

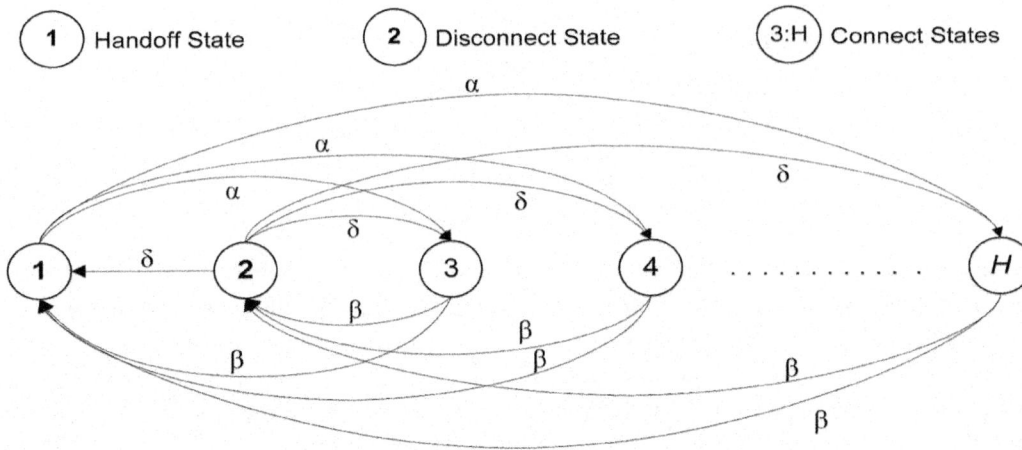

Fig. 9. State transition diagram for the random mobility model.

parameters can be determined as follows: Let T_α, T_δ, and T_β be the mean sojourn time at state $\{1\}, \{2\}$, and $\{3, \cdots, H\}$, respectively. Thus,

$$\alpha = \frac{1}{NT_\alpha} \tag{1}$$

$$\delta = \frac{1}{(N+1)T_\delta} \tag{2}$$

$$\beta = \frac{1}{2T_\beta} \tag{3}$$

The constant values $(N, N+1,$ and $2)$ shown in the denominators of the previous three equations reflect the number of possible destination states from the current state.

The M-matrix (also known as infinitesimal generator matrix) of the space S according to the state transition diagram shown in Fig. 9 is described as follows:

$$M = \begin{bmatrix} -N\alpha & 0 & \alpha & \alpha & \alpha & \cdots & \alpha \\ \delta & -(N+1)\delta & \delta & \delta & \delta & \cdots & \delta \\ \beta & \beta & -2\beta & 0 & 0 & \cdots & 0 \\ \beta & \beta & 0 & -2\beta & 0 & \cdots & 0 \\ \beta & \beta & 0 & 0 & -2\beta & \cdots & 0 \\ \vdots & \vdots & \vdots & \vdots & \vdots & \ddots & \vdots \\ \beta & \beta & 0 & 0 & 0 & \cdots & -2\beta \end{bmatrix}.$$

Let $\pi_i, i = 1, \cdots, H$, be the stationary state probability that the mobile subscriber is in state i. Accordingly, the H-element row vector $\pi = [\pi_1 \quad \pi_2 \quad \pi_3 \quad \cdots \quad \pi_H]$ represents the H stationary state probabilities that satisfy the matrix equation shown in Eq. (4) and their summation is equal to one, i.e., $\sum_{i=1}^{H} \pi_i = 1$.

$$\pi M = 0 \tag{4}$$

As the mobile subscriber moves to one of the connect states $i = 3, \cdots, H$ without any specific preferences and its mobility follows a symmetrical behavior in terms of arrival and departure, the stationary state probabilities $\pi_i, i = 3, \cdots, H$ are all equal. We denote the state probability of being in any connect state by P. Thus, we have $P = \pi_3 = \pi_4 = \cdots = \pi_H$, resulting in

$$\sum_{i=1}^{H} \pi_i = \pi_1 + \pi_2 + NP = 1 \tag{5}$$

By solving matrix Eq. (4), we obtain the state probabilities, π_1 and π_2, of the handoff and disconnect states, respectively.

$$\pi_1 = \frac{(N+2)\beta}{(N+1)\alpha}P \tag{6}$$

$$\pi_2 = \frac{N\beta}{(N+1)\delta}P \tag{7}$$

Substituting π_1 and π_2 into Eq. (4), we obtain the state probability, P, of being in any of the connect states $i, i = 3, \cdots, H$ as follows:

$$P = \left[\frac{(N+1)\alpha\delta}{N\beta\alpha + (N+2)\beta\delta + N(N+1)\alpha\delta} \right] \tag{8}$$

The expected number $E[x_i]$ of subscribers at any connect state $i, i = 3, \cdots, H$, follows the *binomial* distribution, since the presence of each subscriber at state i is a *Bernoulli* experiment with probability of success π_i and probability of failure $(1 - \pi_i)$. Let K be the total number of subscribers in the system. Thus, the expected number of subscribers at any connect state $i, i = 3, \cdots, H$, is given by

$Pr\{x \text{ subscribers are at state } i\} = (\pi_i)^x (1 - \pi_i)^{K-x} \binom{K}{x} = P^x (1 - P)^{K-x} \binom{K}{x}$. This leads to

$$\overline{n}_i = E[x_i] = KP = K \frac{(N+1)\alpha\delta}{N\beta\alpha + (N+2)\beta\delta + N(N+1)\alpha\delta} \tag{9}$$

Similarly, we can obtain the expected number of subscribers at handoff state $\{1\}$, $\overline{n_1}$, and disconnect state $\{2\}$, $\overline{n_2}$. Hence, we have

$$\overline{n_1} = E[x_1] = K\pi_1 = K\left(\frac{(N+2)\beta\delta}{N\beta\alpha + (N+2)\beta\delta + N(N+1)\alpha\delta} \right) \tag{10}$$

$$\overline{n_2} = E[x_2] = K\pi_2 = K\left(\frac{N\beta\alpha}{N\beta\alpha + (N+2)\beta\delta + N(N+1)\alpha\delta} \right) \tag{11}$$

6.2. Modeling neighboring mobility

In the neighboring mobility model, a mobile subscriber, from its current state, can only move to one of its neighbor states, where the selected neighbor is chosen randomly and uniformly. The restriction on the possible brokers the subscriber can move to, from its current broker, is modeled by multiple disconnect and handoff states, a pair for each broker. Thus, the state space S of the neighboring model is defined by

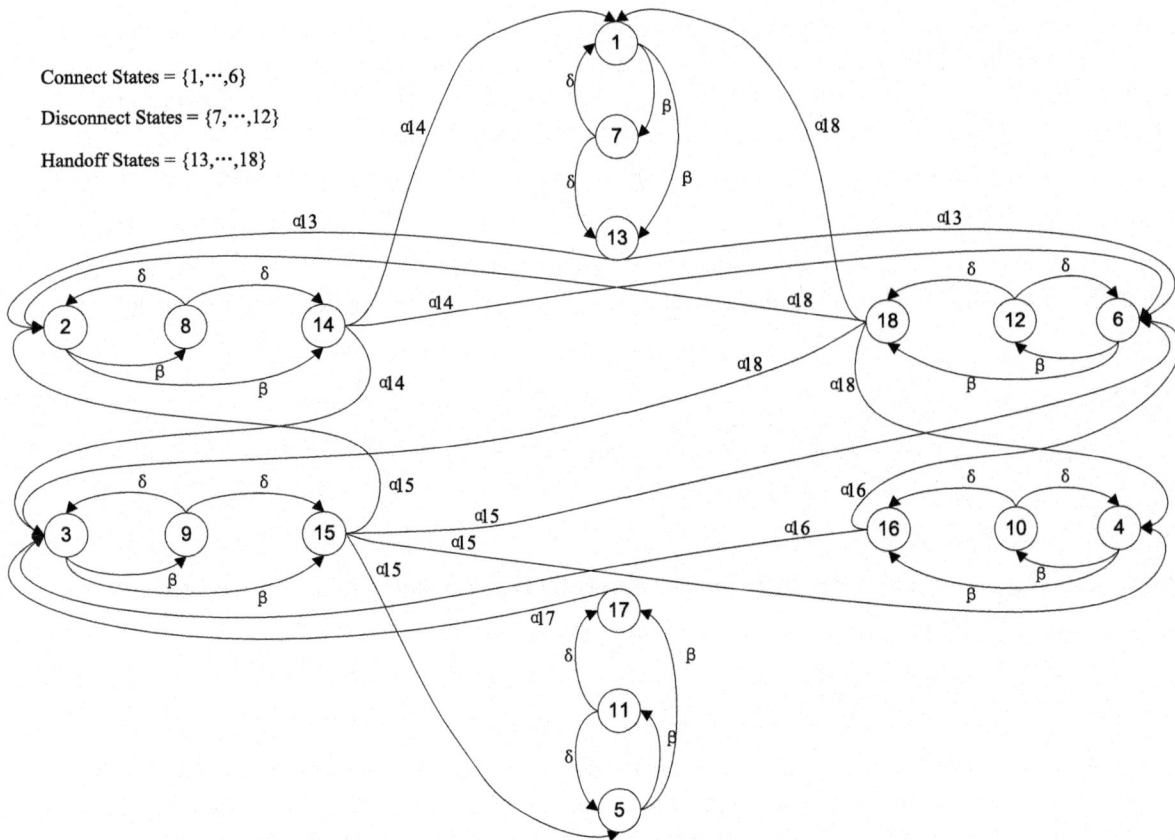

Fig. 10. State transition diagram for neighboring mobility model.

$S = \{1, \cdots, 18\}$, six connect states $\{1, \cdots, 6\}$, six disconnect states $\{7, \cdots, 12\}$, and six handoff states $\{13, \cdots, 18\}$. The transition states of the neighboring mobility model can accordingly be presented by the 18-state Markov chain, as shown in Fig. 10.

The arrows in the figure depict the subscriber mobility between different states while the parameters β, δ, and $\alpha_i, i = 13, \cdots, 18$ correspond to the transition (departure) rates from the connect, disconnect, and handoff states, respectively. Based on the transition state diagram shown in Fig. 10 and the mean sojourn time at each state, we can determine the values of the parameters (β, δ, and α_i) as follows:

$$\beta = \frac{1}{2T_\beta} \tag{12}$$

$$\delta = \frac{1}{2T_\delta} \tag{13}$$

$$\alpha_i = \frac{1}{E_i T_\alpha} \tag{14}$$

In the above three equations, the mean sojourn time at any connect, disconnect, and handoff states is denoted by T_β, T_δ, and T_α, respectively. The number of possible states from any connect or disconnect states is denoted by the constant 2 while the number of next possible neighbor states from any handoff state $i, i = 13, \cdots, 18$ is denoted by E_i.

We next determine the M-matrix of the state space S. As the cardinality of S is 18, the M-matrix has 18×18 entries that denote the transition rates based on the state transition diagram shown in Fig. 10. Thus, the M-matrix of the neighboring model is given by

$$M = \begin{bmatrix}
-2\beta & 0 & 0 & 0 & 0 & 0 & \beta & 0 & 0 & 0 & 0 & 0 & \beta & 0 & 0 & 0 & 0 & 0 \\
0 & -2\beta & 0 & 0 & 0 & 0 & \beta & 0 & 0 & 0 & 0 & 0 & \beta & 0 & 0 & 0 & 0 \\
0 & 0 & -2\beta & 0 & 0 & 0 & 0 & 0 & \beta & 0 & 0 & 0 & 0 & \beta & 0 & 0 & 0 \\
0 & 0 & 0 & -2\beta & 0 & 0 & 0 & 0 & 0 & \beta & 0 & 0 & 0 & 0 & \beta & 0 & 0 \\
0 & 0 & 0 & 0 & -2\beta & 0 & 0 & 0 & 0 & 0 & \beta & 0 & 0 & 0 & 0 & \beta & 0 \\
0 & 0 & 0 & 0 & 0 & -2\beta & 0 & 0 & 0 & 0 & 0 & \beta & 0 & 0 & 0 & 0 & \beta \\
\delta & 0 & 0 & 0 & 0 & 0 & -2\delta & 0 & 0 & 0 & 0 & 0 & \delta & 0 & 0 & 0 & 0 \\
0 & \delta & 0 & 0 & 0 & 0 & 0 & -2\delta & 0 & 0 & 0 & 0 & 0 & \delta & 0 & 0 & 0 \\
0 & 0 & \delta & 0 & 0 & 0 & 0 & 0 & -2\delta & 0 & 0 & 0 & 0 & 0 & \delta & 0 & 0 \\
0 & 0 & 0 & \delta & 0 & 0 & 0 & 0 & 0 & -2\delta & 0 & 0 & 0 & 0 & 0 & \delta & 0 \\
0 & 0 & 0 & 0 & \delta & 0 & 0 & 0 & 0 & 0 & -2\delta & 0 & 0 & 0 & 0 & 0 & \delta \\
0 & 0 & 0 & 0 & 0 & \delta & 0 & 0 & 0 & 0 & 0 & -2\delta & 0 & 0 & 0 & 0 & 0 & \delta \\
0 & \alpha_{13} & 0 & 0 & 0 & \alpha_{13} & 0 & 0 & 0 & 0 & 0 & 0 & -2\alpha_{13} & 0 & 0 & 0 & 0 & 0 \\
\alpha_{14} & 0 & \alpha_{14} & 0 & 0 & \alpha_{14} & 0 & 0 & 0 & 0 & 0 & 0 & 0 & -3\alpha_{14} & 0 & 0 & 0 & 0 \\
0 & \alpha_{15} & 0 & \alpha_{15} & \alpha_{15} & \alpha_{15} & 0 & 0 & 0 & 0 & 0 & 0 & 0 & 0 & -4\alpha_{15} & 0 & 0 & 0 \\
0 & 0 & \alpha_{16} & 0 & 0 & \alpha_{16} & 0 & 0 & 0 & 0 & 0 & 0 & 0 & 0 & 0 & -2\alpha_{16} & 0 & 0 \\
0 & 0 & \alpha_{17} & 0 & 0 & 0 & 0 & 0 & 0 & 0 & 0 & 0 & 0 & 0 & 0 & 0 & -\alpha_{17} & 0 \\
\alpha_{18} & \alpha_{18} & \alpha_{18} & \alpha_{18} & 0 & 0 & 0 & 0 & 0 & 0 & 0 & 0 & 0 & 0 & 0 & 0 & 0 & -4\alpha_{18}
\end{bmatrix}$$

Let $\pi_i, i = 1, \cdots, 18$, be the stationary state probability that the mobile subscriber is in state i. Thus, the 18-element row vector $\pi = \begin{bmatrix} \pi_1 & \pi_2 & \pi_3 & \cdots & \pi_{18} \end{bmatrix}$ represents the 18 stationary state probabilities that satisfy the matrix equation in Eq. (4), and their summation is equal to one, i.e. $\sum_{i=1}^{18} \pi_i = 1$. Note that $\pi_i, i = 1, \cdots, 6, \pi_i, i = 7, \cdots, 12$, and $\pi_i, i = 13, \cdots, 18$ are the state probabilities of the connect, disconnect, and handoff states, respectively. These state probabilities can be obtained by solving the matrix equation indicated in Eq. (4). Now we can readily calculate the expected number of subscribers, $E[x_i]$, at any state $i, i = 1, \cdots, 18$ in a similar way to that described in Section 6.1. Let K be the total number of subscribers in the system. Hence, the expected number of subscribers at any state i is given by

$$Pr\{x \text{ subscribers are at state } i\} = (\pi_i)^x (1 - \pi_i)^{K-x} \binom{K}{x}. \text{ This leads to}$$

$$\overline{n_i} = E[x_i] = K\pi_i \tag{15}$$

6.3. Curve-fitting

A curve-fitting approach is used to relate the average number of subscribers at a broker with a performance metric of interest (here per-broker throughput) using a single function generated from some observed data. Therefore, the generated function can be used to numerically extrapolate near-future outcomes. There are several curve fit forms we could chose from to build a function that gives the "best"

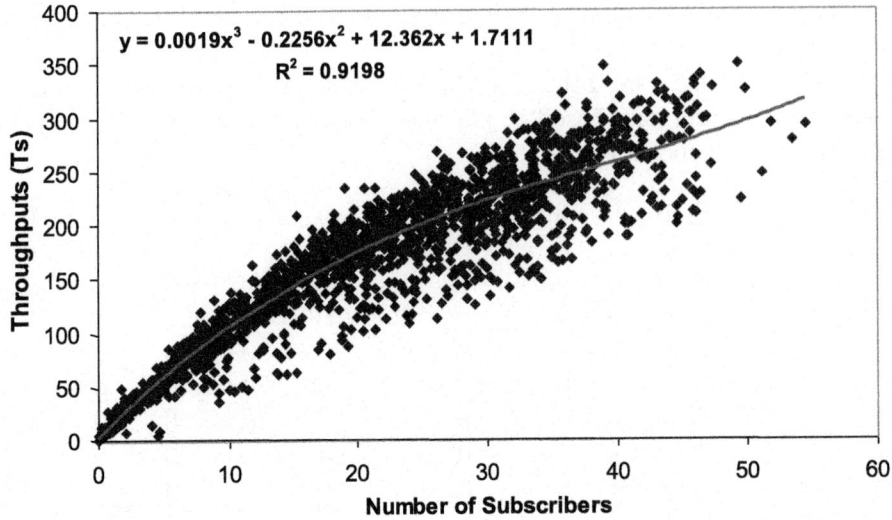

Fig. 11. Polynomial fit.

fit (i.e., the curve with minimum error between the generated curve and data points, usually referred to as *least-square error*). In this paper, we have chosen the *polynomial* curve-fitting approach since it shows the best fit, among the other generic forms we have tried, for our observed data. Using Excel, we fit a 3rd degree polynomial curve. Note that the observed data used for generating the curve is collected from the experiments of our general mobility model (random model). It is also collected from all the brokers used in our experimental setup. Figure 11 depicts the generated polynomial fitting-curve.

From this figure, we obtain the following polynomial equation that can be used to extrapolate the throughput of individual brokers as a function of the expected number of subscribers x.

$$y = (0.0019)x^3 - (0.2256)x^2 + (12.362)x + 1.7111 \qquad (16)$$

The value of $R^2 = 0.9198$, depicted in the graph, is an indicator from 0 to 1 that reveals how closely the estimated values for the fitting-curve correspond to the observed data. A fitting-curve is more reliable when its R^2 (known as *R-squared* or the coefficient of determination) value is at or near 1.

6.4. Comparative study

Next we apply our analytical approach to derive the approximated per-broker throughput results in the *random* and *neighboring* mobility models, and compare them to our experimental results. We first determine the expected number $E(x)$ of subscribers at each broker in both models using CTMC. We then derive the approximated throughput results via Eq. (16) for each individual broker. For all the results reported next, we used the default values shown in Table 1 for T_β, T_δ, and T_α. These values correspond to the used values in our experimental setup. Unless otherwise stated, the total number of subscribers K in the system was set to 200 and the total number of brokers N was set to 6. Thus, the size of the state space S is $H = N + 2 = 8$ in the random model and 18 in the neighboring model.

Random Model Results: Using the default mean sojourn times indicated in Table 1, we can identify the departure rate from each state. From Eqs (1), (2) and (3), we get

$$\alpha = 1/18, \ \delta = 1/84, \ \text{and} \ \beta = 1/120$$

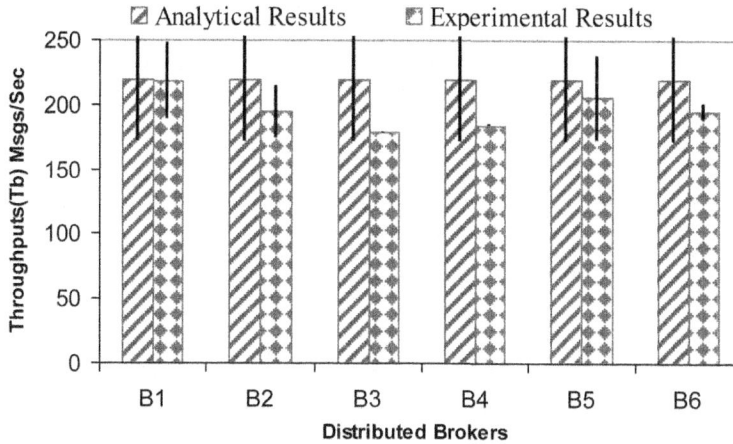

Fig. 12. Per-broker throughput results in the random mobility model.

We now describe the numerical solution for the set of Eqs (9), (10) and (11) using the obtained departure rates $(\alpha, \delta, \text{and} \beta)$. We first use Eq. (11) to obtain the state probability of being in connect state P, which is equal for all connect states as discussed earlier. Thus, we have $P = 0.147679$. Similarly, we can use Eqs (9) and (10) to obtain the state probability of being in the handoff and disconnect states, respectively. Hence, we have $\pi_1 = 0.025316$ and $\pi_2 = 0.088608$. We verify that the obtained state probabilities satisfy Eq. (5), i.e.,

$$\sum_{i=1}^{8} \pi_i = \pi_1 + \pi_2 + NP = 0.025316 + 0.088608 + (6)0.147679 = 1.$$

For a validity check of our analytical model, we have compared the analytical and experimental results in terms of the expected number of subscribers at each broker. Due to space limitations, these results are not shown here but were very close. Further details can be found in [17].

We next describe the numerical results obtained from the random model presented in Section 6.1. Based on Eq. (9), the expected number of subscribers/broker for a total subscriber population of 200 is obtained by $\bar{n}_i = 200 \times 0.147679 = 29.5358$. The expected throughput of each broker is computed by substituting $\bar{n}_i$ into Eq. (16). Thus, the per-broker throughput is given by y $= 218.983$ msgs/sec. We plot y (the expected throughput result) along with the experimental throughput results for the random model in Fig. 12. The lines across the data bars in the next figures represent the upper and lower bound of a 95% confidence interval for throughput results.

Comparing the 95% confidence interval for both sets of results, we note that the confidence intervals of the experimental results overlap all the corresponding intervals of the analytical results. This indicates that the differences between the analytical and experimental results are not statistically significant. From Fig. 12, we also note that the analytical results show relatively higher results compared to the experimental results. This can be attributed to the fact that each broker will have a large number of *proxy* subscribers that buffer messages on behalf of the moving subscribers. As a result, the throughput of individual brokers (especially ones executing on machines with lower hardware configurations such as $B3$ and $B4$) can be noticeably affected due to the overhead imposed by the proxy subscribers. Our analytical model can be refined in the future to take the number of proxy subscribers at each broker into consideration.

Table 5
State probabilities and E(x) of subscribers

State probabilities	Expected number of subscribers
$\pi_1 = 0.1099$	$\overline{n}_1 = E(x_1) = 200 \times 0.1099 = 21.98$
$\pi_2 = 0.1648$	$\overline{n}_2 = E(x_2) = 200 \times 0.1648 = 32.96$
$\pi_3 = 0.2198$	$\overline{n}_3 = E(x_3) = 200 \times 0.2198 = 43.96$
$\pi_4 = 0.1099$	$\overline{n}_4 = E(x_4) = 200 \times 0.1099 = 21.98$
$\pi_5 = 0.0549$	$\overline{n}_5 = E(x_5) = 200 \times 0.0549 = 10.98$
$\pi_6 = 0.2198$	$\overline{n}_6 = E(x_6) = 200 \times 0.2198 = 43.96$
$\pi_{DH} = 0.1209$	$\overline{n}_{DH} = E(x_{DH}) = 200 \times 0.1209 = 24.18$

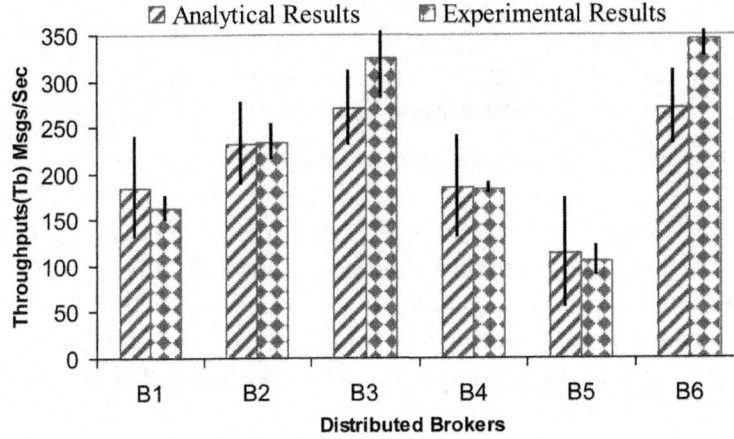

Fig. 13. Per-broker throughput results in the neighboring mobility model.

Neighboring Model Results: Similarly, we use the default mean sojourn times to determine the departure rate from each state. From Eqs (12), (13) and (14), we get

$$\beta = 1/120, \ \delta = 1/24, \ \text{and} \ \alpha_{13 \to 18} = \{1/6, 1/9, 1/12, 1/6, 1/3, 1/12\}$$

We now describe the numerical solution for the matrix equation shown in Eq. (4) using the obtained departure rates (β, δ, and $\alpha_{13 \to 18}$). To solve Eq. (4), we plug in the rate values shown above in the $18 \times 18 M$-matrix presented earlier and then find the solution for Eq. (4). Solving Eq. (4), we obtain the state probability for the different states, and then the expected number of subscribers can be found by multiplying each state probability by the total number of subscribers K. Table 5 presents the state probabilities of the connect states $\pi_i, i = 1, \cdots, 6$ and the total state probability of disconnect and handoff states along with the expected number of subscribers for a total subscriber population of 200. Based on the obtained $\overline{n}_i$, the expected per-broker throughput results can be determined using Eq. (16). Figure 13 depicts a plot of the expected throughput results along with the corresponding experimental results obtained using the neighboring model.

From Fig. 13, we note that the analytical and experimental results are close for most brokers. We also observe that the analytical results show relatively lower throughput results with the central brokers, brokers with a large number of neighbors ($B3$ and $B6$), compared to the experimental results. Generally, the central brokers will be visited by a larger number of subscribers than other brokers, and from the curve shown in Fig. 11, we note that there is a greater variation across the experimental data points with the increase in the average number of subscribers. This may be the reason why brokers $B3$ and $B6$, which serve the highest average number of subscribers, experience the largest deviation between the analytical and experimental results. Another reason can be attributed to the fact that central brokers $B3$ and $B6$ have a larger number of proxy subscribers that buffer messages on behalf of the actual moving subscribers. Thus, the actual subscribers can consume more messages during their connect intervals, resulting in higher throughputs.

7. Conclusions and future work

In this paper, we proposed a novel and efficient *pro-active* mobility management scheme to extend current pub/sub systems to operate in mobile wireless settings. The proposed scheme pro-actively transfers subscriber context to the neighbor brokers prior to its movement to a new broker. The scheme depends largely on the use of a data structure, called *neighbor graph*, which dynamically captures the *set* of next potential brokers where the subscriber context should be transferred. The neighbor graph is automatically built and regularly updated to eliminate the outlier neighbors. We have comprehensively evaluated the performance of our proposed scheme through testbed experiments, comparing it to the durable subscription-based and reactive schemes. Our experimental results demonstrate that the pro-active approach is superior to the alternative solutions with respect to a number of performance metrics such as message loss, message duplication and system throughput. Though its handoff and end-to-end latency may be slightly higher than the durable subscription-base scheme, the pro-active scheme outperforms both alternatives in overall performance. While the performance advantages hold true for both mobility models, more predictable user movements (as in the neighboring mobility model) result in even better performance for our proposed scheme. We conclude our work by introducing a modeling approach to extrapolate the performance of our proposed pro-active approach in a near-size environment (in terms of broker network and/or subscriber population) to our experimental testbed.

A number of directions for future work exist. In our current model, we allow subscribers to choose which broker to attach to. From a system perspective, it may be preferable to let the pub/sub system make this decision, to balance the load between brokers. We are also currently not supporting publisher mobility. Finally, the efficiency of our proposed approach is very dependent on accurately predicting subscriber mobility. In an ideal situation, we would be able to narrow down the potential next-hop neighbors for each subscriber to a small set of target brokers. Newer proposals for handover prediction in cellular networks, such as [5,15,33] may therefore provide ways on further improving the performance of the proposed pro-active mobility management approach.

References

[1] N. Aschenbruck, E. Padilla and P. Martini, A Survey on Mobility Models for Performance Analysis in Tactical Mobile Networks, *Journal of Telecommunications and Information Technology* **2** (2008), 54–61.

[2] R. Baldoni, L. Querzoni, S. Tarkoma and A. Virgillito, Distributed Event Routing in Publish/Subscribe Systems, Middleware for Network Eccentric and Mobile Applications, ISBN 978-3-540-89706-4, Springer Berlin Heidelberg, 2009, 219–244.

[3] G. Banavar, T. Chandra, B. Mukherjee, J. Nagarajarao, R. Strom and D. Sturman, An Efficient Multicast Protocol for Content-Based Publish/Subscribe Systems, In Proceedings of the 19th IEEE International Conference on Distributed Computing Systems (ICDCS'99), Washington, DC, 1999, 262–272.

[4] N. Banerjee, W. Wei and S. Das, Mobility Support in Wireless Internet, *Journal of IEEE Wireless Communications* **10**(5) (2003), 54–61.

[5] L. Barolli, A Speed-Aware Handover System for Wireless Cellular Networks Based on Fuzzy Logic, *Mobile Information Systems* **4**(1) (2008), 1–12.

[6]   I. Burcea, H. Jacobsen, E. Lara, V. Muthusamy and M. Petrovic, Disconnected Operation in Publish/Subscribe Middle-
      ware", In Proceedings of the 2004 IEEE International Conference on Mobile Data Management (MDM'04), Berkeley,
      California, 2004, 39–50.

[7]   M. Caporuscio, A. Carzaniga and A. Wolf, Design and Evaluation of a Support Service for Mobile, Wireless Pub-
      lish/Subscribe Applications, *IEEE Transactions on Software Engineering* **29**(12) (2003), 1059–1071.

[8]   A. Campailla, S. Chaki, E. Clarke, S. Jha and H. Veith, Efficient Filtering in Publish-Subscribe Systems Using Binary
      Decision Diagrams, In Proceedings of the 23rd International Conference on Software Engineering (ICSE'01), Toronto,
      Canada, 2001, 443–452.

[9]   M. Castro, P. Druschel, A. Kermarrec and A. Rowstron, SCRIBE: A Large-Scale and Decentralized Application-Level
      Multicast Infrastructure, *IEEE Journal on Selected Areas in Communications*, *(JSAC'02)* **20**(8) (2002), 100–110.

[10] M. Chelliah, N. Govindaram and N. Gopalan, A Novel Distance Based Relocation Mechanism to Enhance the Performance
      of Proxy Cache in a Cellular Network, *The International Arab Journal of Information Technology* **6**(3) (2009), 258–263.

[11] G. Cugola, E. Nitto and A. Fuggetta, The JEDI Event-Based Infrastructure and its Application to the Development of the
      OPSS WFMS, *IEEE Transactions on Software Engineering* **27**(9) (2001), 827–850.

[12] D. Engelhart, A. Sivasubramaniam, C. Barrett, M. Marathe, J. Smith and M. Morin, *A Spatial Analysis of Mobility
      Models: Application to Wireless Ad Hoc Network Simulation*, In Proceedings of the 37th Annual Simulation Symposium
      (ANSS'04), 2004, 35–43.

[13] P. Eugster, P. Felber, R. Guerraoui and A. Kermarrec, The Many Faces of Publish/Subscribe, *ACM Computing Surveys*
      35(2) (2003), 114–131.

[14] U. Farooq, S. Majumdar and E. Parsons, High Performance Middleware for Mobile Wireless Networks, *In Mobile
      Information Systems Journal* **3**(2) (2007), 107–132.

[15] P. Fülöp, S. Imre, S. Szabó and T. Szálka, Accurate Mobility Modeling and Location Prediction based on Pattern Analysis
      of Handover Series in Mobile Networks, *Mobile Information Systems* **5**(3) (2009), 255–289.

[16] A. Gaddah and T. Kunz, Performance of Pub/Sub Systems in Wired/Wireless Networks, *In Proceedings of the 64th IEEE
      Vehicular Technology Conference (VTC'06)*, Montreal, Canada, 2006, 1–5.

[17] A. Gaddah, *A Pro-Active Mobility Management Scheme for Publish/Subscribe Middleware Systems*, Ph.D. Dissertation,
      Department of Systems and Computer Engineering, Carleton University, Ottawa, Canada, 2008.

[18] S. Gowrishankar, T. Basavaraju and S. Sarkar, Effect of Random Mobility Models Pattern in Mobile Ad hoc Networks,
      International Journal of Computer Science and Network Security (IJCSNS'07) **7**(6) (2007), 160–164.

[19] N. Gupta and P.R. Kumar, A Performance Analysis of the IEEE 802.11 Wireless LAN Medium Access Control,
      Communications in Information and Systems **3**(4) (2003), 279–304.

[20] R. Henjes, D. Schlosser, M. Menth and V. Himmler, Throughput Performance of the ActiveMQ JMS Server, ITG/GI
      Symposium Communication in Distributed Systems (KiVS'07), Bern, Switzerland, 2007, 113–124.

[21] S. Hu, V. Muthusamy, G. Li and H. Jacobsen, Transactional Mobility in Distributed Content-Based Publish/Subscribe
      Systems, In Proceedings of the 29th IEEE International Conference on Distributed Computing Systems (ICDCS'09),
      Montreal, Canada, 2009, 101–110.

[22] Y. Huang and H. Garcia-Molina (2004), "Publish/Subscribe in a Mobile Environment", Wireless Networks Journal,
      Special Issue on Pervasive Computing and Communications, 10(6), pp. 643-652.

[23] H. Jafarpour, S. Mehrotra and N. Venkatasubramanian, A Fast and Robust Content-based Publish/Subscribe Architecture,
      In Proceedings of 7th IEEE International Symposium on Network Computing and Applications, (NCA'08), 2008, 52–59.

[24] A. Jardosh, E. BeldingRoyer, K. Almeroth and S. Suri, Towards Realistic Mobility Models for Mobile Ad hoc Networks,
      In Proceedings of the 9th Annual International Conference on Mobile Computing and Networking (MobiCom'03), San
      Diego, California, 2003, 217–229.

[25] K. Lai, Z. Tari and P. Bertok, Supporting User Mobility through Cache Relocation, *International Journal on Mobile
      Information Systems (MIS'05)* **1**(4) (2005), 275–307.

[26] G. Li, S. Hou and H. Jacobsen, A Unified Approach to Routing, Covering and Merging in Publish/Subscribe Systems
      Based on Modified Binary Decision Diagrams, In Proceedings of the 25th IEEE International Conference on Distributed
      Computing Systems (ICDCS'05), Washington, DC, 2005, 447–457.

[27] G. Li, V. Muthusamy and H. Jacobsen, Adaptive Content-Based Routing in General Overlay Topologies, In Proceedings
      of the 9th ACM/IFIP/USENIX International Conference on Middleware (Middleware'08), Leuven, Belgium, 2008, 1–21.

[28] Y. Liu and G. Maguire, A Predictive Mobility Management Algorithm for Wireless Mobile Computing and Communi-
      cations, In Proceedings of the 4th IEEE International Conference on Universal Personal Communications (ICUPC'95),
      Japan, 1995, 268–272.

[29] N. Liu, M. Liu, J. Zhu and H. Gong, A Community-Based Event Delivery Protocol in Publish/Subscribe Systems for
      Delay Tolerant Sensor Networks, *Journal of Sensors* **9**(10) (2009), 7580–7594.

[30] M. Maier, M. Hein and U. Luxburg, Optimal Construction of K-Nearest-Neighbor Graphs for Identifying Noisy Clusters,
      Journal of Theoretical Computer Science **410**(19) (2009), 1749–1764.

[31] Sun Microsystems, Java Message Service (JMS) API Specification, available on: http://java.sun.com/products/jms, 2009.

[32] M. Menth, R. Henjes, C. Zepfel and S. Gehrsitz, Throughput Performance of Popular JMS Servers, *SIGMETRICS Performance Evaluation Review* **34**(1) (2006), 367–368.

[33] G. Mino, L. Barolli, F. Xhafa, A. Durresi and A. Koyama, Implementation and Performance Evaluation of Two Fuzzy-Based Handover Systems for Wireless Cellular Networks, *Mobile Information Systems* **5**(4) (2009), 339–361.

[34] P. Mogre, M. Hollick, N. d'Heureuse, H. Heckel, T. Krop and R. Steinmetz, A Graph-Based Simple Mobility Model, In Proceedings of the 4th Workshop on Mobile Ad-Hoc Networks (WMAN '07), Bern, Switzerland, 2007, 421–432.

[35] G. Muhl, A. Ulbrich, K. Herrmann and T. Weis, Disseminating Information to Mobile Clients Using Publish-Subscribe, *Journal of IEEE Internet Computing* **8**(3) (2004), 46–53.

[36] M. Musolesi, C. Mascolo and S. Hailes, EMMA: Epidemic Messaging Middleware for Ad Hoc Networks, *Journal of Personal and Ubiquitous Computing* **10**(1) (2006), 28–36.

[37] V. Muthusamy, M. Petrovic, D. Gao and H. Jacobsen, Publisher Mobility in Distributed Publish/Subscribe Systems, In Proceedings of the 4th International Workshop on Distributed Event-Based Systems (ICDCSW'05), Washington, DC, IEEE Computer Society, 2005, 421–427.

[38] National Institute of Standards and Technology, NIST Network Emulation Tool, available on: http://snad.ncsl.nist.gov/itg/nistnet/index.html, 2009.

[39] L. Opyrchal, M. Astley, J. Auerbach, G. Banavar, R. Strom and D. Sturman, Exploiting IP Multicast in Content-Based Publish/Subscribe Systems, In Proceedings of IFIP/ACM International Conference on Distributed Systems Platforms and Open Distributed Processing (Middleware 2000), New York, NY, 2000, 185–207.

[40] P. Pietzuch and J. Bacon, Hermes: A Distributed Event-Based Middleware Architecture, In Proceedings of the 1st International Workshop on Distributed Event-Based Systems (DEBS'02) In Conjunction with the 22nd International Conference on Distributed Computing Systems (ICDCS'02), Vienna, Austria, 2002, 611–618.

[41] I. Podnar and I. Lovrek, Supporting Mobility with Persistent Notifications in Publish/Subscribe Systems, In Proceedings of the 3rd International Workshop on Distributed Event-Based Systems (DEBS'04), Edinburgh, Scotland, UK, 2004, 80–85.

[42] C. Rezende, A. Boukerche, B. Rocha and A. Loureiro, Understanding and Using Mobility on Publish/Subscribe Based Architectures for MANETs, In Proceedings of the 33rd IEEE Conference on Local Computer Networks (LCN'08), Montreal, Canada, 2008, 813–820.

[43] M. Saad and Z. Zukarnain, Performance Analysis of Random-Based Mobility Models in MANET Routing Protocol, *European Journal of Scientific Research (EJSR'09)* **32**(4) (2009), 444–454.

[44] M. Srivatsa and L. Liu, Securing Publish-Subscribe Overlay Services with EventGuard, In Proceedings of the 12th ACM Conference on Computer and Communications Security (CCS'05), Alexandria, VA, USA, 2005, 289–298.

[45] H. Stathes and M. Lazaros, Using Proxy Cache Relocation to Accelerate Web Browsing in Wireless Mobile Communications, In Proceedings of the 10th ACM International Conference on World Wide Web, (WWW'01), Hong Kong, New York, NY, 2001, 26–35.

[46] P. Sutton, R. Arkins and B. Segall, Supporting Disconnectedness-Transparent Information Delivery for Mobile and Invisible Computing, In Proceedings of the 1st International Symposium on Cluster Computing and the Grid(CCGRID'01), Washington, DC, 2001, 277–285.

[47] S. Tarkoma and J. Kangasharju, On the Cost and Safety of Handoffs in Content-Based Routing Systems, *The International Journal of Computer and Telecommunications Networking* **51**(6) (2007), 1459–1482.

[48] C. Wang, A. Carzaniga, D. Evans and A. Wolf, Security Issues and Requirements for Internet-Scale Publish/Subscribe Systems, In Proceedings of the 35th Annual Hawaii International Conference on System Sciences (HICSS'02), Washington, DC, USA, 2002, 303–310.

[49] J. Wang, J. Cao, J. Li and J. Wu, MHH: A Novel Protocol for Mobility Management in Publish/Subscribe Systems, In Proceedings of the 2007 International Conference on Parallel Processing (ICPP'07), IEEE Computer Society, Washington, DC, 2007, 54–61.

[50] B. Wen, F. Liu, Z. Liu and F. Ma, Neighbor Relationship and Optimization in Mesh Overlay Multicast", In Proceedings of the 10th IEEE International Conference on High Performance Computing and Communications (ICHPCC'08), Washington, DC, 2008, 744–749.

[51] J. Yin, X. Wang and D. Agrawal, Modeling and Optimization for Wireless Local Area Network (WLAN), Computer Communications Journal, Special Issue on Performance Issues of Wireless LANs, PANs, and Ad Hoc Networks, **28**(10) (2005), 1204–1213.

[52] S. Yoo, J. Son and M. Kim, A Scalable Publish/Subscribe System for Large Mobile Ad Hoc Networks, *Journal of Systems and Software* **8**(2) (2009), 1152–1162.

[53] A. Zeidler and L. Fiege, Mobility Support with REBECA, In Proceedings of the 23rd International Conference on Distributed Computing Systems (ICDCSW'03), Providence, RI, 2003, 354–360.

[54] Y. Zhao and R. Strom, Exploiting Event Stream Interpretation in Publish-Subscribe Systems, In Proceedings of the 20th ACM Symposium on Principles of Distributed Computing (PODC'01), Newport, RI, 2001, 219–228.

2

BECSI: Bandwidth efficient certificate status information distribution mechanism for VANETs

Carlos Gañán[a,*], Jose L. Muñoz[a], Oscar Esparza[a], Jonathan Loo[b], Jorge Mata-Díaz[a] and Juanjo Alins[a]

[a]*Telematics Department, Universitat Politècnica de Catalunya, Barcelona, Spain*
[b]*Computer Communications Department, Middlesex University, London, UK*

Abstract. Certificate revocation is a challenging task, especially in mobile network environments such as vehicular ad Hoc networks (VANETs). According to the IEEE 1609.2 security standard for VANETs, public key infrastructure (PKI) will provide this functionality by means of certificate revocation lists (CRLs). When a certificate authority (CA) needs to revoke a certificate, it globally distributes CRLs. Transmitting these lists pose a problem as they require high update frequencies and a lot of bandwidth. In this article, we propose BECSI, a Bandwidth Efficient Certificate Status Information mechanism to efficiently distribute certificate status information (CSI) in VANETs. By means of Merkle hash trees (MHT), BECSI allows to retrieve authenticated CSI not only from the infrastructure but also from vehicles acting as mobile repositories. Since these MHTs are significantly smaller than the CRLs, BECSI reduces the load on the CSI repositories and improves the response time for the vehicles. Additionally, BECSI improves the freshness of the CSI by combining the use of delta-CRLs with MHTs. Thus, vehicles that have cached the most current CRL can download delta-CRLs to have a complete list of revoked certificates. Once a vehicle has the whole list of revoked certificates, it can act as mobile repository.

Keywords: PKI, revocation, VANET

1. Introduction

Vehicular ad-hoc networks (VANETs) have recently attracted extensive attentions as a promising technology for revolutionizing the transportation systems. VANETs consist of entities including On-Board Units (OBUs) and infrastructure Road-Side Units (RSUs). Mobile nodes are capable of communicating with each other (i.e. Vehicle to Vehicle Communication -V2V communication) and with the RSUs (i.e. Vehicle to Infrastructure Communication -V2I communication). Multi-hop communication facilitates information exchange among network nodes that are not in direct communication range [3,14], by means of short range wireless technology based on IEEE 802.11p.

Obviously, any malicious behaviors, such as injecting beacons with false information, modifying and replaying the previously disseminated messages, could be fatal to the other users. Thus, identifying the message issuer is mandatory to reduce the risk of such attacks. According to the IEEE 1609.2 standard [13], vehicular networks will rely on the public key infrastructure (PKI). In PKI, a certification

*Corresponding author: Carlos Gañán, Universitat Politècnica de Catalunya (UPC), Jordi Girona 1-3, 08034 Barcelona, Spain. E-mail: carlos.ganan@entel.upc.edu.

authority issues an authentic digital certificate for each node in the network. Due to misbehavior, intentional or otherwise, certificates need to be revoked in order to limit the risk that potential misuse poses to the rest of the network. The IEEE 1609.2 standard [13] states that VANETs will depend on certificate revocation lists (CRLs) to achieve revocation. CRLs are black lists that enumerate revoked certificates along with the date of revocation and, optionally, the reasons for revocation.

As VANETs can have a great amount of nodes (i.e. vehicles), CRLs will be large. Moreover, each vehicle in the network will own many temporary certificates (also called pseudonyms) to protect the users' privacy. Consequently, these lists will require hundreds of Megabytes [12,21,35]. However, distributing and updating CRLs to all vehicles raises a challenge. If there are no more communication media than the own VANET, no trusted-third parties (like the corresponding CA) can be assumed to be permanently available. Thus, online certificate status protocol (OCSP) [28] or, in general, any online solution is not suitable for this context. Several CRLs distribution protocols have been proposed for this purpose. For instance, to distribute these lists efficiently, authors in [26] proposed revocation using compressed CRLs. They divided the CRL into several self-verifiable parts and strongly reduced its size by using Bloom filters. Authors in [12] also propose the use of Bloom filters to store the revoked certificates for increasing the search speed in the CRL. On the other hand, authors in [22] proposed to use regional CAs and short lived certificates to decrease the number of entries in the CRL. We provide more information about these and other similar proposals in Section 2 but as a general conclusion, we could say that most of the research efforts in this context have been put on trying to reduce the size of the CRL, either trying to split it or trying to compress it.

In this article, we address the CRL distribution problem by exploiting the combination of three well-known mechanisms: (1) delta-CRL [1], (2) Merkle hash tree (MHT) [18], and (3) one-way hash chain [16]. By combining these three mechanisms, we design a Bandwidth Efficient Certificate Status Information (BECSI) protocol, that allows increasing the availability and freshness of the certificate status information (CSI) and at the same time reduces the bandwidth necessary to check the validity of a given certificate. BECSI takes advantage of V2V communication to create mobile repositories so that vehicles do not have to rely solely in the RSUs to obtain CSI. We aim to improve the distribution of CSI by transmitting the revocation information that is unknown to a particular user during the validity period of a CRL. The main idea behind BECSI is to allow vehicles requesting for new CSI during the validity period of the current CRL. Thus, revocations that occur during the validity period of the CRL will not be unknown to the vehicles during this whole validity period, reducing the risk of operating with an unknown revoked certificate. Therefore, BECSI reduces the peak bandwidth load associated with the CSI requests as there are more entities in the network that can answer these requests. To achieve, we combine the issuance of delta-CRLs and MHT.

By using the underlying concept of delta-CRLs, we implement a more efficient way of using of distributing CSI inside the VANET. To help minimize frequent downloads of lengthy CRLs, delta-CRLs are published aperiodically. On the other hand, BECSI codes the information included in the CRL and delta-CRLs in different MHTs. Using these MHTs, vehicles are able to act as mobile repositories. To achieve that, we embed some little extra information to the CRL such that allows us to create an efficient and secure request/response protocol. In more detail, we propose a way of efficiently embedding a MHT within the structure of the standard CRL to generate the so-called *extended-CRL* and *extended-delta-CRL*. To create these extended lists, we use an standard way of adding extra information to the CRL. Our extension contains all the necessary information to allow any vehicle or VANET infrastructure element that possesses the *extended-CRL* to build the BECSI tree, i.e., a hash tree with the CSI of the CRL. Using this BECSI tree, any entity possessing the *extended-CRL* can act as repository and efficiently answer to certificate status checking requests of other vehicles. As we will demonstrate by simulation, this makes the distribution of CSI more efficient than distributing complete CRLs (even they are compressed), reducing the data that have to be transmitted over the VANET. We must stress that any entity possessing an *extended-CRL* can act as BECSI repository but that a BECSI repository is not a TTP. In other words, BECSI is offline, which means that no online trusted entity (like a CA) is needed for authenticating the responses produced by BECSI repositories.

The rest of this paper is organized as follows. In Section 2, we present the background related to our mechanism. In Section 3 we describe in depth BECSI. In Section 4, we evaluate the proposed mechanisms. Finally, Section 5 concludes this paper.

2. Background

In this section, first we start describing existing revocation proposals for VANET. Then, we give a brief overview of Merkle Hash Trees (MHT) [18], which is one of the foundations of the proposed certificate validation mechanism. Finally we describe the basics of hash chains.

2.1. VANET revocation mechanisms

2.1.1. Centralized revocation approaches

The IEEE 1609.2 standard [13] proposes an architecture based on the existence of a Trusted Third Party (TTP), which manages the revocation service. In this architecture each vehicle possesses several short-lived certificates (used as pseudonyms), to ensure users' privacy. However, short-lived certificates are not enough as compromised or faulty vehicles could still endanger other vehicles until the end of their certificate lifetimes. Thus, the IEEE 1609.2 promotes the use of CRLs to manage revocation while assuming pervasive roadside architecture.

Other proposals in the literature also assume the existence of a TTP to provide the revocation service. Raya et al. [25] propose the use of a tamper-proof device (TPD) to store the certificates. A TTP is in charge of preloading the cryptographic material in the TPD. Thus, when a vehicle is compromised/misbehaving, it can be removed from the network by just revoking the TPD. To that end, the TTP must include the corresponding revocation information in a CRL. To reduce the bandwidth consumed by the transmission of CRLs, these authors proposed to compress the CRLs by using Bloom filters. However, this method gives rise to false positives which degrades the reliability of the revocation service.

However, even compressed, timely distributing CRLs to all vehicles is not trivial. Some authors [22, 23], instead of using a single central authority, have proposed the use of regional certification authorities which must develop some trust relationships. Papadimitratos et al. [24] suggest restricting the scope of the CRL within a region. Visiting vehicles from other regions require to obtain temporary certificates. Thus, a vehicle will have to acquire temporary certificates if it is traveling outside its registered region. The authors also propose breaking the Certificate Revocation List (CRL) into different pieces, then transmitting these pieces using Fountain or Erasure codes, so that a vehicle can reconstruct the CRL after receiving a certain number of pieces. Similarly, in [32], each CA distributes the CRL to the RSUs in its domain through Ethernet. Then, the RSUs broadcast the new CRL to all the vehicles in that domain. In the case RSUs do not completely cover the domain of a CA, V2V communications are used to distribute the CRL to all the vehicles [15]. This mechanism is also used in [2,7], where it is detailed a public key infrastructure mechanism based on bilinear mapping. Revocation is accomplished through the distribution of CRL that is stored by each user.

$$H_{root} = H_{20} = h(H_{10} \mid H_{11})$$

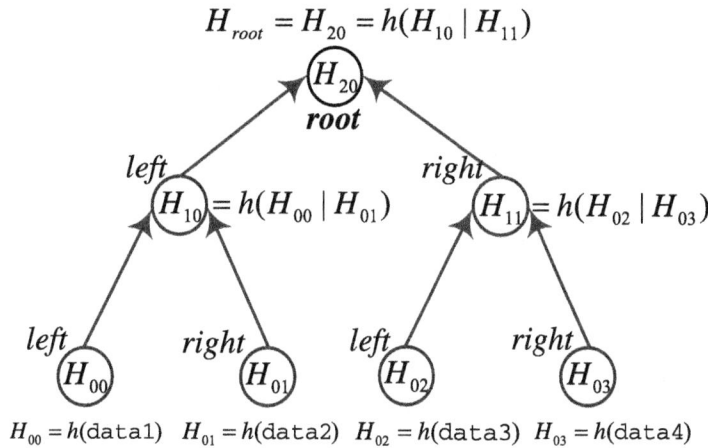

Fig. 1. Sample binary Merkle hash tree.

2.1.2. Decentralized revocation approaches

Decentralized revocation mechanisms provide the revocation service without assuming the existence of a TTP. Some proposals in the literature divert from the IEEE 1609.2 standard and use online status checking protocols instead of CRLs to provide a revocation service in a decentralized manner. This is the case, of the Ad-hoc Distributed OCSP for Trust (ADOPT) [17], which uses cached OCSP responses that are distributed and stored on intermediate nodes. Other group of proposals bases the revocation service on detecting a vehicle to be misbehaving by a set of other vehicles. Then, the detecting set may cooperatively revoke the credential of the misbehaving node from their neighborhood. Moore et al. proposed in [19] a revocation mechanism aiming to prevent an attacker from falsely voting against legitimate nodes. Raya et al. in [25] proposed a mechanism to temporarily revoke an attacker if the CA is unavailable. To do so, the number of accusing neighbor users must exceed a threshold. A similar mechanism based also on vehicle voting is proposed in [34]. Again, by means of a voting scheme, a vehicle can be marked as misbehaving and then be revoked by its neighbors.

Another proposal uses a game-theoretic revocation approach to define the best strategy for each individual vehicle [5,27]. These mechanisms provide incentives to guarantee the successful revocation of the malicious nodes. Moreover, thanks to the records of past behavior, the mechanism is able to dynamically adapt the parameters to nodes' reputations and establish the optimal Nash equilibrium on-the-fly, minimizing the cost of the revocation.

Finally, there are some hybrid approaches that are neither totally centralized nor decentralized [10,33]. For instance, authors in [9] propose the use of authenticated data structures to issue CSI. Using these schemes, the revocation service is decentralized to transmit the CSI but still depends on a CA to decide when a node should be evicted from the VANET.

2.2. The Merkle hash tree

A Merkle hash tree (MHT) [18] is essentially a tree structure that is built with a One Way Hash Function (OWHF). The leaf nodes hold the hash values of the data of interest (data1,data2,...) and the internal nodes hold the hash values that result from applying the OWHF to the concatenation of the hash values of its children nodes. In this way, a large number of separate data can be tied to a single hash value: the hash at the root node of the tree. MHTs can be used to provide an efficient and highly-scalable way to distribute revocation information, as it is described in [8] for MANETs (Mobile Ad Hoc

Networks). A sample MHT is presented in Fig. 1. This hash tree is binary because each node has at most two children or equivalently, two sibling nodes are combined to form a parent node in the next level. We will call these siblings as "left" and "right" and a detailed explanation of how to build the hash tree for BECSI is given in Section 3.3.

A MHT relies on the properties of the One Way Hash Functions (OWHF). It exploits the fact that an OWHF is at least 10,000 times faster to compute than a digital signature, so the majority of the cryptographic operations performed in the revocation system are hash functions instead of digital signatures. In addition, by storing the internal node values, it is possible to verify that any of the leaf nodes is part of the tree without revealing any of the other data.

2.3. Hash chains

The idea of "hash chain" was first proposed by Lamport [16] in 1981 and suggested to be used for safeguarding against password eavesdropping. A hash chain $\mathcal{C}$ is a set of values $s_0, \ldots, s_n$ for $n \in \mathbb{Z}$ such that $s_i = h(s_{i-1})$ for some one-way hash function h, where $i \in [1, n]$ and s_0 is a valid input for h.

Note that hash chains are preimage resistant, i.e., by knowing s_i, s_{i-1} cannot be generated by those who do not know the value s_0, however given s_{i-1}, its correctness can be verified by hashing $h(s_i)$. This property of hash chains has evolved from the property of one-way hash functions. Additionally, hash chains are also second preimage resistant, collision resistant and generate pseudo-random numbers.

In most of the hash-chain applications, first s_n is securely distributed and then the elements of the hash chain are spent (or used) one by one by starting from s_{i-1} and continuing until the value of s_0 is reached. At this point the hash chain is said to be exhausted and the whole process should be repeated again with a different seed to reinitialize the systems.

3. BECSI: Bandwidth efficient certificate status information distribution mechanism

In this Section, we present BECSI, a bandwidth efficient mechanism for certificate status checking over VANETs based on the use of Merkle Hash Trees and hash chains. First we introduce the motivation, goal and security architecture needed to support BECSI, and next we describe the mechanism in depth.

3.1. Motivation and goal

Despite the short-comings related to propagation of revocation information, the need for trusted authorities like CAs to ensure authentication has motivated researchers to propose PKI based security for vehicular networks. Mainly, these mechanisms intend to provide the following set of requirements:

1. *Reliability*: The revocation service must be available at all times.
2. *Memory*: Minimum amount of memory should be required as validation is often carried out in constrained environments.
3. *Bandwidth*: Communication bandwidth should be minimal.
4. *Freshness*: Revocation data should be as updated as possible.

Proposals described in Section 2 mainly deal with the bandwidth requirement. By compressing the CRL, using state-of-the-art coding techniques or partitioning the CRL, these approaches reduce the time required to download CSI. In addition, authors intend to provide a reliable revocation service by decentralizing the CSI distribution points. However, none of these works deals with the freshness of

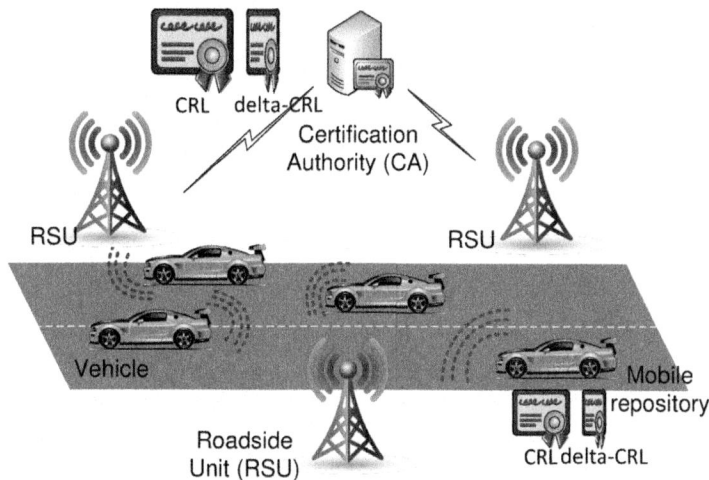

Fig. 2. System architecture.

the CSI. With BECSI we aim not only to reduce the communication overhead but also to increase the availability and freshness of CSI while keeping a reasonable computation cost.

CRLs are normally published in intervals meaning that there will not be any new revocation information available between the issuance and the update of the CRL. Newer revocations will thus be delayed until the next update occurs. High-security applications (e.g. safety applications) cannot cope with this lack of fresh information and render the traditional CRL approach almost useless in VANETs. To solve these problems, BECSI includes an extension to the standard CRL that allows RSUs to act as an offline repositories. Thus vehicles do not have to download the whole extended CRL, and they can just query about the status of a particular certificate.

3.2. Security architecture

The security architecture is an adaptation of a mesh PKI system to a vehicular scenario constructed of peer-to-peer CA relationships. This architecture consists of 4 different types of nodes (see Fig. 2):

1. *Certification authorities*: CAs are responsible for holding and managing the credentials and identities of all the vehicles which are registered under its hood. CAs are responsible for generating the set of certificates that are stored in each OBU. They are also responsible for managing the revocation information and making it accessible to the rest of the entities. By definition of TTP, the CA should be considered fully trusted by all the network entities, so it should be assumed that it cannot be compromised by any attacker. In fact, in our proposal CAs are the only trusted entities within the network.

2. *Road-side units*: RSUs are fixed entities that are fully controlled by the CA. They can access the CA anytime because they are located in the infrastructure-side, which does not suffer from disconnections. If the CA considers that an RSU has been compromised, the CA can revoke it.

3. *Vehicles*: They are the clients of the network. They have their cryptographic material stored in a TPD. Vehicles can check the validity of a certificate using V2I or V2V.

4. *Mobile repositories*: Mobile repositories are vehicles that have previously downloaded the CRL/delta-CRLs and are willing to response to certificate status requests from other vehicles.

$$H_{root} = H_{20} = h(H_{10} | H_{11})$$

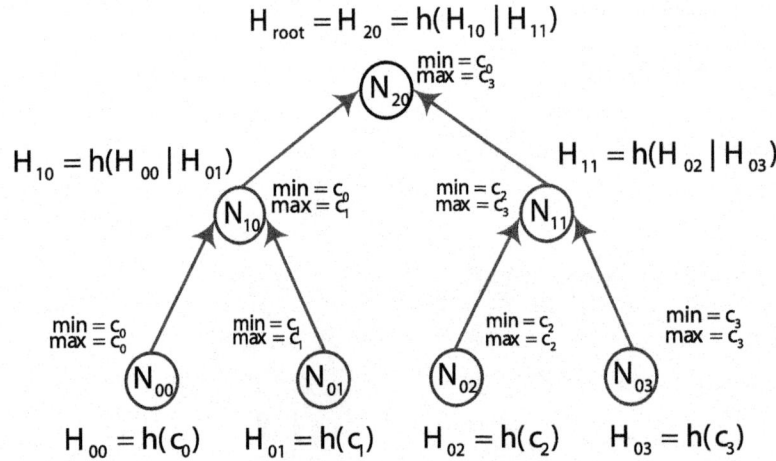

Fig. 3. Sample BECSI *base*-tree.

3.3. BECSI tree

In this section, we introduce the data structure that BECSI uses to handle the revocation service. In this sense, we define the BECSI tree as a composite Merkle Hash Tree (see Section 2). This tree consists of:

- A *base*-tree which is constructed using the serial number of the revoked certificates contained in the base-CRL.
- A set of Δ-trees which are constructed from the serial number of the certificates that are revoked during the validity interval of the base-CRL, i.e, they are constructed from the data contained in the delta-CRLs.

3.3.1. BECSI base-*tree*

The *base*-tree is a binary hash tree where each node represents a revoked certificate that is contained in the base-CRL. We denote by $N_{i,j}$ the nodes within the BECSI *base*-tree, where $i, j \in \{0, 1, 2, \ldots\}$ represent respectively the i-th level and the j-th node in the i-th level. We denote by $H_{i,j}$ the cryptographic (hash) value stored by node $N_{i,j}$ (see Fig. 3).

We denote by $N_{i,j}$ the nodes within the MHT where i and j represent respectively the i-th level and the j-th node. We denote by $H_{i,j}$ the cryptographic variable stored by node $N_{i,j}$.

Nodes at level 0 are called "leaves" and they represent the data stored in the tree. In the case of revocation, leaves represent the set Φ of certificates that have been revoked,

$$\Phi = \{c_0, c_1, \ldots, c_j, \ldots, c_n\}. \tag{1}$$

Where c_j is the data stored by leaf $N_{0,j}$. Then, $H_{0,j}$ is computed as:

$$H_{0,j} = h(c_j), \tag{2}$$

where h is a OWHF.

To build the MHT, a set of t adjacent nodes at a given level i ($N_{i,j}, N_{i,j+1}, \ldots, N_{i,j+t-1}$) are combined into one node in the upper level, which we denote by $N_{i+1,k}$. Then, $H_{i+1,k}$ is obtained by applying h to the concatenation of the t cryptographic variables:

$$H_{i+1,k} = h(H_{i,j} | H_{i,j+1} | \ldots | H_{i,j+t-1}). \tag{3}$$

At the top level there is only one node called the "root". H_{root} is a digest for all the data stored in the MHT.

The sample MHT of Fig. 1 is a binary tree because adjacent nodes are combined in pairs to form a node in the next level ($t = 2$) and $H_{root} = H_{2,0}$.

We define the $\mathcal{D}$igest as the concatenation of the certification authority distinguished number, the root hash and the validity period of the certificate status data. Once created, the $\mathcal{D}$igest is signed by the CA.

$$\mathcal{D}igest_{base} = \{DN_{CA}, H_{root}, \textit{Validity Period}\}_{SIG_{CA}}.$$

We denote as the $\mathcal{P}ath_{c_j}$ as the set of cryptographic values necessary to compute H_{root} from the leaf c_j.

It is worth noting that $\mathcal{D}$igest is trusted data because it is signed by the CA and it is unique within the tree while $\mathcal{P}$ath is different for each leaf. Thus, If the MHT provides a response with the proper $\mathcal{P}ath_{c_j}$ and the MHT $\mathcal{D}$igest, any vehicle can verify whether $c_j \in \Phi$.

For instance, let us suppose that a certain user wants to find out whether c_1 belongs to the sample MHT of Fig. 1. Then,

$$\mathcal{P}ath_{c1} = \{H_{0,0}, H_{1,1}\},$$

$$\mathcal{D}igest = \{DN_{CA}, H_{2,0}, \textit{Validity Period}\}_{SIG_{CA}}.$$

The response verification consists in checking that $H_{2,0}$ computed from the $\mathcal{P}_{c_1}$ matches $H_{2,0}$ included in the $\mathcal{D}$igest:

$$H_{root} = H_{2,0} = h(h(h(c_1)|H_{0,0})|H_{1,1}).$$

Note that the BECSI *base*-tree can be built by a trusted third party (e.g. a CA) and distributed to a non-TTP because a leaf cannot be added or deleted to Φ without modifying H_{root}, which is included in the $\mathcal{D}$igest and as the $\mathcal{D}$igest is signed, it cannot be forged by a non-TTP. To do such a thing, an attacker would need to find a pre-image of a OWHF which is computationally infeasible by definition.

3.3.2. BECSI Δ-trees

BECSI Δ-trees are constructed in the same way that the *base*-tree. However they present two differences with respect to the *base*-tree:

- Each leaf of the Δ-trees refers to certificates that were revoked during the validity interval of the base-CRL.
- The root of the Δ-tree is calculated by hashing the top-hash of the tree with the corresponding value of a hash chain. For more details about the construction of the hash chain see Section 3.4.3.

Figure 4 shows the simplest possible Δ-tree which contains only two revoked certificates.

Note that these Δ-trees have the same properties that the *base*-tree, so that the root node is unique and cannot be forged. Thus, the $\mathcal{D}$igest is composed as:

$$\mathcal{D}igest_{\Delta_i} = \{DN_{CA}, H^i_{\Delta root}, \textit{Validity Period}\}_{SIG_{CA}}.$$

Similarly, the $\mathcal{P}$ath consist of the set of cryptographic values necessary to compute $H^i_{\Delta root}$ from the leaf c_j. Note that this path is shorter in the case of the leafs of the Δ-trees because they contain less revoked certificates than the *base*-tree. The length of the Δ-trees is fixed as they are constructed from the delta-CRLs that have fixed size.

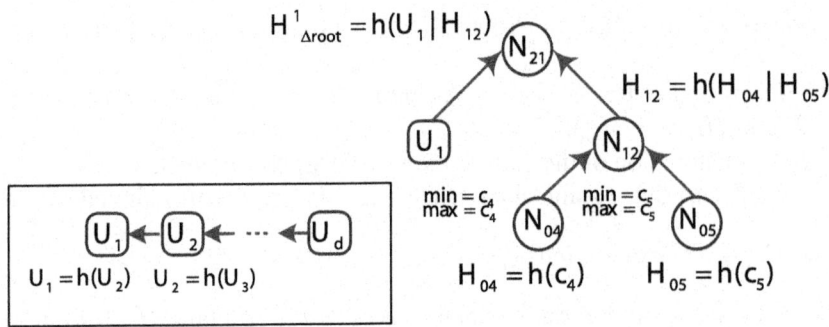

Fig. 4. Sample BECSI Δ-tree.

3.4. Operating mode

BECSI consists in four phases. During the first phase of *Bootstraping*, the CA creates the "extended-CRL", that is, a CRL in which a signed extension is appended. This extension will allow non-trusted third parties (non-TTP) to answer CSI requests in an off-line way when required. Once this extended-CRL has been constructed, it is distributed to the RSUs. In the second phase of *Repository Creation*, a non-trusted entity (i.e. a RSU or a vehicle) gets the extended-CRL and becomes a CSI repository for other VANET entities. Next, during the third phase *CSI Update*, the CA creates "extended-delta-CRLs" of fixed size. A delta-CRL is a time-stamped digitally signed revocation list containing information about new revocations that occurred since the issuance of a prior base-CRL.[1] To construct these fixed-size delta-CRLs, the CA has to wait to have enough new revoked certificates. Therefore, the issuance of these delta-CRLs is aperiodic. Once, the delta-CRL is constructed, the CA appends a signed extension corresponding to the root node of the Δ-tree. The extended-delta-CRLs are distributed to the RSUs and mobile repositories. Moreover, the CA broadcasts the number of issued extended-delta-CRLs to avoid CSI suppression attacks. Finally, in the fourth stage of *Certificate Status Checking*, vehicles can use an efficient protocol to obtain the CSI from any available VANET repository. Henceforward, we give a more detailed description of these three stages.

3.4.1. Bootstraping

In this first phase, the CA creates the *extended-CRL* and delivers it to the RSUs. An *extended-CRL* is basically a standard CRL with an appended extension. This extension can be used by non-TTP (e.g. RSUs and vehicles inside the VANET) to act as repositories and answer to CSI requests. All the tasks of this system initialization are performed in the CA locally.

These are the steps that the CA must carry out:

1. The CA creates a *tbs-CRL* (to be signed CRL), i.e., a list including the serial number of the certificates that have been revoked (along with the date of revocation), the identity of the CA, some time-stamps to establish the validity period, etc.

2. The CA creates the BECSI *base*-tree, i.e., a MHT constructed with the serial numbers within the previous *tbs-CRL* as leaves of the tree. The BECSI *base*-tree is a binary tree, and is constructed following the methodology explained in Section 3.3. The leaves of the *base*-tree are ordered by

---

[1] A base-CRL is a complete CRL that contains a complete list of revoked certificates, to which the revocation list in the delta-CRL needs to be applied to produce the latest list of revoked certificates. Base and delta CRLs have similar data structures.

increasing serial number. Therefore, the bottom left leaf stores the revoked certificate with lowest serial number. Note that if the BECSI base-tree is formed by an odd number n of leaves, there is a leaf $N_{0,n-1}$ that does not have a pair. Then the single node is simply carried forward to the upper level by hashing its $H_{0,n-1}$ value. We proceed in the same way if any i-th level is formed by an odd number n of nodes. Once created the MHT, the CA obtains the root hash.

3. The CA calculates the extension, which consists basically of the $\mathcal{D}$igest and the last value U_0 of a hash chain. The hash chain will be used to make users aware of the number of issued delta-CRLs. Recall that the $\mathcal{D}$igest is calculated as the concatenation of the CA distinguished number, the root hash of the base-tree and the validity period of the CSI, and after that signed by the CA. Obviously, the distinguished number and the validity period should be the same than the ones contained in the tbs-CRL. In fact, the BECSI base-tree is just a different way of representing the CSI, but the hash tree will be valid during the same time and will provide the same information than the CRL. Once calculated, this $\mathcal{D}$igest is appended to the tbs-CRL.

4. The CA creates the hash chain. To that end, the CA picks a random value for U_d. By hashing this value iteratively, the CA forms a one-way chain of self-authenticating values, and assigns the values sequentially to the time intervals (one value per delta-CRL). The last value of the chain U_0 is appended to the tbs-CRL along with the $\mathcal{D}$igest, generating the *tbs-extended-CRL*.

5. The CA signs the tbs-extended-CRL, generating the *extended-CRL*. Notice that this second overall signature not only authenticates all the CSI, but also binds this CSI to the $\mathcal{D}$igest. The *extended-CRL* is only slightly larger than the standard CRL, as we will show later in Section 4.

6. Finally, the CA distributes copies of the *extended-CRL* to the designated RSUs, which will act as the typical PKI repositories, in the same manner as they would do with a standard CRL.

After this first phase, the RSUs have a copy of the *extended-CRL*, which contains exactly the same CSI than a standard CRL and it is valid for the same time. The advantage of an *extended-CRL* is that any non-TTP in possession of it can generate again the BECSI base-tree locally, and obtain the root hash. As the *extended-CRL* also includes the $\mathcal{D}$igest, which is signed by the CA, this entity has an authenticated version of the BECSI base-tree and can answer to CSI requests in an off-line way.

3.4.2. CSI repositories creation

In this phase, RSUs and freewill vehicles become new CSI repositories of the VANET. Vehicles that become mobile repositories allow the distribution of CSI in areas with poor coverage. To become a repository an entity must follow the following steps:

1. The entity obtains the *extended-CRL* (and *extended-delta-CRLs*) either from the CA or from another entity that has an up-to-date copy of the *extended-CRL* (and *extended-delta-CRLs*) in its cache. Notice that the CA uses a secure wireline to communicate with the RSUs, while the RSUs use a wireless link to communicate with the vehicles.

2. Once the *extended-CRL* (and *extended-delta-CRLs*) has been downloaded, the entity verifies that the signature of the *extended-CRL* (and *extended-delta-CRLs*) is valid and corresponds to the CA. If so, the entity generates locally the BECSI base-tree (and Δ-trees) using the serial numbers within the *extended-CRL* (and *extended-delta-CRLs*) and following the same algorithm than the CA (as explained in Section 3.3). The root hash of the tree created from the *extended-CRL* (and *extended-delta-CRLs*) entries must match the signed root value contained in the $\mathcal{D}$igest$_{base}$ (and $\mathcal{D}$igest$_{\Delta_i}$).

3. At this moment, the entity can respond to any status checking request from any vehicle until the corresponding $\mathcal{D}$igest expires.

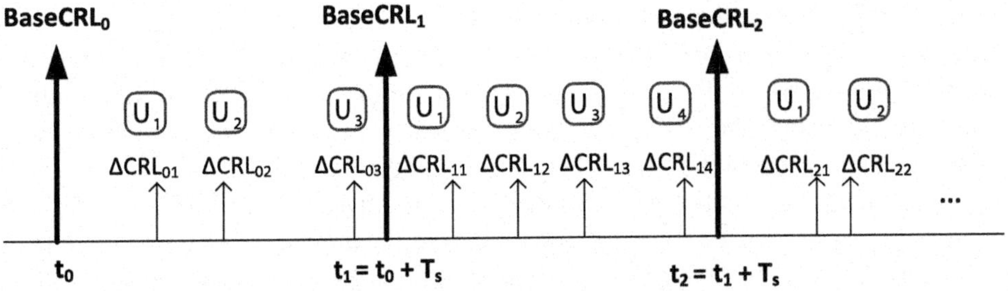

Fig. 5. Delta-CRLs issuance scheduling.

Fig. 6. CCH/SCH timing.

3.4.3. CSI update

After the first two phases, any entity of the VANET is capable of downloading the CRL from a repository or it can just check the status of a given certificate using the capabilities of a MHT. However, in order to improve the freshness of the revocation information and avoid potential bottlenecks when obtaining new CSI, BECSI also provides CSI updates during the validity interval of the CRL.

To alleviate high CRL distribution costs, BECSI uses a hybrid delta-CRL scheme. BECSI issues a variable number of delta-CRLs during the validity interval of the base CRLs (as shown in Fig. 5), reducing the total bandwidth load on the CRL distribution points (RSU and mobile repositories). The size of these delta-CRLs is fixed a priori by the CA. Consequently, the number of delta-CRLs issued during the validity interval of the base-CRL depends on the number of revoked certificates during this interval.

To ensure that any vehicle entity is aware of the number of issued delta-CRLs, the CA discloses a value U_i of a hash chain each time a new delta-CRL is issued. This hash value allows users to make sure of how many Δ-trees have been published by the CA. Thus a non-TTP cannot lie about the amount of revocation information that has been published. Note that the corresponding value U_i is used to calculate the root value of the Δ-tree, binding the Δ-tree to the U_i. BECSI takes advantage of the physical layer used in VANETs to transmit the hash value U_i to vehicles.

The physical layer in VANETs is based on the Dedicated short range communication (DSRC) protocol [14]. DSRC is a 75 MHz band in the 5.9 GHz frequency range with seven non-overlapping channels. Two different channel types are described for use in DSRC. The first type is the control channel, referred to as CCH, which is a single channel reserved for short, high-priority application and system control messages [13]. During the CCH, every node broadcasts a beacon that provides trajectory and other information about the vehicle. The other type of channel is the service channel, or SCH, which has six different 10 MHz channels that support a wider range of applications and data transfer. During CCH time channel activities on SCH are suspended and vice versa. BECSI uses the CCH to transmit the corresponding U_i. Each node in the VANET monitors the CCH during time periods designated as control channel intervals. The time period for an entire CCH Interval and SCH Interval is called a Sync Interval (see Fig. 6). Between CCH intervals, nodes may switch to participate on a SCH for applications such as file downloads [21].

Each regional CA sends to RSUs an authenticated message M containing the corresponding U_i and a time-stamp.

$$CA \rightarrow RSUs : M = [U_i, \text{Time Stamp}]_{Sign_{CA}}$$

Note that regional CAs are expected to have a wireline to communicate with their corresponding RSUs. The time stamp included in the message allows vehicles to verify the freshness of the message. Thus, it is avoided potential forgery or replay attacks. The size of this message is 72 bytes:

- 64 bytes for the ECDSA-256 CA's signature.
- 4 bytes for the timestamp representing seconds UTC since the epoch ('1970-01-01 00:00:00' UTC).
- 4 bytes for representing the U_i value.

During the CCH interval, RSUs broadcast this message to OBUs in range. However, not in every CCH interval M is sent. Depending on the certificate revocation rate, each regional CA will choose the rate at which they have to transmit the U_i to the vehicles. Normally, vehicles will remain under the coverage of an RSU for more than 100 ms. Therefore, CAs have to adjust the frequency at which M is sent to avoid vehicles receiving multiple copies of the same message. Notice that as M is signed by the CA, any vehicle can act as repository and transmit this message without being able to modify it. The hash chain is initialized with secret nonce that the CA generates (U_d) and includes in the *extended*-CRL. By hashing U_d, the other U_i nodes of the chain are calculated. As the validity interval of the CRL is finite, the length of the chain is also finite, i.e., U_0 is the last node of the chain that is calculated after hashing d times U_d. Thus each value can be calculated applying a hash function h to the previous value, and the first value of the hash chain is the secret nonce U_d.

$$U_d \xrightarrow{h} U_{d-1} \xrightarrow{h} \ldots \xrightarrow{h} U_i \xrightarrow{h} U_2 \xrightarrow{h} U_1 \xrightarrow{h} U_0$$

On the other hand, BECSI not only issues delta-CRLs aperiodically, it also includes an extension in the delta-CRLs as it does with the base-CRL. Thus, any VANET entity that has cached the extended-delta-CRL can construct the BECSI Δ-tree (as shown in Section 2.2). With this MHT, a non-TTP can respond to any entity requesting the status of a particular certificate. The response can be authenticated by the requesting party by means of the extension.

3.4.4. Certificate status checking

After the third phase, RSUs and some vehicles will be able to act as repositories. The last stage of the mechanism consists in providing the certificate status information to any vehicle that needs to validate the status of a certificate. Under BECSI, vehicles have to option to check the status of a certificate:

- Downloading the standard CRL and the available delta-CRLs from any repository. This option is desirable when the connectivity to the infrastructure is high and the network congestion is low. For instance, in a urban scenario during non-rush hours where the deployed infrastructure should be enough to serve the CSI.
- Requesting the status of a particular certificate to any repository. With this option, the requesting vehicle only gets the status of a single certificate, so the bandwidth load is low. Vehicles should use this option when they need a quick response (e.g. authenticating safety messages), or when the VANET conditions are not good enough to download the whole CRL and delta-CRLs.

Independently of the option, vehicles that need to check the status of a certificate must locate a valid repository. To do so, vehicles use a Service Discovery Protocol (SDP) to find a RSU or a vehicle that is acting as repository. Once the repository has been located, the vehicles can query for the CRL or query for the status of a particular certificate using the following status checking protocol.

The protocol for status information exchange is based on the hash tree structure and it allows checking the integrity of a single *extended-CRL* or *extended-delta-CRL* entry with only some hash material plus the $\mathcal{D}$igest (included in the extension) and the corresponding U_i. On the one hand, this is much more efficient than broadcasting the entire *extended-CRL* and the *extended-delta-CRLs*. On the other hand, the mechanism is fully offline (the only trusted authority is the CA), which is a very good feature because sometimes it may be impossible for vehicles to reach the CA due to lack of coverage.

Hence, a vehicle that needs to check the status of a certificate must follow the next steps:

1. The vehicle uses a service discovery protocol to find either a RSU or a mobile repository inside its coverage range for status checking.
2. The vehicle sends the serial number of the certificate that is going to be verified to the repository. The repository searches the target certificate in the base-tree and the Δ-trees. In the case the certificate is found, the repository sends the $\mathcal{P}$ath, i.e., the hash values of the nodes of the base-tree (or Δ-tree) which are needed to calculate the signed root. To calculate the path, the repository follows a recursive algorithm that starts from the root and goes across the MHT until the target leaf is reached (see Algorithm 1).

Algorithm 1: Algorithm to calculate the $\mathcal{P}$ath of a given certificate.

Input: SN_{target}
Output: $\mathcal{P}$ath
foreach *base-tree and Δ-tree$_i$* **do**
    if $SN_{target} \in \Delta$-tree$_i$ **then**
      | $k = i$
    else
      └ $k = 0$
$N_{ij} = root_k$;
while $N_{ij}.max \neq N_{ij}.min$ **do**
    $i = i - 1$
    $j = 2 \cdot j$
    if $N_{ij}.max < SN_{target}$ **then**
      $\mathcal{P}$ath.add(N_{ij})
      $j = j + 1$
    else
      └ $\mathcal{P}$ath.add$(N_{i,j+1})$
return $\mathcal{P}$ath,k

---

3. The vehicle verifies that the H_{root} (or the H^i_{root}) calculated from the $\mathcal{P}$ath matches the H_{root} (or the H^i_{root}) contained in the $\mathcal{D}$igest$_{base}$ (or the $\mathcal{D}$igest$_{\Delta_i}$).

Notice that as all H_{root} and all $\mathcal{D}$igests are signed by the CA, it is just as impractical to create falsified values of the $\mathcal{P}$ath as it is to break a strong hash function. In case the certificate is not revoked, the repository sends the adjacent leaves to the requested certificate. To this respect, the repository has to prove that a certain certificate (SN_{target}) does not belong to the set of revoked certificates (Φ). To prove that $SN_{target} \notin \Phi$, as the leaves are ordered, it is enough to demonstrate the existence of two leaves, a minor adjacent (SN_{minor}) and a major adjacent (SN_{major}) for the base-tree and each Δ-tree that fulfill:

1. $SN_{major} \in \Phi$.
2. $SN_{minor} \in \Phi$.
3. $SN_{minor} < SN_{target} < SN_{major}$.
4. SN_{minor} and SN_{major} are adjacent nodes.

So in the worst case, where d delta-CRLs have been published, the repository will have to send $2d + 1$ $\mathcal{P}$aths to proof that a certificate is not revoked, i.e, the serial number is not contained neither in the base-CRL nor in the delta-CRLs. Note that in any case, the amount of data necessary to proof that is smaller than the whole CRL. Therefore, checking vehicles have to know exactly the number of published Δ-trees, i.e., to check that a certificate is not revoked it must corroborate that it does not belong to any MHT. In any case, the data that the repository needs to send to a node to perform the status checking can be placed in a single UDP datagram using 802.11p link-layer.

4. Performance evaluation

In this section, we evaluate the efficiency of the proposed status checking protocol and we compare it with other certificates status management protocols designed for VANET. First, we define a set of metrics to compare the performance of revocation schemes. Then, BECSI is evaluated through simulation using NCTUns [31]. NCTUns was chosen for its advanced IEEE 802.11 model library and ability to integrate with any Linux networking tools.

4.1. Comparison criteria

– *Query Cost (Q_{cost})*: This criterion measures the cost of certificate validity checking. The cost represents bandwidth requirement from repositories to vehicles. Therefore, we calculate this cost as the size of a CSI query (s_q) plus the size of its response (s_r):

$$Q_{cost} = s_q + s_r. \tag{4}$$

– *Request Ratio*: This metric captures the amount of requests that the VANET entities perform to update the CSI. If client validation requests arrive independent of each other, an exponential inter-arrival probability density function can be used to derive the request rate (R) for downloaded CRLs as in [6]:

$$R_t = N_{veh}\lambda e^{\lambda t}, \tag{5}$$

where N_{veh} is the total number of vehicles in the VANET and λ is the ratio of certificates validated per day by each vehicle.

– *Window of vulnerability (WOV)*: This criterion captures the risk of operating with cached CSI. It indicates how long the new revocation data might be held by CAs before being distributed to vehicles. In this paper, WOV is measured in number of hours, which is reasonable because typically CRLs are normally updated every day. We estimate the WOV not only taking into account the

Table 1

Comparison of the overhead introduced by BECSI and other certificate validation mechanisms

Mechanism	Request size	Response size	Query Cost
CRL	73 bytes	145 Mbytes	~145 Mbytes
Compressed CRL (Bloom Filter-2% false positives)	73 bytes	10 Mbytes	~10 Mbytes
ADOPT	66 bytes	586 bytes	652 bytes
BECSI tree	73 bytes	725 bytes	778 bytes

CRL Header ($\sim$ 50 bytes)

– Issuer's name: 32 bytes (if X.500 name used)
– CRL issuance time (thisUpdate): 6 bytes
– Next CRL issuance time (nextUpdate): 6 bytes

List of revoked certificates (9 bytes per revoked certificate)

– Serial number : 3 bytes
– Revocation date : 6 bytes
– CRL entry extensions (e.g. revocation reason)

CRL general extensions (e.g. CRL Number)

Signature of CRL issuer (64 bytes for ECDSA-256 bit)

Fig. 7. Key elements of X.509 v2 CRL.

validity interval of issued CSI but also the ratio of unknown revoked certificates during this interval as in [20]. Thus,

$$WOV(t) = \frac{\rho(t - t_0)}{(1 - \rho)T_c + \rho(t - t_0)},$$ (6)

where T_c is the mean certificate lifetime, ρ is the revocation ratio of revoked certificates, and t_0 is the issuing instant of the CSI.

– *Scalability*: This criterion shows how a revocation mechanism scales in large VANETs, measured as the ratio of increased costs (in terms of update and query costs) over increased size of the vehicles (measured in the number of certificates, queries and revoked certificates). If we assume a stable certificate revocation rate and query rate, a larger VANET typically indicates more revoked certificates and queries in unit time.

4.2. Analytical evaluation

In this section, we compare analytically the performance of BECSI to other certificate status validation mechanisms. To that end, we compare BECSI with these mechanisms in terms of aforementioned metrics.

4.2.1. Query cost analysis

First of all, we start estimating the size of a CRL in a vehicular environment.

Figure 7 describes the size of each of the elements that compose a CRL. Note that, in a VANET, the size of the CRL will depend mainly on the number of revoked certificates, so that the size of the CRL header is negligible compared to the total size of the CRL. Let N_{veh} be the total number of vehicles in the region that the CRL needs to cover, ρ the average percentage of certificates revoked, L_f the lifetime of a certificate, and $\bar{s}$ the mean number of pseudonyms of a vehicle. Additionally, let N_{rev} be the number of non-expired certificates that were revoked, i.e., the number of certificates that the CRL contains. According to [30], the probability of a certificate being revoked follows an exponential distribution. Then, the probability of a given certificate to become revoked at any time period of its lifetime $i \in [0, \ldots, L_f]$ can be expressed as:

$$P_{rev}(i) = L_f e^{-i \cdot L_f}.$$

When a certificate is revoked at time period i of its lifetime, it stays in the CRL for $L_f - i$ time periods. Thus, the expected time a revoked certificate stays in the CRL can be estimated as:

$$E(L_f - i) = E(L_f) - E(i) = L_f - \frac{1}{L_f} = \frac{L_f^2 - 1}{L_f} \simeq L_f.$$

Then, we can estimate the mean number of revoked certificates in a CRL as:

$$\overline{N_{rev}} = N_{veh} \cdot \rho \cdot \bar{s} \cdot L_f.$$

Finally, we estimate the size of a CRL in a VANET. As shown in Fig. 7, CRL entries will have varying sizes, but according to 1609.2 standard [13], 14 bytes per entry is a realistic figure, i.e, $s_e = 14$ bytes. The size of the CRL header is negligible compared to the total size of the CRL. According to NIST statistics [4], 10% of the certificates need to be revoked during a year, i.e., $\rho = 0.1$. Recall that in a VANET, each vehicle owes not only an identity certificate, but also several pseudonyms. The number of pseudonyms may vary depending on the degree of privacy and anonymity that it must be guaranteed. According to Raya, Papadimitratos, and Hubaux in [26] the OBU must store enough pseudonyms to change pseudonyms about every minute while driving. This equates to about 43,800 pseudonyms per year for an average of two hours of driving per day [21]. Haas, Hu, and Laberteaux in [11] recommend changing pseudonyms every 10 minutes, and driving 15 hours per week. This equates to 4,660 pseudonyms per year, but they recommend storing five years of pseudonyms for a total of about 25,000 pseudonyms per OBU. Therefore, we set $\bar{s} = 25,000$. Regarding to the certificate lifetime, according to [30], it ranges from 26 to 37 days. In this manner, we set the lifetime to 1 month. Therefore, the expected CRL size is:

$$CRL_{size} = \overline{N_{rev}} \cdot s_e = N_{veh} \cdot \rho \cdot \bar{s} \cdot L_f \cdot s_e$$

Assuming that a regional certification authority manages a very short population of around 50,000 vehicles, the expected CRL size is $CRL_{size} \simeq 145$ Mbytes.

On the other hand, the response size of BECSI (when using the MHTs to generate authenticated responses) is much smaller than a CRL as it consists only of the $\mathcal{D}$igest and the $\mathcal{P}$ath for a given certificate. Using the SHA-1 algorithm (hash size of 160 bits), and ECDSA-256 the size of the response of BECSI for 10,000,000 revoked certificates (including pseudonyms) is of approximately 725 bytes.

In terms of the total overhead introduced to the network, Table 1 shows the *Query Cost* for current proposed certificate validation mechanisms. Note that the request size is very similar for all the mechanisms. However, the size of the response varies significantly, e.g., BECSI and ADOPT response sizes are six orders of magnitude smaller than conventional CRL. Figure 8 shows the size of the response for CRL, Compressed CRL, ADOPT [17] and BECSI depending on the number of revoked vehicles in the network. While ADOPT response size is constant, the size of the response when using CRL or a compressed version of the CRL increments with the number of revoked certificates. Notice that, the CRL size grows linearly with the number of revoked certificates, while BECSI response sizes describe a logarithmic growth. Therefore, in terms of Query Cost, BECSI is more efficient than CRL and the compressed CRL. Regarding ADOPT, its response size is slightly smaller than in BECSI, but it lacks of the benefits that BECSI provides to operate during disconnections. ADOPT relies on the fact that any vehicle stores the previously received CSI responses. In vehicular scenarios the number of cached responses could be huge, and therefore, also a huge storage capacity is required in the vehicle. In addition, ADOPT does not guarantee that a vehicle obtains the status of a given certificate when needed. So, ADOPT has smaller responses, but it does not provide as fresh information as BECSI, it forces VANET nodes to store a large amount of CSI data and finally, it makes the network more vulnerable due to the potential unavailability of required CSI.

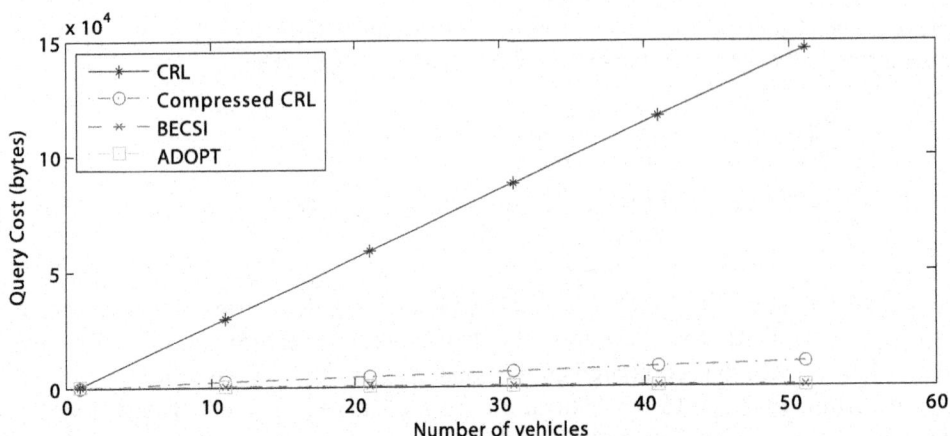

Fig. 8. Response size vs. number of vehicles.

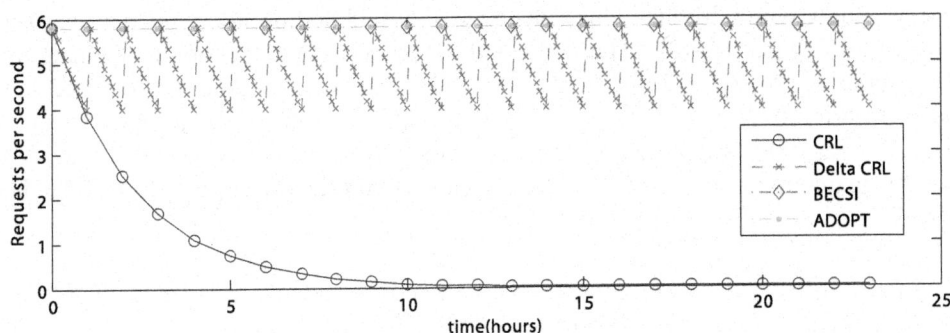

Fig. 9. Request rate for different revocation mechanisms.

4.2.2. *Request ratio analysis*

In the traditional method of certificate revocation, each CRL includes a *nextUpdate* field that specifies the time at which the next CRL will be issued. Thus, once a relying party has obtained a CRL in order to perform a validation, it will not need to request any further information from the repository to perform future validations until the time specified in the *nextUpdate* field of the CRL in its cache has been reached. So, during the period of time in which a CRL is valid (i.e., the most current), each relying party will make at most one request to the repository for revocation information. This request will be made the first time after the current CRL is issued that the relying party performs a validation. Thus, the request ratio of the CRL decreases during the validity interval of the CRL following an exponential function (see Fig. 9). Figure 9 shows the request rate for a CRL, issued using the traditional method, over the course of 24 hours. The graph in this figure was drawn assuming that a CRL was issued at time 0 and that no other CRLs were issued during the period of time shown in the graph. It was also assumed that there are 50,000 vehicles each validating an average of 10 certificates per day.

Figure 9 shows also an example of delta-CRLs issued in the traditional manner. In this example, vehicles download base CRLs at most once every 24 hours. Delta-CRLs are then obtained to ensure that validations are based on certificate status information that is at most 1 hour old. Each validation will require access to a delta-CRL and its corresponding base CRL (either downloaded from the repository or generated locally from a delta-CRL and a previous base CRL). So, the request rate for delta-CRLs will be the same as the request rate for full CRLs in a system that does not use delta-CRLs. Base CRLs, on the other hand, will be downloaded less frequently.

Finally, regarding the cases of BECSI and ADOPT, the request rate is almost constant, i.e., every time a vehicle needs to check the status of a particular certificate they must query a repository. Note that this rate decreases with time, as vehicles also have the ability to store previously queried CSI. However, as the number of valid certificates is so large in VANETs, this decrement is imperceptible.

4.2.3. WOV analysis

The window of vulnerability (WOV) affects update and query bandwidth requirements and/or repositories processing loads directly, while these two factors are two major features determining scalability of CSI issuing mechanisms. WOV presents a direct tradeoff between the security/ timeliness and system scalability. No window of vulnerability means high security and is thus desirable; however, it requires either timely certificate status update from CAs that can force a high update cost and incur security risk.

The traditional way of issuing CRL is the worst mechanism in terms of WOV. During the whole validity of the CRL, vehicles are unaware of new revoked certificates. Therefore, the WOV will increase during the validity of the CRL as there will be more unknown revoked certificates as times goes by. Figure 10 shows the WOV for a CRL issued periodically each 24 hours, and with a constant revocation rate $\rho = 0.1$. Note that the revocation rate determines the slope of the function, i.e., higher revocation rate will give higher WOV.

Compressed CRLs have the same WOV that the standard CRL as they are just a compressed version of the CRL issued with the same lifetime. In the same way, ADOPT also has the same WOV that a CRL. ADOPT presents a distributed mechanism that takes advantages of V2V communications to issue cached CSI. However, the CSI source of this cached revocation information is a CRL. Therefore, the validity period of the cached information in ADOPT is the same that the validity period of the source CRL, i.e., the WOV is the same.

Figure 10 also shows the WOV for BECSI and delta-CRLs. Note that both mechanisms improve the WOV. Traditional delta-CRLs reduce the WOV as they are issued periodically during the validity of the base-CRL. Therefore, the interval of vulnerability of the base-CRL is reduced as many times as delta-CRLs are issued during the lifetime of the base-CRL. In the example shown in the figure, for each base-CRL issued each day, a delta-CRL is issued each 2.4 hours. Thus, 10 delta-CRLs are issued during the validity interval of the base-CRL, reducing the WOV ten times.

In the same way, BECSI also uses delta-CRLs to construct the tree structure. Therefore, the tree structures used in such scheme reduces not only update or query costs, but also the WOV. Recall that BECSI does not issue delta-CRLs periodically, but these are issued with a fixed size. In this sense, despite that fact that the WOV could be higher than with the traditional delta-CRL issuing mechanism, the maximum WOV is always constant. Thus CAs can manage the WOV by selecting the size of the delta-CRLs. In the example shown in the Fig. 10, the number of delta-CRLs issued during the validity interval of the base-CRL is reduced compared to the traditional delta-CRL issuing mechanisms. Note that in this example, BECSI's WOV is never higher than 0.0005.

4.2.4. Scalability analysis

When the vehicular population is large, CRLs tend to become large imposing high bandwidth costs on the CRL distribution points. Hence traditional CRL-based schemes do not scale well. If clients have limited bandwidth capability as is the case of the 802.11p, downloading large CRLs will be user-unfriendly.

With traditional delta-CRLs, the base-CRLs are issued less frequently (as shown in Fig. 5), this reducing the total bandwidth load on the CRL distribution points. However, that use of the traditional

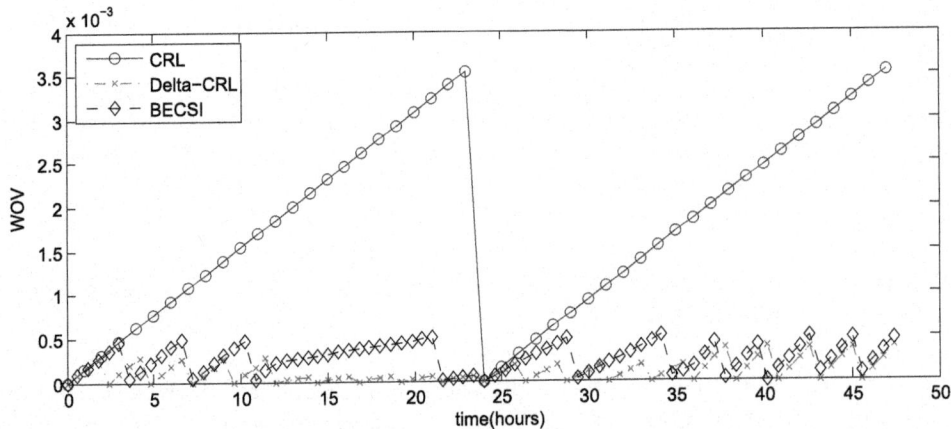

Fig. 10. WOV for different revocation mechanisms.

delta-CRL does not lead to a significant reduction in bandwidth as one would expect. If delta-CRLs are issued very frequently, there is no advantage in using traditional delta-CRLs. Therefore, although the scalability improves compared to simple CRLs mechanisms, traditional delta-CRLs scalability depends on the issuing periodicity of the delta-CRLs.

BECSI takes advantage of the delta-CRLs and optimize the issuing interval so that delta-CRLs remain constant in size. With BECSI, delta-CRLs always have the same size, but they are issued aperiodically. Thus, BECSI becomes more scalable than traditional delta-CRL where depending on the revocation rate the issuing period of the delta-CRLs could be bandwidth-inefficient. Moreover, BECSI also takes advantage of the capabilities of the V2V communication, allowing any vehicle in the network to become a mobile repository. In this sense, BECSI (as ADOPT), multiplies the number of potential repositories, and, therefore, its scalability is also increased.

4.3. Simulation

In the previous section, we have seen analytically that BECSI mechanisms outperform CRL in terms of Query Cost, WOV and scalability. Moreover, BECSI also improves other revocation mechanisms such as ADOPT when analyzing the availability of fresh CSI. In this section, we evaluate the proposed mechanism in a VANET scenario taking into account the specific characteristics of these networks. Using the simulator NCTUns [31], BECSI is evaluated.

The reference scenario is shown in Fig. 11. This scenario consists of 4 two-lane roads forming a 1000 × 500 m rectangle. Three RSUs are placed every 300 meters. Note that there are some areas of the highway that are not covered.

Table 2 summarizes the values of the configuration parameters used in the reference scenario. Note that we have configured our simulation to use the Nakagami propagation model. We choose this propagation model because empirical research studies have shown that a fading radio propagation model, such as the Nakagami model, is best for simulation of a vehicular environment [29].

Using this scenario as reference, firstly, we compare the CSI validation delay of the BECSI scheme with that of the classical PKI [13] under a well-deployed VANET. In the conducted simulation, we consider the cryptography delay only due to hashing operations and point multiplication operations on an elliptic curve, as they are the most time-consuming operations in the proposed protocol. Let T_{hash}

Table 2
Parameter values for the reference scenario

Parameter	Value
Area	1000×500 m
Number of RSUs	3
Number of OBUs	100
RSU Transmission range	300 m
MAC	IEEE 802.11p
Propagation model	Nakagami
Number of caching nodes	20
Maximum speed	120 km/h

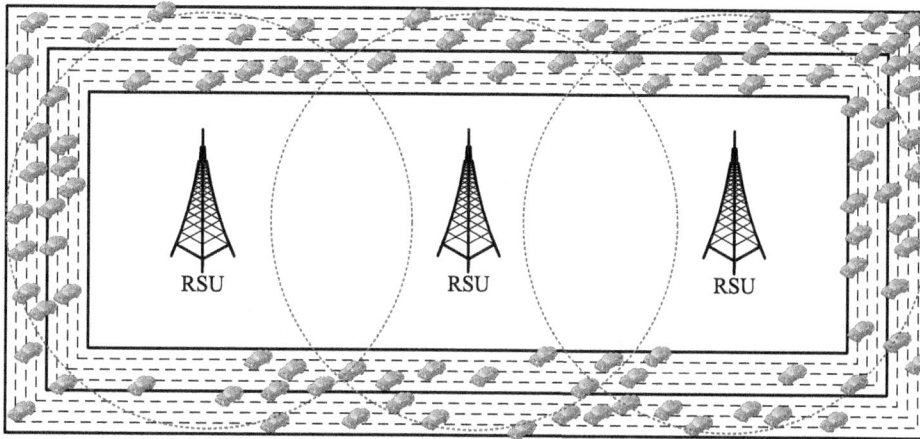

Fig. 11. Simulation scenario.

and T_{mul} denote the time required to perform a pairing operation and a point multiplication, respectively, respectively. Elliptic curve digital signature algorithm is the digital signature method chosen by the VANET standard IEEE1609.2, where a certificate and signature verification takes $4T_{mul}$, and a signature generation takes T_{mul}. To verify a credential in the basic scheme described in Section 3.4.4, a vehicles must perform a hash operation to compute the current contents of leaf node in the BECSI-tree corresponding to SN_i. Finally, it performs $logN$ hash operations to compute the root of the BECSI-tree using the $\mathcal{P}$ath. Therefore, the total computation overhead when checking the status of a certificate is $T_{hash}(logN + 1) + 4T_{mul}$. In [36], T_{mul} are found for an MNT curve with embedding degree $k = 6$ that is equal to 0.6 ms. In our simulation, we use an Intel Core i7 950 (at 3.07 GHz) which is able to perform 1015952 SHA-1 Hashes per second, i.e, $T_{hash} = 0,98\mu$s. Therfore the expected time to check the validity of a $\mathcal{P}$ath in BECSI with is 2.4 ms.

In VANETs, the most important issue in any revocation method is the delay of delivering the CSI to the vehicles to prevent that misbehaving vehicles from jeopardizing the safety of its neighbors. Consequently, we measure the revocation delay as delay from the moment a vehicle issues a CSI request until the moment the new CSI is received. Table 3 shows the average time spent by a vehicle to retrieve CSI from a repository.

It is worth noting that the worst mechanisms in terms of delay are the traditional CRL and delta-CRL as requesting entities are downloading all the available CSI. However, the delay of the conventional CRL compared with the proposed BECSI protocol decreases with the number of CSI requests. The variations in time to download the CRL are due to the number of intermediate RSUs existing in the connection

Table 3
Time required to retrieve CSI

Revocation mechanism	Average Time	Standard Deviation
CRL (300 KB)	2,23 min	0,51 min
Compressed-CRL (20 KB)	7,01 sec	1,12 sec
Traditional Delta CRL (2.5-15 KB)	4,47 sec	2,12 sec
ADOPT (652 B)	705,06 ms	200,81 ms
BECSI Delta CRL (8 KB)	6,02 sec	0,05 sec
BECSI MHT (778 B-912 B)	483,02 ms	20,31 ms

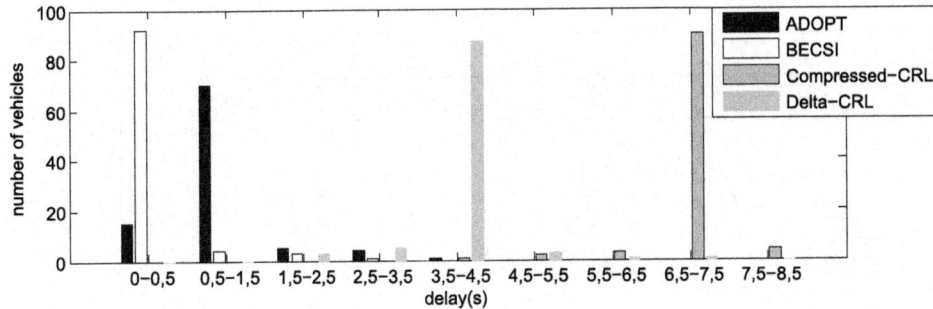

Fig. 12. Histogram plot of the CSI delay depending on the revocation mechanism.

between the CA and the vehicle sending the revocation request. The average time to validate the status of a certificate in ADOPT is lower than BECSI because of the number of hops that are necessary to retrieve the cached CSI. BECSI in its MHT mode of operation is the fastest in average when validating the status of a certificate. However, this mode of operation has a also a notable deviation. While in ADOPT the high deviation is due to the number of hops, in BECSI this deviation is mainly due to the number of Δ-trees that a vehicle has to check when a certificate is not revoked. Note also, that there are also some deviations from the theoretical expected results. This is due to several reasons such as the non-uniform distribution of the mobile repositories, the distance to the repositories or the congestion of the channel. Figure 12 shows the number of vehicles that are able to download the CSI in a particular range time depending on the revocation mechanisms. As expected, with BECSI and ADOPT almost all the 100 vehicles are able to download and process the CSI in less than 1,5 seconds. However, with Delta-CRLs and compressed-CRLs it takes from 4 to 8 seconds to retrieve the CSI.

Finally, we also evaluate the overhead introduced by BECSI. BECSI introduces overhead due to the transmission of the value of the hash chain in the control channel. To evaluate this in the CCH channel, we configure the RSUs to transmit this message every second. As expected, the vehicle is receiving messages from the RSU in range every 100 ms; and every second it receives the message M that involves an increase of the incoming throughput of 72 bytes. In this sense, the overhead introduced by the BECSI mechanism is 4% of the toal capacity of the CCH channel.

5. Conclusions

The revocation service is critical to permit efficient authentication in VANETs. Decentralized approaches based on reputation and voting schemes provide mechanisms for revocation management inside the VANET. However, the local validity of the CSI and the lack of support for extending its validity to the global VANET restrain their utilization in the real scenarios. The IEEE 1609.2 standard suggest the use of CRLs to manage the revocation data. However, the tradicional way of issuing CRLs do not fit well in a VANET where huge number of nodes are involved and where several pseudonym certificates are assigned in addition to vehicle identity certificates.

In this paper, we have presented BECSI, a bandwidth efficient certificate status checking mechanism based on the use of a hybrid delta-CRL scheme and MHTs. BECSI introduces an extension to both base-CRL and delta-CRL allowing any non-TTP to act as repository. The main advantage of this *extended-CRL* and *extended-delta-CRL* is that the road-side units and vehicles can build an efficient structure based on an authenticated hash tree to respond to status checking requests inside the VANET, saving time and bandwidth. Thus, vehicles do not have to download the whole CRL but query for the status of the certificate they need to operate with. Moreover, as *extended-delta-CRLs* have a fixed size, BECSI avoids the traditional problem of optimizing the validity windows of delta-CRLs. Thus, the risk of operating with unknown revoked certificates remains constant during the validity interval of the base-CRL, and CAs have the ability to manage this risk by setting the size of the delta-CRLs.

Analytical and simulation results show that allocating a small bandwidth is enough to ensure that vehicles receive CSI responses within few seconds. The performance improvement is obtained at expenses of adding the signed hash tree extension to the standard-CRL. BECSI evaluation shows that not only improves in terms of bandwidth but also in terms of scalability (increase in the number of available repositories) and vulnerability (controlled WOV). In this way, BECSI becomes an offline certificate status validation mechanism as it does not need trusted responders to operate. Therefore, BECSI significantly achieves great efficiency and scalability, especially when deployed in heterogeneous vehicular networks.

Acknowledgments

This work is funded by the Spanish Ministry of Science and Education under the projects CONSOLIDER-ARES (CSD2007-00004) and TEC2011-26452 "SERVET", FPU grant AP2010-0244, and by the Government of Catalonia under grant 2009 SGR 1362.

References

[1] ITU-T X.509: Information technology – Open Systems Interconnection – The Directory: Public-key and attribute certificate frameworks, 2005.

[2] F. Armknecht, A. Festag, D. Westhoff and K. Zeng, Cross-layer privacy enhancement and non-repudiation in vehicular communication. In *4th Workshop on Mobile Ad-Hoc Networks (WMAN'07)*, 2007.

[3] R. Bera, J. Bera, S. Sil, S. Dogra, N.B. Sinha and D. Mondal, Dedicated short range communications (DSRC) for intelligent transport system. In *Wireless and Optical Communications Networks, 2006 IFIP International Conference on*, 2006, pp. 1–5.

[4] S. Berkovits, S. Chokhani, J. Furlong, J. Geiter and J. Guild, Public key infrastructure study: Final report. Technical report, *MITRE Corporation for NIST*, 1995.

[5] I. Bilogrevic, M. Manshaei, M. Raya and J. Hubaux, Optimal Revocations in Ephemeral Networks: A Game-Theoretic Framework. In *Proceedings of the 8th International Symposium on Modeling and Optimization in Mobile, Ad Hoc, and Wireless Networks (WiOpt 2010)*, IEEE, 2010, pp. 184–193.

[6] D.A. Cooper, A model of certificate revocation. In *Fifteenth Annual Computer Security Applications Conference*, 1999, pp. 256–264.

[7] C. Fan, R. Hsu and C. Tseng, Pairing-based message authentication scheme with privacy protection in vehicular ad hoc networks. In *Proceedings of the International Conference on Mobile Technology, Applications, and Systems*, Mobility '08, 2008, pp. 82:1–82:7.

[8]   J. Forné, J.L. Muñoz, O. Esparza and F. Hinarejos, Certificate status validation in mobile ad hoc networks, *Wireless Commun* **16** (February 2009), 55–62.

[9]   C. Gañán, J.L. Muñoz, O. Esparza, J. Mata-Díaz and J. Alins, Toward revocation data handling efficiency in vanets. In *Proceedings of the 4th international conference on Communication Technologies for Vehicles*, Nets4Cars/ Nets4Trains'12, Berlin, Heidelberg, 2012. Springer-Verlag, pp. 80–90.

[10] C. Gañán, J.L. Muñoz, O. Esparza, J. Mata, J. Hernández-Serrano and J. Alins, COACH: Collaborative certificate status checking mechanism for vanets. *Journal of Network and Computer Applications*, 2012.

[11] J.J. Haas, Y. Hu and K.P. Laberteaux, Design and analysis of a lightweight certificate revocation mechanism for VANET. In *Proceedings of the sixth ACM international workshop on VehiculAr InterNETworking*, VANET '09, New York, NY, USA, 2009, pp. 89–98. ACM.

[12] J.J. Haas, Y. Hu and K.P. Laberteaux, Efficient certificate revocation list organization and distribution. *Selected Areas in Communications, IEEE Journal on* **29**(3) (March 2011), 595–604.

[13] IEEE, IEEE trial-use standard for wireless access in vehicular environments – security services for applications and management messages. *IEEE Std 1609.2-2006*, 2006, pp. 1–117.

[14] D. Jiang and L. Delgrossi, IEEE 802.11p: Towards an International Standard for Wireless Access in Vehicular Environments. In *Vehicular Technology Conference, 2008. VTC Spring 2008. IEEE*, May 2008, pp. 2036–2040.

[15] K.P. Laberteaux, J.J. Haas and Y. Hu, Security certificate revocation list distribution for vanet. In *Proceedings of the fifth ACM international workshop on VehiculAr Inter-NETworking*, VANET '08, 2008, pp. 88–89.

[16] L. Lamport, Password authentication with insecure communication. *Commun ACM* **24** (November 1981), 770–772.

[17] G.F. Marias, K. Papapanagiotou and P. Georgiadis, ADOPT: a distributed ocsp for trust establishment in MANETs. *11th European Wireless Conference 2005*, 2005.

[18] R.C. Merkle, A certified digital signature. In *Advances in Cryptology (CRYPTO89). Lecture Notes in Computer Science*, number 435, Springer-Verlag, 1989, pp. 234–246.

[19] T. Moore, J. Clulow, S. Nagaraja and R. Anderson, New strategies for revocation in ad-hoc networks. In *Proceedings of the 4th European conference on Security and privacy in ad-hoc and sensor networks*, ESAS'07, 2007, pp. 232–246.

[20] J.L. Muñoz, O. Esparza, C. Gañán and J. Parra-Arnau, PKIX certificate status in hybrid manets. In *WISTP*, volume 5746 of *Lecture Notes in Computer Science*, Springer, 2009, pp. 153–166.

[21] M.E. Nowatkowski, Certificate Revocation List Distribution in Vehicular Ad-hoc Networks. PhD dissertation, Georgia Institute of Technology, May 2010.

[22] P. Papadimitratos, L. Buttyan, T. Holczer, E. Schoch, J. Freudiger, M. Raya, Zhendong Ma, F. Kargl, A. Kung and J.-P. Hubaux, Secure vehicular communication systems: design and architecture, *Communications Magazine, IEEE* **46**(11) (November 2008), 100–109.

[23] P. Papadimitratos, L. Buttyan, J.-P. Hubaux, F. Kargl, A. Kung and M. Raya, Architecture for secure and private vehicular communications. In *Telecommunications, 2007. ITST '07. 7th International Conference on ITS*, June 2007, pp. 1–6.

[24] P. Papadimitratos, G. Mezzour and J.P. Hubaux, Certificate revocation list distribution in vehicular communication systems. In *Proceedings of the fifth ACM international workshop on VehiculAr Inter-NETworking*, VANET '08, 2008, pp. 86–87.

[25] M. Raya and J.P. Hubaux, The security of vehicular ad hoc networks. In *Proceedings of the 3rd ACM workshop on Security of ad hoc and sensor networks*, SASN '05, 2005, pp. 11–21.

[26] M. Raya, D. Jungels, P. Papadimitratos, I. Aad and J.-P. Hubaux, Certificate revocation in vehicular networks. Technical report, EPFL, 2006.

[27] M. Raya, M. Manshaei, M. Félegyhazi and J.P. Hubaux, Revocation games in ephemeral networks. In *Proceedings of the 15th ACM conference on Computer and communications security*, CCS '08, 2008, pp. 199–210.

[28] S. Santesson and P. Hallam-Baker, Online Certificate Status Protocol Algorithm Agility. RFC 6277 (Proposed Standard), June 2011.

[29] V. Taliwal, D. Jiang, H. Mangold, C. Chen and R. Sengupta, Empirical determination of channel characteristics for dsrc vehicle-to-vehicle communication. In *Proceedings of the 1st ACM international workshop on Vehicular ad hoc networks*, VANET '04, New York, NY, USA, 2004, pp. 88–88. ACM.

[30] D. Walleck, Y. Li and S. Xu, Empirical analysis of certificate revocation lists. In *Proceeedings of the 22nd annual IFIP WG 11.3 working conference on Data and Applications Security*, 2008, pp. 159–174.

[31] S.Y. Wang and C.L. Chou, Nctuns tool for wireless vehicular communication network researches, *Simulation Practice and Theory* **17** (2009), 1211–1226.

[32] A. Wasef, Y. Jiang and X. Shen, DCS: An Efficient Distributed-Certificate-Service Scheme for Vehicular Networks, *Vehicular Technology, IEEE Transactions on* **59**(2) (Feb 2010), 533–549.

[33] A. Wasef and X. Shen, EMAP: Expedite Message Authentication Protocol for vehicular ad hoc networks, *Mobile Computing, IEEE Transactions on*, (99) (2011), 1.

[34] A. Wasef and X. Shen, EDR: Efficient Decentralized Revocation Protocol for Vehicular Ad Hoc Networks, *Vehicular Technology, IEEE Transactions on* **58**(9) (Nov 2009), 5214–5224.

[35] A. Wasef and X. Shen, MAAC: Message authentication acceleration protocol for vehicular ad hoc networks. In *Global Telecommunications Conference, 2009. GLOBECOM 2009. IEEE*, 30 2009–Dec. 4 2009, pp. 1–6.

[36] C. Zhang, R. Lu, X. Lin, P. Ho and X. Shen, An efficient identity-based batch verification scheme for vehicular sensor networks. In *INFOCOM 2008. The 27th Conference on Computer Communications. IEEE*, April 2008, pp. 246–250.

3

Performance evaluation of acoustic underwater data broadcasting exploiting the bandwidth-distance relationship

P. Nicopolitidis[a,*], K. Christidis[a], G.I. Papadimitriou[a], P.G. Sarigiannidis[b] and
A.S. Pomportsis[a]

[a]*Department of Informatics, Aristotle University of Thessaloniki, Box 888, 54124, Thessaloniki, Greece*
[b]*Department of Engineering Informatics and Telecommunications, University of Western Macedonia,
50100, Kozani, Greece*

Abstract. Despite being a fundamental networking primitive, data broadcasting has so far received little attention in the context of underwater networks. This paper proposes an adaptive push system for data broadcasting in underwater acoustic wireless networks with locality of client demands. The proposed system exploits the characteristic relationship between the bandwidth of an underwater acoustic link and the transmitter-receiver distance in order to improve performance in environments with locality of client demands. Simulation results show superior performance of the proposed approach in the underwater environment compared to existing systems.

Keywords: Underwater networks, acoustic communications, data broadcasting, locality of demand

1. Introduction

The underwater environment poses a number of unique challenges for implementing communication. Due to the fact that radio waves and optical beams cannot easily be applied in underwater, the transmission method of choice for the physical layer of underwater environment is the use of acoustic waves. Underwater acoustic communications entails a number of interesting characteristics, with the most important ones being the Bit Error Rate, the propagation delay and the typical Bandwidth-distance relationship exhibited by such links. Specifically, a) the Bit Error Rate of underwater acoustic links is higher than that of terrestrial radio-based ones, b) the low speed of sound in the water (1.5 Km/sec) gives a propagation delay that is five orders of magnitude larger than that of terrestrial radio-based links of the same length and c) there exists a relationship between the maximum coverage area of an underwater acoustic transmitter and both the centre frequency of the channel available to the transmitter and the channel's bandwidth. As the transmission distance increases, the operating frequency band must be shifted toward lower frequencies and its available bandwidth reduces. Typical values for this bandwidth range from several tens of KHz for areas of a few kilometres to several Hz for areas that span tens of kilometres.

---

*Corresponding author. E-mail: petros@csd.auth.gr.

Despite being a popular networking primitive, data broadcasting [16,18,19,25–30] has so far received little attention in the context of underwater networks with the only relevant work presented in [20, 21,23]. [20] proposes three reliable broadcast protocols based on the capabilities of Forward Error Correction (FEC). The Bandwidth-distance relationship of the underwater acoustic environments allows clients to forward the server's broadcast to clients further away. More concretely, this is accomplished by employing specific frequency bands where the signals are not expected to travel long distances. This way a considerable reduction of the number of transmissions required to complete the broadcast is achieved, which in turn reduces both the overall energy consumption and the total time it takes to complete the broadcast. A quite different approach is presented in [21] which makes use of Fountain codes to enhance the efficiency of the data dissemination process in the face of poor channel conditions. The proposed broadcasting scheme can provide a trade-off between different performance metrics such as delay, advancement per hop and transmission power.

Data broadcasting in the underwater acoustic environment to a number of clients with overlapping demands for data items, has been addressed in [23]. Apart from achieving adaptation of its broadcast schedule according to the a-priori unknown needs of the clients, the proposed system also efficiently combats the problem of high latency of the underwater acoustic wireless environment. Simulation results in [23] show superior performance of the proposed system in the underwater environment compared to adaptations of existing terrestrial push systems.

This paper proposes an adaptive push-based data broadcasting system that extends the above-mentioned method in environments where there is a need for transmission of data items to multiple underwater clients that exhibit locality of demand [27–29]. The latter means that clients are grouped into groups, each one located at a different place with members of each group having similar demands, different from those of clients at other groups. Examples of such scenarios include situations where underwater groups of clients (e.g. submarines or divers or even sensors) are positioned at different locations inside a server's coverage area. These groups demand specific information from a Base Station, which is essentially the Broadcast Server. The terrain map for the group's location for example, is information, which differs according to the location of the group, and even if it is crucial and popular for one group it remains completely useless for another.

The topology of the proposed system comprises groups of clients that exhibit locality of demand for items and a Base Station with acoustic transmission capabilities, all with full-duplex acoustic modems as already proposed in other acoustic underwater protocols [11,17]. The Base Station from its side undertakes the communication with the terrestrial data repository either via a cable or a radio-wireless link. A Learning Automaton [14] is being used by the Base Station in order to estimate the overall transmission probability of each data item, by continuously adapting to the demand pattern of the client population via simple feedback received from the clients. Compared to its predecessor [23], the proposed system increases performance due to exploitation of locality of demand via the bandwidth-distance relationship.

Underwater acoustic wireless networks can support a large variety of applications. The proposed system can provide support for several situations. The Base Station can disseminate data items that contain graphical and contextual information (e.g. map, objects and temperature currents respectively). Some examples are maritime communications, environmental and equipment monitoring (e.g. tsunami and earthquake detection), manned missions such as monitoring of stations or divers who are equipped with underwater GPS, (or another positioning method [3–5,15] and an acoustic navigation system aiming at the mapping and positioning of the diver. Having knowledge of its current position, the equipment (e.g. [13]) of each diver will expect to receive and thus acknowledge only data items that contain useful information regarding its present position.

The context of this paper is organized as follows. Section 2 describes the proposed underwater data broadcasting system for environments with locality of client demands. Simulation results that reveal the superiority of the proposed system are presented in Section 3. Finally, Section 4 concludes this paper.

2. The proposed push system

2.1. The broadcasting algorithm-basics

The broadcasting algorithm operates as follows: The Broadcast Server is equipped with an S-model Linear Reward-Inaction Learning Automaton (LA) that contains the server's estimate p_i of the actual demand probability d_i of the client population for each data item i among the set of the items the server broadcasts. Learning Automata [7,8,14] are mechanisms that can be applied to learn the characteristics of a system's environment. In the area of data networking, Learning Automata have been applied to several problems, including the design of self-adaptive MAC protocols for wired and wireless platforms (e.g. [9,10,24]) and routing (e.g. [1,2]).

Clearly $\sum_{i=1}^{N} p_i = \sum_{i=1}^{N} d_i = 1$, where N is the number of items in the server's database. At each cycle the server will transmit the item i that maximizes the cost function $G(i) = (T - R(i))^2 \frac{p_i}{l_i} \left(\frac{1+E(l_i)}{1-E(l_i)} \right)$ [16]. In the above Equation T is the current time, $R(i)$ the time when item i was last broadcast, l_i is the length of item i and $E(l_i)$ is the probability that an item of length l_i is erroneously received. For items that haven't been previously broadcast $R(i)$ is set to -1. If the maximum value of $G(i)$ is shared by more than one item, the algorithm selects one of them randomly.

The server will probe the data items in order to receive feedback from them so as to estimate the demand probability for each data item. For the feedback transmission, CDMA, which has also been used in the underwater acoustic environment [6,12], is chosen. On the other hand, data item are broadcast via narrowband modulation over the entire available bandwidth for data broadcasting. The number of mobile clients that can be supported in the system is given by the capacity of CDMA, which is given by $\frac{W_{FB}/U_{FB}}{E_b/N_o} + 1$. W_{FB} represents the transmission bandwidth for the feedback, U_{FB} represents the feedback transmission bit rate and E_b/N_o is the bit energy-to-noise power spectral density.

To combat the problem of increased latency of the underwater links, which are underutilized by application of terrestrial stop-and-wait push systems [25], where link latency is insignificant, [23] proposes that a continuous data broadcasting approach is employed. To this end, the length of each client feedback will comprise a $1 + \lceil \log_2 N \rceil$ number of bits so as to code the id of the item is acknowledges. To collect feedback from the clients the server uses an in-band channel over the transmitted data items. This channel is realized by piggybacking two binary numbers, A and B on the of each item i, $1 \leqslant i \leqslant N$. A is a one-bit number, while B comprises a $\lceil \log_2 N \rceil$ number of bits. Two cases can be distinguished: a) the server does not request feedback after the broadcast of item i, it notifies the clients for this by setting $A = 0$ and b) the server wants to collect feedback by the clients that are currently waiting to receive item j, it sets on the header of the broadcast item i, $A = 1$ and $B = $ decimal j in order to probe item j. In order to select which item to probe, the cost function G is separately applied to produce the sequence of probed items, assuming that time elapses only when transmissions of items with probing requests are made. This means that the server needs to store a separate vector R' for keeping the time when each item was last probed by the server.

After the transmission of an item that requests feedback response, the server will continuously broadcast items for a time interval equal to the duration of the feedback transmission. For this interval, clients who successfully received data items, will read the binary flag A = 0 at the item's header and will not respond with a feedback.

When feedback arrives at the server, the latter will read the item id on the incoming feedback and uses it to update the estimation of the corresponding items. The Learning Automaton, which maintains the probability distribution vector p, estimates the demand probability d_i of each information item i. Assuming item j is the server's k^{th} transmission, the following Linear Reward–Inaction (L_{R-I}) probability updating scheme [14] is used whenever a feedback for item j is received at the server:

$$p_m(k + 1) = p_m(k) - L\left(\frac{p_m(k) - a}{ClNum}\right), \forall j \neq m$$
$$p_j(k + 1) = p_j(k) + \left(\frac{L}{ClNum}\right)\sum_{m \neq j}(p_m(k) - a)$$

(1)

where L, a take values in $(0, 1)$ and $p(k)$ takes values in $(a, 1)$. Thus, the probabilities of the items do not change when the probing of an item j does not satisfy any waiting client, whereas following a broadcast that satisfies clients, the probability of item j is increased. To detect the actual number of clients, $ClNum$, the server can broadcast a control item that forces every client to respond with a feedback and then waits for feedbacks for a time period equal to the time needed by a client a the maximum coverage distance from the server to receive the control packet and send its feedback to the server. L defines the speed of convergence. The lower the value of L, the more accurate the estimation made by the automaton, a fact that comes at expense over convergence speed. Parameter a prevents the probabilities of less popular items from taking values close to zero increasing the adaptability of the automaton. Using this scheme, convergence of the item probabilities estimated by the Automaton to the actual overall demand pattern of the client population is achieved [23,25–29].

2.2. Exploiting the bandwidth-distance relationship

As already mentioned above, there exists a fundamental relationship between the distance of the transmitter and the receiver and both the centre frequency of the channel available to the transmitter and the channel's bandwidth. This relationship between bandwidth and distance of an underwater acoustic channel has been extensively examined in [21] and is summarized in Fig. 1 for transmitter-receiver distances that range between 0 and 5 Km. The curve represents the channel's server frequency in KHz while the vertical bars represent the available bandwidth for the respective transmitter-receiver distances.

The above-mentioned relationship can be exploited in order to provide a performance increase to the system. In order to quantify this, we present the following example. Consider a system which does not employ the bandwidth-distance relationship, equipped with a server that acoustically transmits items to underwater clients located inside the server's coverage area of a 5 Km radius. According to Fig. 1, in order to provide coverage to all clients inside the 5 Km range, the server will use a channel with bandwidth about 10 KHz, which we split into 8 KHz for item transmission and 2 KHz for feedback transmission in order to support a certain number of clients according to the capacity equation of CDMA. So, the transmission to each client is made via the same bandwidth and as a result with the same data rate.

To further increase performance, we exploit the bandwidth-distance relationship of underwater acoustic links in order to decrease the item transmission time by transmitting at higher rates to clients located

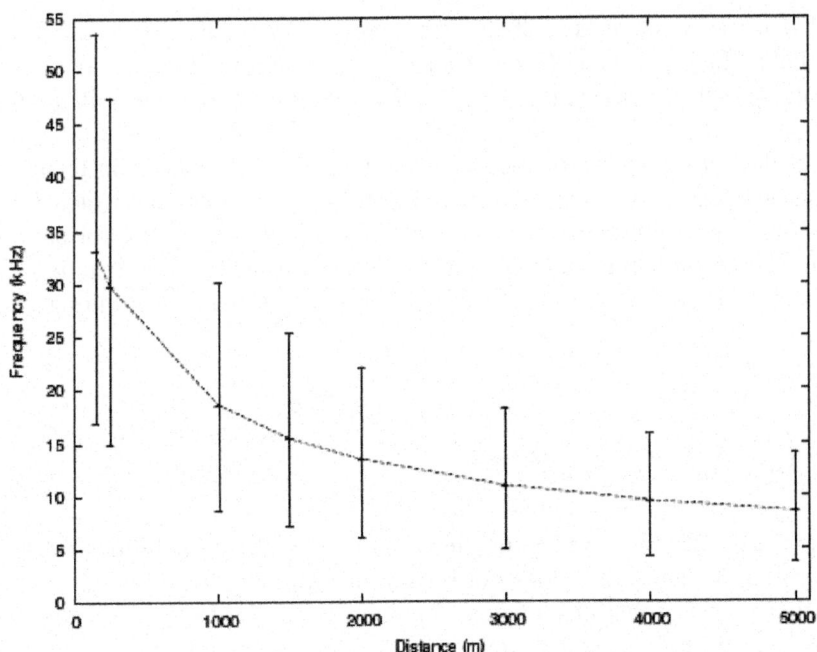

Fig. 1. The effect of distance on the channels center frequency and its available bandwidth (reported in [21]).

closer to the Broadcast Server. Reverting to the previous example and taking into consideration the effect of distance on the available bandwidth (Fig. 1), the server transmission channel for a group of clients located at a distance of 1Km will be set at around 22 KHz. If we again use 2 KHz to support the same number of clients over the feedback channel, the remaining 20 KHz for data item transmissions provide a two-fold increase in the transmission rate. The closer a client is located to the server, the higher the data rate that will be used.

By using the data from Fig. 1 and application of interpolation for distances up to 5 Km and extrapolation for longer distances, we can obtain the corresponding channel bandwidths for several transmitter-receiver distances. Since the server a) possesses knowledge of its position, b) it broadcasts data items that concern the environmental surroundings at certain distances and c) it knows which items contain information that concern specific positions at specific distances, it will be able to broadcast data items to different areas at different channels that are determined from the server's distance to that area. The underwater clients on the other hand can pinpoint their position via their on-board location mechanism [3–5,22] and since the position of the server is known, each client can compute its distance from the server and thus determine the channel (center frequency and bandwidth) over which it will receive data from the server.

3. Performance evaluation

In order to assess the performance increase of the proposed approach, we used an event-driven simulator coded in C++ to compare it to the underwater data broadcasting system of [23] and to the stop-and-wait approach of [25] adapted to the underwater environment. We consider a broadcast server having a database of N equally-sized items. The server is initially unaware of the demand for each item, so in the beginning of simulation process, every item has a probability estimate p_i of $1/N$. In the systems of [23,

Table 1
Group positions uniformly distributed

	Group 1 distance (Km)	Group 2 distance (Km)	Group 3 distance (Km)	Group 4 distance (Km)	Group 5 distance (Km)
MAX_DISTANCE 2 Km	0.4	0.8	1.2	1.6	2.0
MAX_DISTANCE 5 Km	1	2	3	4	5

25], a fixed bit-rate is used, so the server broadcasts all items with the same bit rate. The proposed system with exploitation of the bandwidth-distance relationship is a variable bit-rate system because the server determines the bit rate to use according to the server-client distance. As far as the number of feedback bits are concerned, one-bit feedback is used for the method of [25] while $1 + \lceil \log_2 N \rceil$ bits are needed for the feedback of the proposed approach and that of [23].

We consider $ClNum$ underwater clients that due to the time volatile nature of item contents, have no cache memory – an assumption also made in other similar research both for terrestrial and underwater environments (e.g. [16,23,25–29]). Clients are grouped into G groups each one of which is located at a different distance from the antenna. Any client belonging to group g, $1 \leqslant g \leqslant G$ is interested in the same subset Sec_g of the server's database. All items outside this subset have a zero demand probability at the clients of the group. Finally, all subsets are of the same size and $Sec_i \neq Sec_j, \forall i, j \in [1, \ldots, G], i \neq j$.

To simulate some disagreement in the client demand patterns, we introduce the parameters Dev and $Noise$. A coin toss weighted by Dev, is made for all clients. If the outcome of the toss states that the client will deviate from the initial demand pattern, then a new pattern for this client is created. This pattern is produced as follows: with probability $Noise$ the demand probability of each item in the client's demand pattern database is swapped with another item that is selected in a uniform manner from the interval $[1 \ldots N]$.

The simulation is carried out until at least $Numreq$ requests are satisfied at each client, meaning that overall, at least $Numreq \times ClNum$ requests have been served. The following constant parameter values were used for the simulation: $N = 500$, $G = 5$, $Numreq = 5000$, $L = 0.15$, $a = 10^{-4}$. Dev and $Noise$ parameter when set to non-zero take values of 0.3 and 0.5 respectively. Item size is set to 10^3 bits, $ClNum = 81$ and $E_b/N_0 = 5$. Finally, the speed of sound in the water is set to 1.5 Km/sec.

The Database subsets accessed by the different client groups are chosen as follows: Each of the 5 groups demands the following sections of items respectively: $Sec_1 = [1 \ldots 200]$, $Sec_2 = [201 \ldots 350]$, $Sec_3 = [351 \ldots 400]$, $Sec_4 = [401 \ldots 450]$, $Sec_5 = [451 \ldots 500]$. Each group g is placed at x kilometers from the server, and we assume that all clients of this group have this same distance. Table 1 shows the distance of each of the 5 groups from the Broadcast Server. Assuming that each Sec_i subset comprises Num items, the demand probability $d(i)$ of an item in region i in that subset is calculated according to the following Zipf distribution as $c \left(\frac{1}{i}\right)^\theta$, where $c = 1/\sum_k \left(\frac{1}{k}\right)^\theta, k \in [1 \ldots Num]$. The Zipf distribution, which is used in other relative papers as well [16,23,25–29] can efficiently model applications that are characterized by a certain amount of commonality in client demands. θ is a parameter named access skew coefficient. For $\theta = 0$, the Zipf distribution reduces to a uniform distribution of demand for the items in a subset, while as the value of θ increases, increasingly skewed demand patterns are produced. Finally, we did not take into consideration reception errors as the goal is to present the relative performance of the compared systems.

For the system of [23] we used an available bandwidth for broadcasting and feedback channel implementation of $W_{DATA} = 8$ KHz and $W_{FB} = 2$ KHz respectively. Consequently, the broadcasting rate and the feedback transmission rate were set to $U_{DATA} = 1900$ bps and $U_{FB} = 5$ bps respectively. These values correspond to a distance of 10 Km. For server coverage areas of 5 and 2 Km radius the

Table 2
Parameters of simulation environments

	Clients in Group 1	Clients in Group 2	Clients in Group 3	Clients in Group 4	Clients in Group 5
N1	1...16	17...32	33...48	64...49	65...81
N2	1...32	33...56	57...64	65...72	73...81
N3	1...9	10...17	18...25	26...49	50...81

MAX_DISTANCE 2Km Solid plots: Dev=0.0, Dashed plots: Dev=0.3
Client groups are in positions 0.4, 0.8, 1, 1.6, 2 Km.

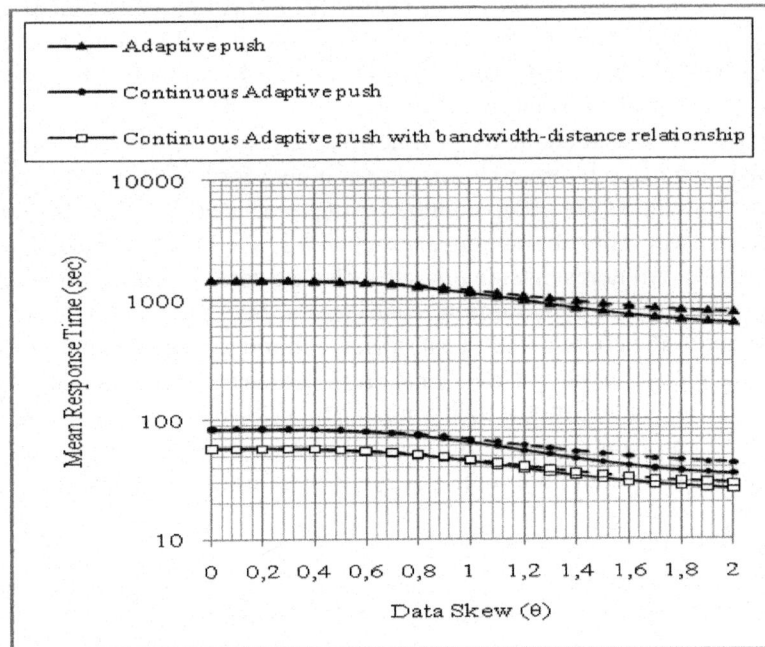

Fig. 2. Mean response time versus data access skew coefficient for the three systems in Network N2 for a server coverage area of 2 Km radius.

server uses W_{DATA} of 9 KHz and 10 KHz respectively, with respective proportional increases in the item transmission speed. The above parameters via Eq. (2) allow for $ClNum = 81$. For the system of [25], which utilizes 1-bit feedback, for the same feedback duration, 81 clients can be supported with $W_{FB} = 200$ Hz and $U_{FB} = 0.5$ bps. Thus for this system $W_{DATA} = 9.8$ KHz and consequently $U_{DATA} = 2352$ bps, with respective proportional increases in the item transmission speed for server coverage areas of 5 and 2 Km radius.

Finally, for the proposed system with exploitation of the bandwidth-distance relationship, the values of W_{DATA}, U_{DATA}, W_{FB} and U_{FB} for broadcasting to a client located at the border of the coverage area are the same with the ones used for [23,25]. However for client groups closer to the Broadcast Server W_{DATA} and U_{DATA} result from Fig. 1 according to the group's distance from the server.

We compared the performances of the three approaches in the following three environments, whose characteristics are shown in Table 2. We can see that in Network N1 client groups equally sized, in Network N2 the bigger groups are positioned closer to the server and in Network N3 the smaller groups are positioned closer to the server.

Figures 2–9 present the simulation results for the proposed approach and those of [23,25]. In these Figures, the maximum coverage area of the server, (marked as $MAX\_DISTANCE$ in the Figures) is

MAX_DISTANCE 5Km Solid plots: Dev=0.0, Dashed plots: Dev=0.3
Client groups are in positions 1, 2, 3, 4, 5 Km.

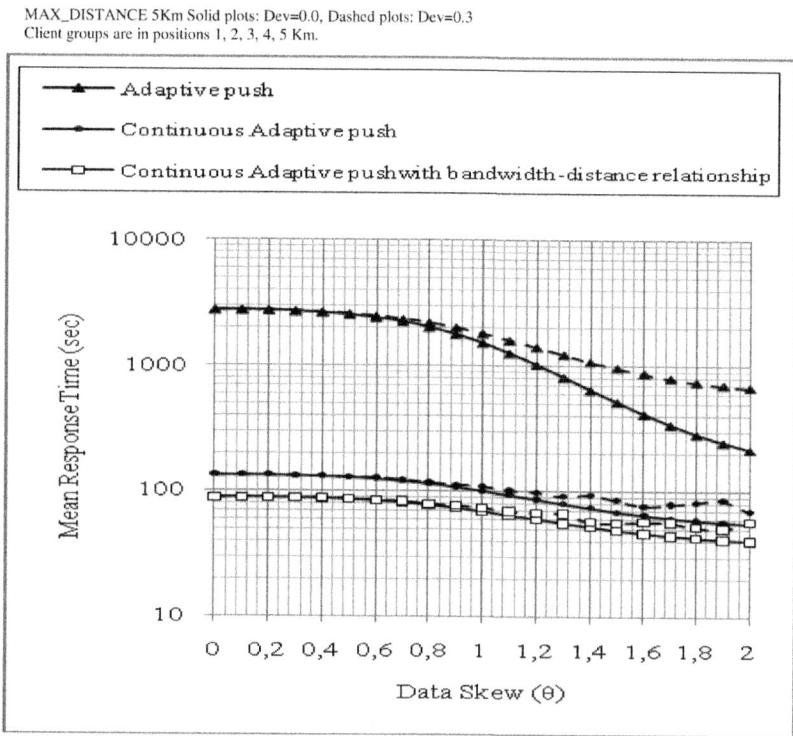

Fig. 3. Mean response time versus data access skew coefficient for the three systems in Network N2 for a server coverage area of 5 Km radius.

MAX_DISTANCE 2Km Solid plots: Dev=0.0, Dashed plots: Dev=0.3
Client groups are in positions 0.4, 0.8, 1, 1.6, 2 Km.

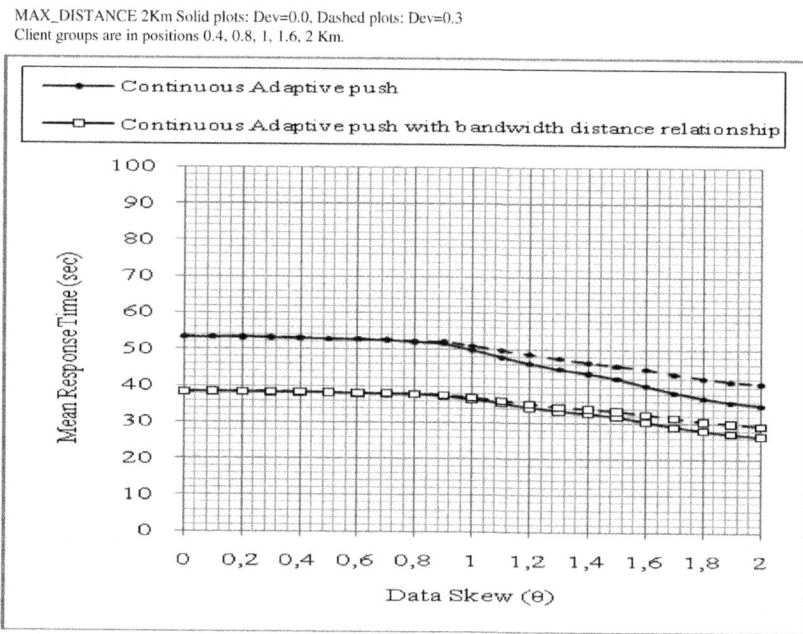

Fig. 4. Mean response time versus data access skew coefficient for the proposed and the continuous adaptive push system of [3] in Network N1 for a server coverage area of 2 Km radius.

MAX_DISTANCE 5Km Solid plots: Dev=0.0, Dashed plots: Dev=0.3
Client groups are in positions 1, 2, 3, 4, 5 Km.

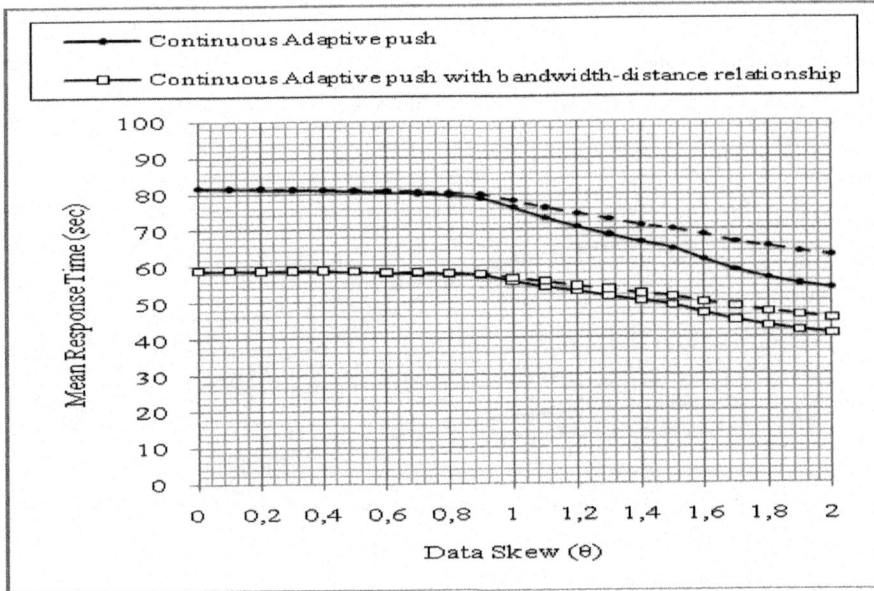

Fig. 5. Mean response time versus data access skew coefficient for the proposed and the continuous adaptive push system of [3] in Network N1 for a server coverage area of 5 Km radius.

MAX_DISTANCE 2Km Solid plots: Dev=0.0, Dashed plots: Dev=0.3
Client groups are in positions 0.4, 0.8, 1, 1.6, 2 Km.

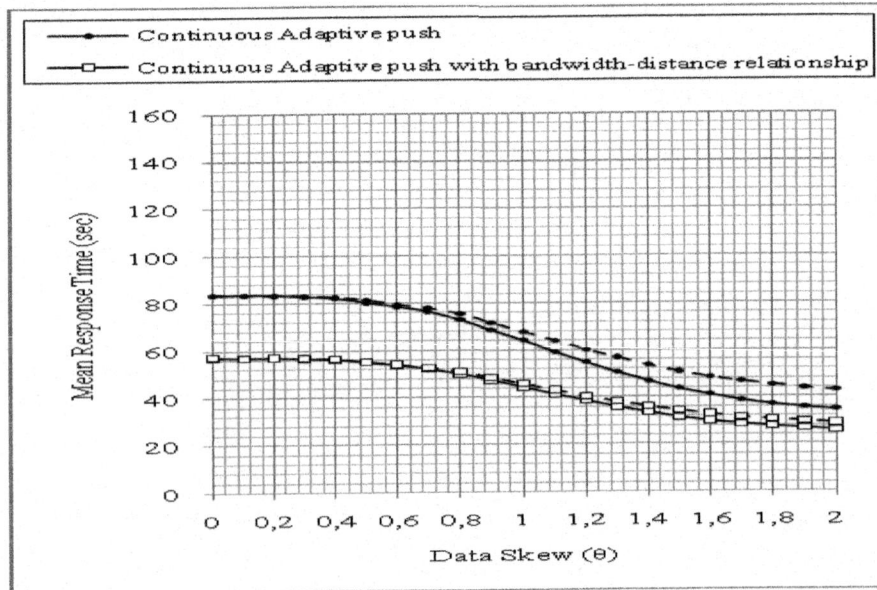

Fig. 6. Mean response time versus data access skew coefficient for the proposed and the continuous adaptive push system of [3] in Network N2 for a server coverage area of 2 Km radius.

MAX_DISTANCE 5Km Solid plots: Dev=0.0, Dashed plots: Dev=0.3
Client groups are in positions 1, 2, 3, 4, 5 Km.

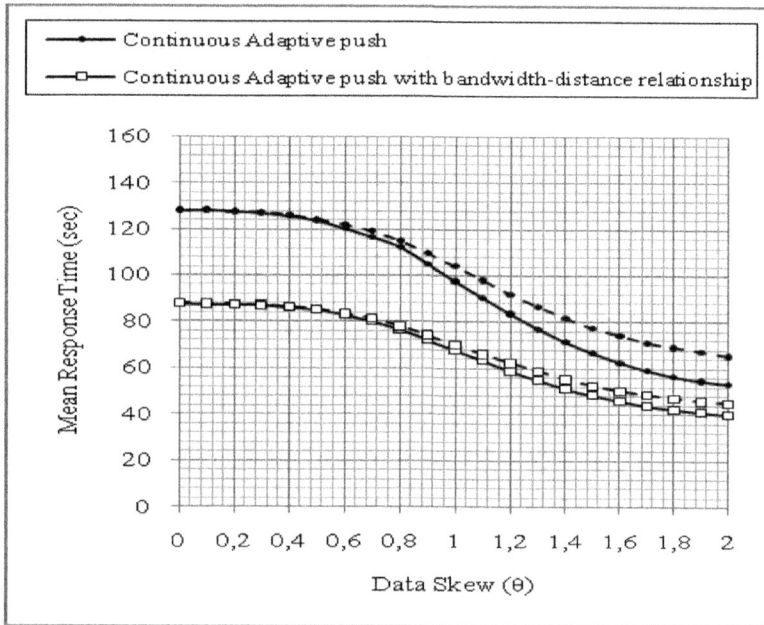

Fig. 7. Mean response time versus data access skew coefficient for the proposed and the continuous adaptive push system of [3] in Network N2 for a server coverage area of 5 Km radius.

MAX_DISTANCE 5Km Solid plots: Dev=0.0, Dashed plots: Dev=0.3
Client groups are in positions 1, 2, 3, 4, 5 Km.

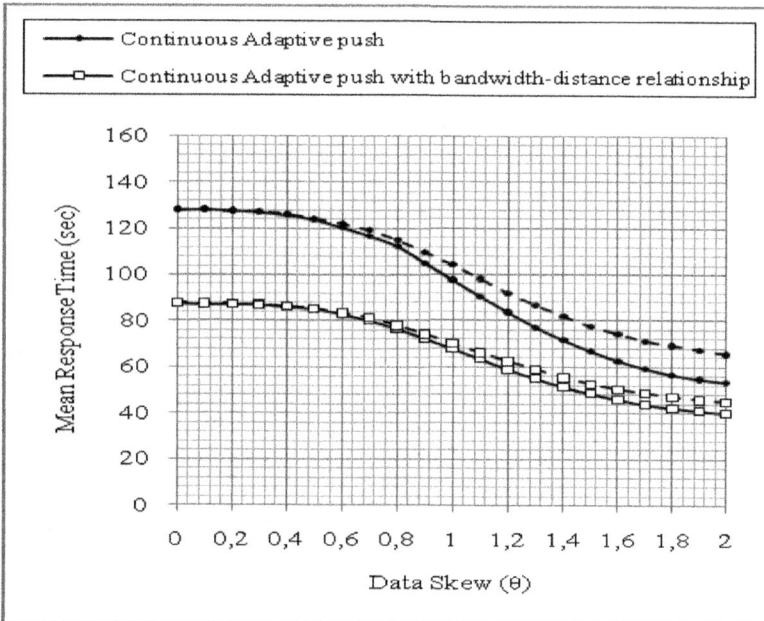

Fig. 8. Mean response time versus data access skew coefficient for the proposed and the continuous adaptive push system of [3] in Network N3 for a server coverage area of 2 Km radius.

MAX_DISTANCE 5Km Solid plots: Dev=0.0, Dashed plots: Dev=0.3
Client groups are in positions 1, 2, 3, 4, 5 Km.

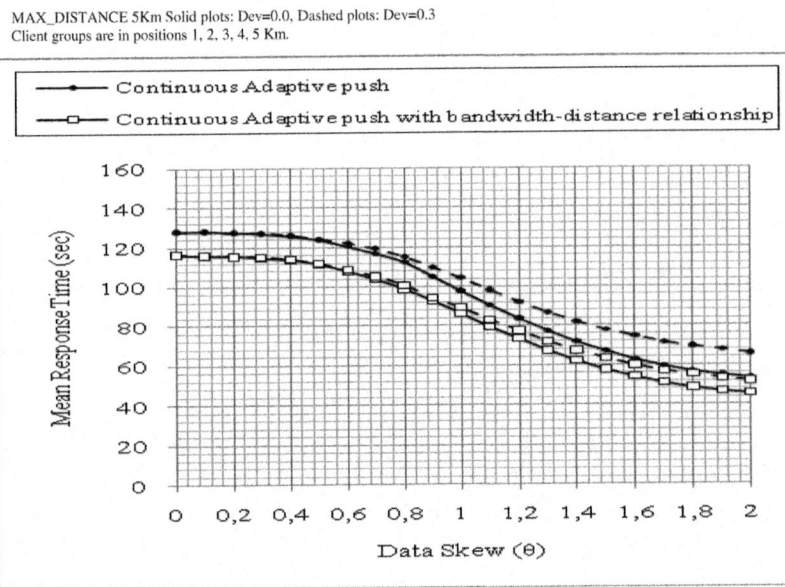

Fig. 9. Mean response time versus data access skew coefficient for the proposed and the continuous adaptive push system of [3] in Network N3 for a server coverage area of 5 Km radius.

2 or 5 Km. Specifically, Figs 2, 3 compare the performances of the proposed approach to that of [23, 25] in Network N2 with a logarithmic (base 10) scale used for the Y-axis. The pairs of Figs 4, 5, 6, 7 and 8, 9 compare the performance of the proposed approach to that of [23] for Networks N1, N2 and N3 respectively. All these Figures plot the performance of the compared systems versus the data access skew coefficient θ. The used metric is the clients mean response time, a performance metric widely used in data broadcasting (e.g. [16,23,25–29]). Client mean response time is the mean time a client waits in order to receive a requested item. In all Figures, the plots termed "Adaptive push", "Continuous adaptive push", "Continuous Adaptive push with bandwidth-distance relationship" correspond to the stop-and-wait approach of [25], to the underwater push system of [23] and the proposed approach with exploitation of the bandwidth-distance relationship respectively. In all Figures, dashed plots correspond to the performance of the systems that make use of disagreement in the client demands with $Dev = 0.3$ and $Noise = 0.5$, whereas the solid ones correspond to the performances of the systems with $Dev = 0$. The main conclusions that can be drawn from the Figures are the following:

- The performance of all schemes improves for increasing values of the data skew parameter. This is expected behavior [16,23,25–29] as the Learning-Automaton adaptation mechanism manages to learn the actual demand probabilities of the various information items and use these values on the selection of the item to broadcast. Moreover the performances of all approaches decrease for $Dev = 0.3$ due to the less commonality in client demands [23,25–28].
- The approach of [25] has performance that is worse by orders of magnitude compared to that of the proposed one. This is because the adaptive push system of [25] waits after each item transmission for the sum of twice the maximum propagation delay plus the item transmission time plus the feedback duration in order to receive a feedback, which obviously explains the huge mean response time rates. Contrarily, the two other methods allow a continuous transmission of data items for feedback collection, which in turn hides the increased transmission latency that burdens the system for the

first item transmission. The same result appeared in simulations for environments N1 and N3 and is the reason that the performance of [25] is omitted from results in Figs 4–9.

– The proposed system that exploits the bandwidth-distance relationship results in a significant performance increase over that of [23]. These gains are larger in environments, where the bigger groups are located closer to the Broadcast Server. However, even in environments like N3, where the biggest groups are furthest away for the Server, the proposed system achieves a significant performance increase via the exploitation of the bandwidth-distance relationship which yields higher data rates for clients located closer to the Server.

4. Conclusion

As the popularity of underwater wireless networks increases, data broadcasting has emerged as an efficient way of disseminating information to underwater clients, with a high degree of commonality in their demand patterns. In many cases, clients are grouped into several groups, each one in a different position, with the members of each group having similar demands. This paper proposes an adaptive push system for data broadcasting in underwater acoustic wireless networks with locality of client demands. It exploits the characteristic relationship between bandwidth of an underwater acoustic link and transmitter-receiver distance in order to provide increased performance. Simulation results show superior performance of the proposed approach in the underwater environment compared existing push systems.

References

[1]   A.A. Economides, Learning Automata Routeing in Connection-oriented Networks, *International Journal of Communication Systems* **8**(4) (July–August 1995), 225–237.

[2]   A.A. Economides, P.A. Ioannou and J.A. Silvester, Decentralized Adaptive Routing for Virtual Circuit Networks Using Stochastic Learning Automata, in: *Proceedings of IEEE INFOCOM 1988*, New Orleans, USA, March 27–31 1988, pp. 613–622.

[3]   A. Mahajian and M. Walworth, 3-D position sensing using the differences in the Timeof-Flights from a wave source to various receivers, *IEEE Transactions on Robotics and Automation* **17** (February 2001), 91–94.

[4]   D. Moore, J. Leonard, D. Rus and S. Teller, Robust distributed network localization with noisy range measurements, in: *Proceedings of Sensys*, November 3–5, 2004, Baltimore, U.S.A., pp. 50–61.

[5]   D.Niculescu and B.Nathi, "Ad hoc positioning system (APS)", *In Proceedings of IEEE GLOBECOM 2001*, San Antonio, U.S.A, November 25–29, 2001, pp.2926–2931.

[6]   D. Pompili, T. Melodia and I.F. Akyildiz, A CDMA-Based Medium Access Control for Underwater Acoustic Sensor Networks, *IEEE Transactions on Wireless Communications* **8**(4) (April 2009), 1899–1909.

[7]   G.I. Papadimitriou, A New Approach to the Design of Reinforcement Schemes for Learning Automata: Stochastic Estimator Learning Algorithms, *IEEE Transactions on Knowledge and Data Engineering* **6**(4) (August 1994), 649–654.

[8]   G.I. Papadimitriou, Hierarchical Discretized Pursuit Nonlinear Learning Automata with Rapid Convergence and High Accuracy, *IEEE Transactions on Knowledge and Data Engineering* **6**(4) (August 1994), 654–659.

[9]   G.I. Papadimitriou and A.S Pomportsis, Learning Automata-Based TDMA protocols for Broadcast Communication Systems with Bursty Traffic, *IEEE Communication Letters* **4**(3) (March 2000), 107–109.

[10]   G.I. Papadimitriou and A.S. Pomportsis, Self-Adaptive TDMA Protocols for WDM Star Networks: A Learning-Automata-Based Approach, *IEEE Photonics Technology Letters* **11**(10) (October 1999), 1322–1324,

[11]   H.-H. Ng, W.-S. Soh and M. Motani, MACA-U: A Media Access Protocol for Underwater Acoustic Networks, in: *Proceedings of IEEE GLOBECOM*, New Orleans, USA, December 2008.

[12]   H.X. Tan and W.K.G. Seah, Distributed CDMA-based MAC Protocol for Underwater Sensor Networks, in: *Proceedings of IEEE LCN 2007*, 15–18 October, 2007 Dublin Ireland, pp. 26–36.

[13]   http://wetpc.com.au/html/technology/wearable.htm.

[14]   K.S. Narendra and M.A.L. Thathachar, Learning Automata: An Introduction, Prentice Hall, 1989.

[15] Kenneth and D. Frampton, Acoustic self-localization in a distributed sensor network, *IEEE Sensors Journal* **6** (2006), 166–172.

[16] N.H. Vaidya and S. Hameed, Scheduling Data Broadcast In Asymmetric Communication Environments, *Wireless Networks* **5**(3), 171–182.

[17] N. Chirdchoo, W.-S. Soh and K.-C. Chua, Aloha-Based MAC Protocols with Collision Avoidance for Underwater Acoustic Networks, in: *Proceedings of IEEE INFOCOM*, Anchorage, USA, May 2007.

[18] S. Acharya, M. Franklin and S. Zdonik, Dissemination-based Data Delivery Using Broadcast Disks, *IEEE Personal Communications* **2**(6) (December 1995), 50–60.

[19] S.S. Manvi, M.S. Kakkasageri and J. Pitt, Multiagent Based Information Dissemination in Vehicular Ad-hoc Networks, *Mobile Information Systems* **5**(4) (2009), 363–389.

[20] P. Casari and A.F. Harris III, Energy-efficient Reliable Broadcast in Underwater Acoustic Networks, *in Proceedings of ACM WUWNet*, Montreal, Canada, September 2007.

[21] P. Casari, M. Rossi and M. Zorzi, Towards Optimal Broadcasting Policies for HARQ based on Fountain Codes in Underwater Networks, *in Proceedings of IEEE/IFIP WONS*, Garmisch-Partenkirchen, Germany, January 2008.

[22] P. Fülöp, S. Imre, S. Szabó and T. Szálka, Accurate Mobility Modeling and Location Prediction Based on Pattern Analysis of Handover Series in Mobile Networks, *Mobile Information Systems* **5**(3) (2009), 255–289.

[23] P. Nicopolitidis, G.I. Papadimitriou and A.S. Pomportsis, Adaptive Data Broadcasting in Underwater Wireless Networks, *IEEE Journal of Oceanic Engineering* **35**(3) (July 2010), 623–634.

[24] P. Nicopolitidis, G.I. Papadimitriou and A.S. Pomportsis, Learning-Automata-Based Polling Protocols for Wireless LANs, *IEEE Transactions on Communications* **51**(3) (March 2003), 453–463.

[25] P. Nicopolitidis, G.I. Papadimitriou and A.S. Pomportsis, Using Learning Automata for Adaptive Push-Based Data Broadcasting in Asymmetric Wireless Environments, *IEEE Transactions on Vehicular Technology* **51**(6) (November 2002), 1652–1660.

[26] P. Nicopolitidis, G.I. Papadimitriou and A.S. Pomportsis, Continuous-flow Wireless Data Broadcasting, *IEEE Transactions on Broadcasting* **55**(2) (June 2009), 260–269.

[27] P. Nicopolitidis, G.I. Papadimitriou and A.S. Pomportsis, Exploiting Locality of Demand to Improve the Performance of Wireless Data Broadcasting, *IEEE Transactions on Vehicular Technology* **55**(4) (July 2006), 1347–1361.

[28] P. Nicopolitidis, G.I. Papadimitriou and A.S. Pomportsis, Multiple Antenna Data Broadcasting for Environments with Locality of Demand, *IEEE Transactions on Vehicular Technology* **56**(5) (September 2007), 2807–2816.

[29] V. Kakali, G.I. Papadimitriou, P. Nicopolitidis and A.S. Pomportsis, A New Class of Wireless Push Systems, *IEEE Transactions on Vehicular Technology* **58**(8) (October 2009), 2529–4539.

[30] Z. Mammeri, F. Morvan, A. Hameurlain and N. Marsit, Location-dependent Query Processing Under Soft Real-time Constraints, *Mobile Information Systems* **5**(3) (2009), 205–232.

4

Sharing with caution: Managing parking spaces in vehicular networks

Thierry Delot[a,c,*], Sergio Ilarri[b], Sylvain Lecomte[a] and Nicolas Cenerario[a]

[a]*University Lille North of France, Valenciennes, France*
[b]*IIS Department, University of Zaragoza, Zaragoza, Spain*
[c]*INRIA Lille Nord Europe, Villeneuve d'Ascq, France*

Abstract. By exchanging events in a vehicular ad hoc network (VANET), drivers can receive interesting information while driving. For example, they can be informed of available parking spaces in their vicinity. A suitable protocol is needed to disseminate the events efficiently within the area where they are relevant. Moreover, in such a competitive context where each vehicle may be interested in a resource, it is crucial not to communicate that resource to each driver in the vicinity. Otherwise, those drivers would waste time trying to reach a parking space and only one of them would be fulfilled, which would lead to a poor satisfaction in the system.

To solve this problem, we detail in this paper a reservation protocol that efficiently allocates parking spaces in vehicular ad hoc networks and avoids the competition among the vehicles. We have integrated our protocol within VESPA, a system that we have designed for vehicles to share information in VANETs. An experimental evaluation is provided, which proves the usefulness and benefits of our reservation protocol in both parking lots and urban scenarios. Besides, we present an in-depth study of the state of the art on this topic, that shows the interest and the originality of our approach.

Keywords: Vehicular networks, data sharing, parking space allocation

1. Introduction

Nowadays, there is a great interest in developing systems to assist drivers on the road, providing them with different types of relevant information. VANETs rely on the use of short-range networks (a few hundred meters), like IEEE 802.11, Ultra Wide Band (UWB), or WAVE (IEEE 802.11p, IEEE 1609), for vehicles to communicate [27] and provide bandwidth in the range of Mbps. Using such communication networks, the driver of a car can receive information from its neighbors. Many pieces of information can be exchanged in the context of inter-vehicle communications, for instance to warn drivers when a potentially dangerous event arises (an accident, an emergency braking, an obstacle on the road, etc.) or to try to assist them (with information about traffic congestions, real-time traffic conditions, etc.). As opposed to communication approaches based on a support infrastructure or a wide-area network such as a 3G cellular network, the use of ad hoc communications can facilitate a quick exchange of data with neighboring nodes at no cost, which will encourage the cooperation among close vehicles. VANETs have attracted an intensive research attention (for some recent examples, see [19,35,44]).

*Corresponding author: Thierry Delot, University Lille North of France, LAMIH UMR UVHC/CNRS and INRIA, Le Mont Houy, 59313 Valenciennes, France. E-mail: Thierry.Delot@univ-valenciennes.fr.

Among the different types of information that a driver may find interesting, information about available parking spaces is one of the most valuable. Finding an available parking space is indeed stressful, time-consuming and contributes to increasing traffic; so, according to [23], searching for parking spaces is a basic component of urban traffic congestion (between 5% and 10% of the traffic in cities and up to 60% in small streets). Besides, it leads to fuel consumption and environment pollution due to the emission of gases. Some studies emphasize the costs of searching for parking spaces. For example [10], indicates that nearly one out of two vehicles on the move are searching for a parking space. Similarly, from a study by the Imperial College in London (Imperial College Urban Energy Systems Project) it is known that, during congested hours, more than 40% of the total fuel consumption is spent while looking for a parking space. As a final example, according to a study by Donald Shoup [41], in Westwood Village (a commercial district next to the campus of the University of California, Los Angeles) parking searching leads every year to about 47000 gallons of gasoline, 730 tons of CO_2 emissions and 95000 hours (eleven years) of drivers' time. Moreover, the Directive 2010/40/EU of the European Parliament sets among the priority areas for Intelligent Transport Systems the provision of both information and reservation services for (safe and secure) parking places for trucks and commercial vehicles.

Vehicular networks bring new opportunities but also significant challenges from the data management point of view [19]. The work presented in this paper is an extension of VESPA [12,16,18,20] (Vehicular Event Sharing with a mobile P2P Architecture, http://www.univ-valenciennes.fr/ROI/SID/tdelot/vespa/), which is a system developed to share information about events in inter-vehicle ad hoc networks. In such environments, both push-based (data dissemination) and pull-based (query dissemination) approaches [46] have been proposed. On the one hand, with a push-based approach the vehicles receive data from their neighbors and then it is possible to process queries locally over the data received in order to provide interesting information to the driver (relevant for that time and location). On the other hand, with a pull-based approach a query is actually disseminated over the vehicular network in order to collect the relevant data from the vehicles. Push-based solutions are more frequently used. So, data about the events occurring on the road or available resources such as parking spaces may have to be communicated to a potentially large set of vehicles, depending on the relevance of the data to the drivers. As opposed to other proposals, VESPA aims at supporting all types of events. Thus, VESPA proposes a dissemination protocol [12] and a relevance estimation mechanism [16,20] not only suitable for stationary events (e.g., an emergency braking, an accident, etc.) but also for mobile events (e.g., an emergency vehicle asking preceding vehicles to yield the right of way, a vehicle with a non-functioning brake light, etc.). When supporting such mobile events, the set of vehicles for which the event information is relevant evolves according to both the movements of the vehicle generating the event (e.g., a vehicle that brakes suddenly) and the other vehicles involved (in the previous example, the vehicles behind). Besides, the direction of traffic is also of major importance in establishing the relevance of shared information, even for non-mobile events (e.g., consider a traffic jam affecting only the vehicles moving in one direction).

In this paper, we focus on the exchange of information about available parking spaces using VESPA. As opposed to other types of events, it is not enough to indicate the presence of the event to the driver. Indeed, if the same information (i.e., the same available parking space) is presented to several interested drivers, this will lead to a competition between the vehicles and only one of them will be able to take that parking space. It is so crucial to propose a solution for parking spaces to be "reserved". By reservation we understand the fact that the parking space should be communicated to only one driver, as there is no way to physically prevent other vehicles from taking a public parking space; so, thanks to the reservation protocol, the information about an available parking space is disclosed to a single driver. The fully decentralized environment imposed by vehicular networks makes that reservation process particularly challenging since vehicles keep moving and no reliable link or central server can be used. Although other solutions have been proposed to disseminate information about available parking spaces using short range communications (e.g., [49]), to the best of our knowledge, no other work has tackled the problem of parking space reservation in VANETs. This paper extends [17]; among the extensions, we could mention an extensive study of related proposals, an experimental evaluation in urban environments, and an analysis of reliability issues.

The structure of the rest of this paper is as follows. In Section 2, we introduce the representation of events in VESPA and describe how the relevance of parking spaces received by vehicles is evaluated using the concept of Encounter Probability. In Section 3, we present our reservation protocol. In Section 4 we discuss some reliability issues. In Section 5, we evaluate experimentally our solution. In Section 6, we compare our work with other approaches by presenting a detailed study of the state of the art (including both research proposals and existing systems). Finally, in Section 7 we summarize our conclusions and indicate some ideas for future research.

2. Relevance of events

One of the major problems in inter-vehicle applications is how to estimate the relevance of the events received. The traditional approach is to define a relevance function used to determine whether an event received should be considered or ignored. The relevance function used in VESPA, called *Encounter Probability* (*EP*), which estimates the likelihood that a vehicle is going to meet a certain event, is used to verify whether an event is relevant for a vehicle or not. Thus, when an event is received by a vehicle, the EP for this event is computed. If the EP computed is higher than a certain *storage threshold* (ST) then the event is stored in a local data cache, as it is considered *relevant for the vehicle*. If the EP is also higher than a certain *relevance threshold* (RT, with RT ⩾ ST) and besides the driver is interested in that type of event (because the event is interesting for any driver – e.g., an event indicating an accident – or because the driver submitted a query to retrieve events of that type), then the event is *relevant for the driver* and has to be communicated to the driver. Finally, a *dissemination threshold* (DT) is also used to decide whether the event should be communicated to other neighboring vehicles or not.

In the following, we first describe how events are represented and then we briefly summarize the method used to compute the *Encounter Probability* (*EP*).

2.1. Representation of events in VESPA

VESPA relies on a generic structure to represent events whatever their type (both *stationary* vs. *mobile* and *direction-dependent* vs. *direction-independent* events are considered). Each event is described using several attributes: 1) a unique *Key* set by the event source, 2) a *Version* number allows to distinguish between different updates of the same event (e.g., used to indicate that a mobile event has changed its location), 3) an *Importance* value helps to determine the urgency of presenting that information to the driver (e.g., accidents that may affect the driver have a high importance), 4) a *CurrentPosition* indicates the time and place for the data generated, 5) a *DirectionRefPosition* and a *MobilityRefPosition* store preceding reference positions used to compute an Encounter Probability, 6) a *LastDiffuserPosition* and a *HopNumber* contain the location of the last vehicle which relayed the message and the number of rediffusions of the event (for purposes of the dissemination protocol, as explained in [12]), and 7) a *Description* field contains additional information for the driver.

Table 1
Example of dissemination of an available parking space

Key	Vrs	Imp	CP	LDP	Hop	Desc
$v1e1$	1	1	50° 19'15.91 N 3°30'51.11 E 10h25m17s	–	0	Available parking space
$v1e1$	1	1	50° 19'15.91 N 3°30'51.11 E 10h25m17s	50° 19'15.91 N 3°30'51.11 E 10h25m17s	1	Available parking space
$v1e1$	1	1	50° 19'15.91 N 3°30'51.11 E 10h25m17s	50° 19'17.51 N 3°30'54.71 E 10h25m18s	2	Available parking space

The type of the event (stationary or mobile, direction-dependent or not) is not explicitly represented as an attribute of the event, as it can be inferred from some of the other message fields. Thus, when dealing with a stationary object/event, the *MobilityRefPosition* will always be set to *null*. Similarly, for non-direction-dependent events the value of *DirectionRefPosition* will be set to *null* to allow the identification of such type of event.

2.1.1. Example: Disseminating the availability of parking spaces

The protocol used in VESPA to disseminate information about available parking spaces and other types of events is described in detail in [12]. It aims at delivering events to potentially interested vehicles and relies on the concept of EP. Indeed, the probability that an event relevant for a vehicle is also relevant for its neighbors is high. So, when the EP computed for the event reaches the dissemination threshold, the dissemination of the event has to be continued in the network. Such an EP-based dissemination protocol ensures the adaptivity of the dissemination of an event according to its type. This is indeed crucial when several types of events are considered (e.g., the information about an available parking space has to be broadcasted to all surrounding vehicles whereas the information about an emergency braking should only be diffused to the vehicles behind).

As an example, we illustrate in Table 1 several steps of the diffusion of an event representing an available parking space using the dissemination protocol that we introduced in [12]. In the table, *Vrs* represents the version number, *Imp* the importance of the event, *CP* the current position, *LDP* the last diffuser's position, and *Desc* the description of the event. The first row represents the message emitted by a vehicle leaving a parking space or by a static sensor detecting the car leaving. None of the fields corresponding to the reference positions are filled here because of the type of event (*DirectionRefPosition* and *MobilityRefPosition* are both *null* for stationary non-direction-dependent events like available parking spaces), and so they are not shown in Table 1. The second row presents the message generated by a vehicle relaying the message presented in the first row of the table; the only attributes modified are the *HopNumber*, increased because the message is relayed, and the *LastDiffuserPosition*, corresponding to the position of the previous sender of the message (i.e., the generator of the event). The third row shows a next rediffusion of the message. For simplicity, only the longitude and latitude (and not the altitude) of GPS positions are shown.

2.2. Computation of the EP

In this section, we briefly summarize the way the EP is computed in VESPA. Actually, there are two different methods considered. The first method does not require any road maps or other information about the environment and estimates future locations of vehicles and events based on geographic vectors, in order to estimate the probability that they will meet [16]. The second method attempts to achieve a higher accuracy by exploiting the information available in real road maps [20]. For more details about the computation of the Encounter Probability, along with an evaluation of its benefits, we refer the interesting reader to [12,16,20].

2.2.1. Computation of the EP based on geographic vectors

In order to compute the Encounter Probability between a vehicle and an event, two movement vectors are defined for a vehicle (computed by sampling the vehicle's locations periodically): the *direction vector* and the *mobility vector*. The *direction vector* allows estimating future positions of the vehicle on a short term, whereas the *mobility vector* captures an overall impression of the vehicle's direction and allows to estimate future positions on the long term. Each vehicle can compute its direction vector and its mobility vector easily. Similarly, each vehicle can compute the *mobility and direction vectors of the events* it receives. For that purpose, it uses the data associated to the events, and more precisely the *CurrentPosition* attribute and either the *DirectionRefPosition* or the *MobilityRefPosition* attribute, respectively. Finally, for each event, a *vehicle's mobility vector (and direction vector) in relation to the event* is computed by the vehicle, so that managing a single mobility and direction vector for each couple <vehicle, event> is enough.

Then, the computation of the EP is based on four factors: the minimal geographical distance between the vehicle and the event over time (Δd), the difference between the current time and the time when the vehicle will be closest to the event (Δt), the difference between the time when the event is generated and the moment when the vehicle will be closest to the event (Δg), and the angle between the direction vector of the vehicle and the direction vector of the event (represented by a colinearity coefficient c). These elements are weighted by considering different *penalty coefficients* (α, β, γ and ζ) and aggregated to compute a value for the EP:

$$EP = \frac{100}{\alpha \times \Delta d + \beta \times \Delta t + \gamma \times \Delta g + \zeta \times c + 1}$$

For more details about the computation of the EP with geographic vectors, please see [16].

2.2.2. Computation of the EP based on maps

The information used in the previous method can be noisy, as for example the vehicle could abruptly change its direction at any time due to the constraints imposed by the road network infrastructure. So, we have proposed a second method to compute the EP when digital road maps are available. The goal is to try to compute the EP in a more precise way. This method is based on the estimation of the *TTL (Time to Live)* of an event, which is the time interval during which the event will be valid, and on the distinction between two different types of events: *attraction events* and *repulsion events*. *Attraction events* represent events that the driver would like to reach (e.g., in this paper parking spaces); for them, a *Reachability Probability (ReachP)* is defined based on the TTL and the time needed to reach the event by following the shortest path between the vehicle and the event (*TTR*). *Repulsion events* are events that the driver would like to avoid (e.g., traffic congestions or accidents); for them, a *Need to Escape Probability (NeedEsP)* is defined based on the TTL and the time needed by the vehicle to reach the last intersection that offers the vehicle an alternative route to avoid the repulsion event (*TTE*). The value of the EP is computed as *ReachP* or *NeedEsP* depending on the type of event:

$$ReachP = \begin{cases} 100 & \text{if } TTL > TTR \\ 0 & \text{otherwise} \end{cases}$$

$$NeedEsP = \begin{cases} 100 & if\ TTL > TTE \\ 0 & otherwise \end{cases}$$

According to experiments performed in the context of VESPA, this method to compute the EP can increase the accuracy. Nevertheless, the approach based on geographic vectors will also behave well in scenarios where the information about road maps may be unavailable. For more details, see [20].

3. Reservation protocol for parking spaces

In Section 2, we have described how to estimate the relevance of events. For most events, these mechanisms can be used efficiently to disseminate the events in the vehicular network to warn drivers. However, disseminating information about an available parking space with the basic VESPA approach is not enough, as this would lead to competition between vehicles to try to take that space. So, in this section, we present a solution to allocate available parking spaces in VANETs.

Coordinating different vehicles in vehicular ad hoc networks for them to choose one vehicle to which the parking space will be allocated is not an easy task. Indeed, no centralized server is available in environments where vehicles only communicate through short range communication networks. Moreover, all the vehicles have the same importance/role. So, we propose in the following a protocol in which a *coordinator* vehicle is chosen for each parking space. The role of such a coordinator is to collect, from the neighbors interested in finding an available parking space, the necessary information to decide to which vehicle the resource will be allocated. In the following, we will call "reservation" the process consisting of allocating parking spaces to vehicles. Our goal is to ensure that the information about a parking space is shown only to the driver of the vehicle that is chosen to occupy it, in order to minimize competition.

3.1. Basics of the protocol

Using vehicle-to-vehicle communications, a vehicle can inform the other surrounding ones when it is about to leave a parking space. Therefore, a message describing the event "available parking space" is generated and broadcasted using VESPA, as explained before. To avoid causing an unnecessary competition among vehicles, a suitable reservation protocol should aim at:

- Indicating to the driver an available parking space reserved for his/her vehicle (and not a list of all the received available parking spaces).
- Maximizing the probability that the parking space will still be available when the driver arrives there.
- Maximizing the use of resources (by avoiding that a parking space remains available if there are vehicles searching in its vicinity).
- Ensuring a fair use of resources (i.e., it should be equitable and avoid situations where a vehicle has consistently higher priority than others).
- Minimizing the actions that the driver has to perform, with two goals: 1) not to disturb her/him while driving, and 2) to prevent the drivers from disseminating themselves misleading information that they could use to their own advantage (i.e., to obtain a parking space before the others).
- Avoiding network congestion.

Our solution to reserve parking spaces relies on the election of one *coordinator* per available parking space (i.e., a vehicle with a temporary particular role in the allocation of an available space). The coordinator is responsible for the allocation of the parking space to a vehicle according to a predefined policy.

It should be noted that if several coordinators were in charge of the same parking space, the information could be disseminated faster, but this would lead to the competition situations that we want to avoid. In this section, we detail how our protocol works for the case where there are vehicles interested in the resource in the communication range of the vehicle diffusing it, and in Section 3.2 we explain how the advertisement range can be extended (if needed) by switching the coordinator.

At first, the coordinator is the vehicle that leaves the parking space.[1] It sends a message to inform the vehicles in its communication range that a parking space is available. Then it waits for potential answers. Among the vehicles receiving the information, only those interested in the parking space answer to the coordinator. So, each vehicle receiving a notification about an available parking space has to verify if that information is relevant. For this purpose, the vehicle computes the Encounter Probability (presented in Section 2) for the event received, to estimate if it can reach the space while it is still estimated available. In that case, it declares its interest in the parking space. Each interested vehicle v_i provides to the coordinator its vehicle's identifier and the information necessary for the coordinator to choose the vehicle to which the resource has to be allocated, such as (depending on the allocation policy) the time t_i, corresponding to the current duration of the search of a parking space for that vehicle.

After a period of time T (amount of time during which the coordinator waits for answers from interested vehicles), the coordinator chooses, among the vehicles that answered, the one for which the parking space is "reserved" (i.e., the one whose driver will receive the information about the parking space, including its geographic coordinates). Different policies may be applied to choose the vehicle to which the parking space should be allocated. The choice of an appropriate allocation policy will be further discussed in Section 5, in the context of a experimental evaluation; summing up, we consider that an appropriate policy could be to assign the parking space to the vehicle with the highest EP (to maximize the probability to find it free when it reaches it). Finally, the vehicle sends a message to the coordinator to confirm that it will take the parking space. This confirmation message avoids losing parking spaces: in case the chosen vehicle does not accept the parking space (or if it does not receive the allocation message), another vehicle is chosen by the coordinator. This exchange will be performed in a short time.

The flow charts of the algorithms executed by the vehicles searching for a parking space and the coordinator are detailed in Figs 1 and 2, respectively. Regarding Fig. 1, the method *declareInterest* is executed when a vehicle searching for a parking space receives an available parking space event (*apse*) from a coordinator and *confirmInterest* is executed when a vehicle is informed by a coordinator that it has been allocated the parking space; regarding the latter, it should be noticed that a vehicle may send an abort message if it has already found another parking space in the meanwhile or if the driver changed his/her mind and does not want to park anymore. Regarding Fig. 2, the method *advertiseParking (Event apse)* is executed when a vehicle leaves a parking space (or when it becomes a coordinator through delegation, as explained in Section 3.2), *sort(answers)* sorts the answers received according to the allocation policy used and the information communicated by the interested vehicles, and *notification* becomes true when the winning vehicle confirms the allocation.

3.2. Extending the notification range

We have described so far the interactions between the coordinator and the interested vehicles to allocate an available parking space. However, we have considered that at least one of the vehicles within the

---

[1]If the event is generated by a fixed sensor in the parking space, an initial coordinator is chosen between the vehicles receiving this event.

declareInterest(Event apse, Coordinator c)

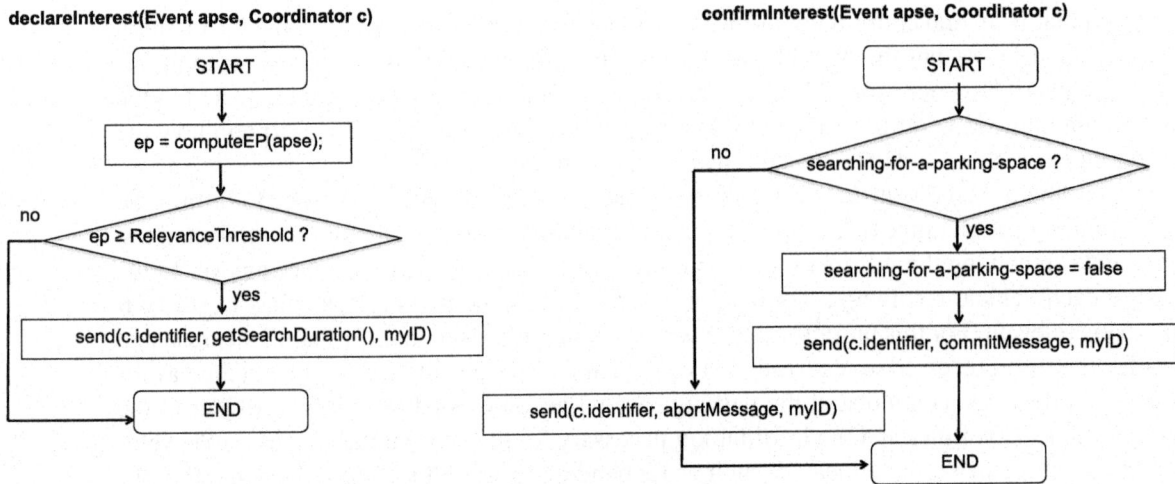

confirmInterest(Event apse, Coordinator c)

Fig. 1. Behavior of the receiving side.

advertiseParking(Event apse)

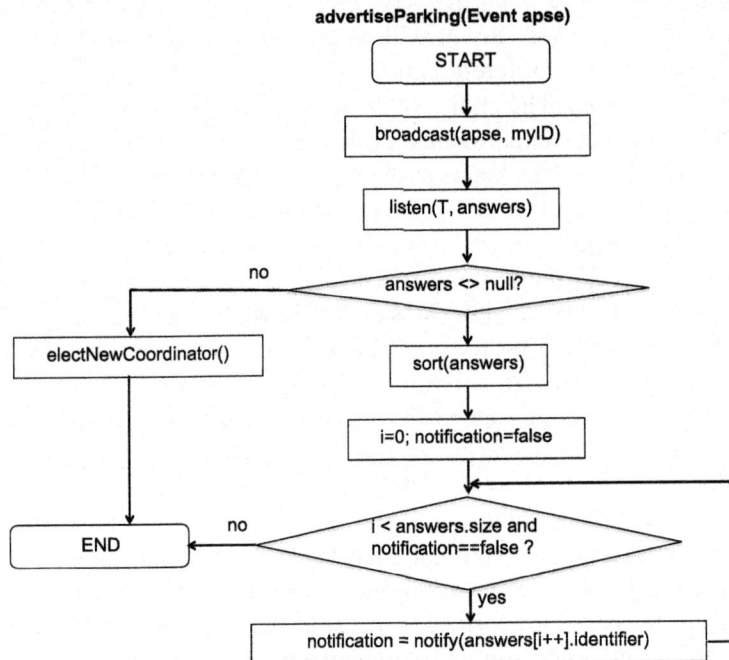

Fig. 2. Behavior of the coordinator in the normal process.

communication range of the coordinator was interested in finding an available parking space. If this is not the case, the information has to be relayed farther to try to find a vehicle interested in the resource. Anyway, it is not possible anymore to interact with the same coordinator using multi-hop techniques. Indeed, due to the use of short range communication networks and the absence of any fixed support infrastructure, we have no way to ensure that the coordinator would still be reachable when using multi-hop relaying (all vehicles are highly mobile). Thus, the decision process could not be guaranteed.

Instead, we rather try to choose a new coordinator. The new coordinator has to be farther from the

electNewCoordinator()

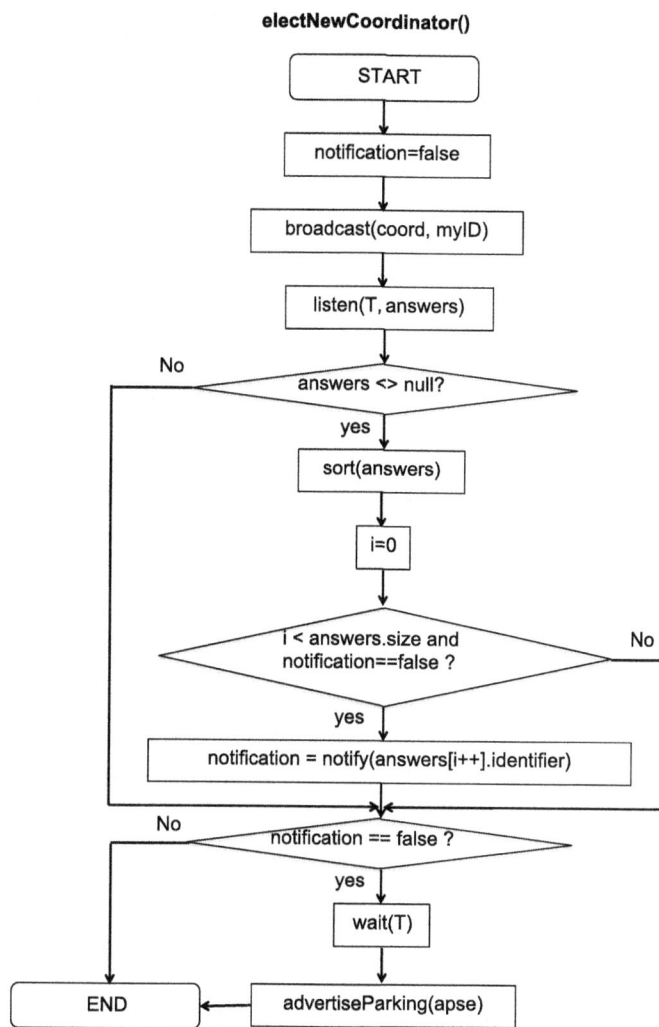

Fig. 3. Behavior of the coordinator when it needs to be replaced.

resource to increase the probability to reach new potentially interested vehicles. In the case of parking spaces, the coordinator should however not be selected too far away from the available slot. Indeed this would increase the probability that another vehicle arrives to park in that parking space in the meanwhile. Instead, we choose the new coordinator according to the demand in terms of available parking spaces in its vicinity.

Our goal is to find an interested vehicle as quickly and as close to the parking space as possible. Therefore, each vehicle periodically broadcasts to its neighbors its "state" (i.e., whether it is searching for a parking space or not). This allows each vehicle to estimate the approximate number of nearby vehicles searching for a parking space. So, when a new coordinator has to be elected, the former one broadcasts a message to its neighbors. The neighboring vehicles receiving that message (and not already coordinators for another parking space) reply to the coordinator by indicating their estimations. The coordinator then sorts the candidates in increasing number of these estimations and contacts the vehicles in that list following that order until one vehicle confirms the reception of the proposal and so becomes the new coordinator. In case no candidate coordinator answers, the former one keeps its role and, after a while, broadcasts again the message about the available parking space. By then, new vehicles interested can now be in its neighborhood. If not, a process to switch the coordinator is initiated again. The flow chart of the algorithm executed by the coordinator when it needs to find a replacement is shown in Fig. 3; in this case, *notification* becomes true when a vehicle accepts to become the new coordinator. The flow chart of the corresponding algorithm for the other vehicles is trivial from the descriptions in the text.

Thus, vehicles interested in finding an available parking space can be located even if they are further from the available parking space than the communication range of the wireless network used.

3.3. Some remarks

In the following, we discuss some situations that may arise during the allocation protocol:

- *Several advertisements of different parking spaces.* Obviously, several available parking spaces can be communicated to the same vehicle at the same moment (e.g., two close vehicles can leave their parking space or become coordinators at the same time). Thus, a vehicle can receive different messages, issued by different coordinators, indicating an available parking space. Then, a difficulty arises for that vehicle to choose to which coordinator it should send a positive answer. This is not easy because the vehicle does not know by which coordinator it could be elected since it does not even know which other vehicles will answer. So, it is not possible for the vehicle to take an informed decision.

  Consequently, we choose to let the vehicle answer to all the available coordinators. Once again, the confirmation message used in the protocol helps to avoid losing a parking space in case the same vehicle is elected twice by two different coordinators. At the coordinator side, if one vehicle does not confirm its interest in the parking space, the coordinator tries to contact the second vehicle in the list of candidates, and so on. Notice that an alternative solution could be to reply only to one coordinator (chosen randomly), but in this case the probability to receive a parking space would decrease.

- *Vehicles not compliant with the proposed protocol.* It should be noted that we do not assume that all the vehicles will have the proposed system installed. There may be other vehicles not participating in the data sharing or even vehicles that participate but whose drivers follow a behavior different from the one expected:

  * *Choice of a parking space different from the one assigned.* A vehicle that has been allocated a parking space could see a different parking space available, before reaching the one assigned, and take it. In that case, it could advertise again its allocated parking space (that will remain available), acting as a coordinator for it. In this way, the information about its availability would not get lost.

  * *Occupation of the parking space by other vehicles.* In the previous case, the vehicle could actually end up occupying a parking space reserved for another vehicle. Similarly, vehicles not equipped with the proposed system could find and occupy any parking space. So, even if we try to help the elected vehicle to reach the parking space while it is still available, we cannot ensure that no other driver will see and use that space before. This is obviously unavoidable. What our protocol eventually ensures is an effective dissemination of information about available parking spaces, without leading to a competition.

It is also important to emphasize that the proposed protocol relies in the concept of Encounter Probability explained in Section 2. Therefore, the information about available parking spaces is not propagated randomly, but within the areas where the information could reach interesting vehicles.

4. Reliability issues

In this section, we discuss some aspects about the reliability of the proposed protocol. In particular, we analyze what may happen if some message gets lost or if duplicates arise during the advertisement and allocation process, as well as potential solutions and their limitations.

In the vehicular network, a message is communicated to other nearby vehicles by broadcasting. When a vehicle tries to communicate a message to another vehicle but the message does not reach the intended destination (because the target vehicle has moved out of the communication range), we say that *the message has been lost*. We distinguish between this kind of problem and the case when the broadcast itself fails (e.g., due to interferences/collisions in the channel, fading, obstructions, reflections, or other propagation effects). Thus, in the case of *broadcasting failure* the communicating vehicle will be able to detect the problem by overhearing the communication channel; in case it detects a failure, it will retry the communication. The situation is different when a message gets lost, as the communicating vehicle may be unaware of this situation unless acknowledgement messages are included in the protocol to ensure the reception. However, it should be noticed that the acknowledgement itself may get lost. Therefore, receiving it is a guarantee that the vehicle received the message but the destination may have received the message even if no acknowledgment is received. This is consistent with what is mentioned in [11] (in relation to the WAVE protocol): "Unlike unicast traffic, broadcast frames are never acknowledged by the receivers; therefore, failed transmissions cannot be detected by the sender, and broadcast frames cannot be retransmitted."

Now, let us analyze all the communication messages exchanged in the protocol described in Section 3:

1. *Announcement of the availability of a parking space by the coordinator*. First, the coordinator advertises the parking space in its vicinity. The advertisement is disseminated by broadcast in the area within the communication range of the coordinator. As it is not aimed at a specific vehicle, it cannot get lost. Of course, if there is no vehicle within range, then the message will not be received and the protocol to choose a new coordinator and extend the notification range will be started. This will lead to a slight increase in the delay of the allocation protocol, but it is very unlikely that the coordinator will change again without need.

2. *Communication of interest by the interested vehicles*. Each vehicle receiving the advertisement and searching for parking communicates a message indicating its interest in the parking space.

   If the communication of interest from a vehicle gets lost, then such a vehicle will not be among the candidates to receive the parking space. Therefore, that potential candidate will not be chosen by the coordinator and the allocation decision may be sub-optimal. In the worst case, if that was the only interested vehicle or all the messages from the interested vehicles get lost, the coordinator may end up assuming that there is no interested vehicle in the vicinity. So, a new coordinator may be chosen unnecessarily.

3. *Communication of a decision by the coordinator*. The coordinator decides to which vehicle the parking space should be allocated based on certain spatio-temporal criteria. Once it has decided, it communicates the assignment to the chosen vehicle.

   If the assignment message gets lost, the allocation will fail. The coordinator will not receive the confirmation from the chosen vehicle and so it will select another interested vehicle. As a consequence, the final assignment may be sub-optimal. However, selecting the second best candidate should not be a major problem.

4. *Confirmation of the parking space assignment by the chosen vehicle*. The vehicle chosen receives the information from the coordinator and sends back a confirmation to indicate that it has successfully received the information and accepts the parking space.

If the confirmation message is not received by the coordinator due to a communication problem, the coordinator will assume that the allocation was not successful. So, it will start the allocation process again to find another vehicle that may be interested in the parking space. In this way, the parking space will be allocated to another vehicle. However, the previously-chosen vehicle will be unaware of this situation and the driver will also make use of the information to try to occupy the space. As a result, some competition will be generated between the first vehicle and the second one. This is an important problem, but the situation is still better than in the case of a pure data sharing approach where the information about parking spaces is communicated to all the vehicles, which would generate a higher competition for the spots.

5. *Announcement of the need of a new coordinator.* When the advertisement ranges needs to be extended, the coordinator sends a message requesting a new coordinator.

   The announcement is disseminated within the communication range of the coordinator, but it is not sent to any specific vehicle. Therefore, it cannot get lost. If there is no vehicle within range, the coordinator will try again by repeating the dissemination.

6. *Communication of answers from the potential candidates for new coordinator.* When the need of a coordinator is advertised, the vehicles receiving the message answer with their estimations about nearby vehicles searching for an available parking space.

   If one of these answers gets lost, then that vehicle will not be considered as a candidate to take on the role of coordinator. This may lead to a sub-optimal election of the new coordinator, as that vehicle could have been the best choice. However, as all the vehicles are within the communication range of the coordinator, we do not expect very big differences in terms of the number of neighboring vehicle searching for a parking space; therefore, the impact of this communication failure will be small. Of course, if all the answers get lost, then the process of searching a new coordinator has to be restarted. Whereas this will lead to a slight increase in the delay of the protocol used to choose a new coordinator, it is very unlikely that the problem will happen again next time.

7. *Acceptance of the coordinator's role.* The process of change of coordinator concludes when the vehicle selected as a new coordinator sends a confirmation to the previous coordinator.

   However, if the message indicating the acceptance gets lost, then the coordinator will try to find another target to pass on its responsibility to allocate the parking space. As a result, a second coordinator will be chosen. This will lead to the existence of several coordinators for the same parking space. Consequently, the parking space may be allocated to more than one vehicle (as each coordinator will perform one allocation of the parking space). This situation is quite similar to the one discussed in 4), but in this case from the point of view of the coordinator.

So, the two most important problems occur: 1) when the confirmation from a chosen vehicle to the coordinator gets lost in the assignment process, and 2) when the acceptance of the vehicle chosen as a new coordinator gets lost in the process of election of a new coordinator. Both cases will probably lead to a situation where the assignment of the parking space is communicated to several vehicles (even if from the point of view of a coordinator the parking space is assigned to only one vehicle), and therefore to some competition.

The possibility to lose those types of messages in unavoidable unless unicast messages and an appropriate multi-hop routing protocol (e.g., [31]) are used. However, broadcasting/geocasting is the predominant communication mode in vehicular networks [12,24,32,40] and avoids the overhead of routing messages to specific vehicles. Therefore, the only alternative that would completely avoid the possibility of some competition introduced by the data sharing system would be to remove the two types of messages mentioned. In that case, we would be implicitly assuming: for the first type of message, that

Table 2
Summary of potential problems due to message losses

Message type	Problem if lost	Impact	
Availability of parking space	–	–	
Communication of interest	Pb. 1: Sub-optimal allocation	Very low	Process of allocation of parking spaces
	Pb. 2: Unnecessary election of a new coordinator (if all the communications of interest get lost) → increase of delay	Low	
Parking assignment	Pb. 3: Sub-optimal allocation	Very low	
Confirmation of the parking assignment	Pb. 4: Multiple assignment of the parking space	Medium	
Request for new coordinator	–	–	Process of selection of a new coordinator
Answer to the request for new coordinator	Pb. 5: Sub-optimal allocation	Very low	
	Pb. 6: Restart of the process to elect a new coordinator (if all the communications of interest get lost) → increase of delay	Low	
Acceptance of the coordinator's role	Pb. 7: Multiple coordinators → Multiple assignment of the parking space	Medium	

the vehicle that is assigned the parking space will receive the message and accept it; and for the second type of message, that the chosen coordinator will receive the message and accept the role. If this is not the case, the information about the available parking space would get lost and not communicated to any vehicle. Therefore, we advocate keeping these confirmation messages to ensure that the information will be successfully delivered to some vehicle, while at the same time avoiding the competition that would arise if all the vehicles receive the information.

Table 2 shows a summary of the types of messages used in the proposed protocol, the problems that may arise if they get lost, and their impact.

5. Experimental evaluation

In this section, we describe our experimental evaluation. In particular, we evaluated different strategies to advertise available parking spaces in two different scenarios: parking lots and urban environments.

We have developed a prototype of VESPA for smartphones (see http://www.univ-valenciennes.fr/ROI/SID/tdelot/vespa/prototype.html), which can be used in a real scenario. However, due to obvious scalability reasons (it is difficult to perform repeatable scenarios with a high number of vehicles in a real environment), we use a simulator that we have developed to evaluate our system. Tests in a real environment are thus used mostly for verification, to calibrate our simulations, and for demonstration purposes.

We notably observed that the time elapsed since a vehicle started searching for a parking space could not be the only criterion for the allocation of parking spaces. Indeed, if a driver is close to an available parking space and sees it, he/she will park his/her vehicle if he/she is searching for such a resource (i.e., the space will be taken even if the event reporting the availability of that parking space is not communicated to that driver). Thus, it seems interesting to consider the value of the Encounter Probability of the resource for the vehicles searching; for example, we need to take into account that nearby vehicles are more likely to occupy the space.

5.1. Brief description of the simulator

In order to evaluate our solution with an important number of vehicles, a simulator has been designed. We needed a testing system that could simulate realistic vehicles' movements, wireless exchanges, the generation of events, etc. Moreover, whereas the traffic naturally strongly impacts the dissemination of resources among vehicles, the information exchanged among those vehicles also influences the way the vehicles move. For example, if an interested driver receives the position of an available parking space, he/she will move in that direction. This is why we decided to develop our own simulator, since no existing one could fit our needs.

Our simulator allows simulating vehicles on real road networks stored on digital road maps and also creating artificial roads and parking lot structures by hand. In the simulated scenario, the vehicles drive from a random departure point to a destination point through roads defined according to real maps. The choice of roads used for each vehicle to reach its destination is computed using Dijkstra's shortest path algorithm. When a vehicle does not have any destination point because it is looking for a parking space but does not know any available slot, a random road is chosen (simulating a driver that is looking around for a place to park). A communication range of 200 m is considered by default.

It should be emphasized that the behavior of the vehicles during the simulations depends on both the information provided by the inter-vehicle communication system (i.e., the positions of available parking spaces communicated to the driver) and the information that the drivers observe. For example, to simulate a real environment, a driver searching for a parking space will take an available one once he/she sees it, even if a farthest one has been allocated to her/him.

5.2. Relevance evaluation and reservation policies

As explained in Section 3.1, different allocation policies could be considered. Moreover, the relevance of parking spaces could be evaluated using different strategies, such as the EP-based approach discussed in Section 2. In this section, we briefly mention some strategies that could be considered for relevance evaluation and parking allocation. Thus, some strategies that could be considered for relevance evaluation are:

- One with no inter-vehicle communication (i.e. drivers searching only park their car when they physically see an available space), called *View Only*.
- One for which the relevance is evaluated using our Encounter Probability, called *VESPA only* (see Section 2). We will use the geographic-based method to consider the worst-case where a road map is not available.

  The penalty coefficients used to compute the Encounter Probability for parking spaces in parking lots are: $\alpha^{-1} = 50$, $\beta^{-1} = 30000$, $\gamma^{-1} = 600$ and $\zeta = 0$, in order to prefer free parking spaces located in the same row even if there are closer ones (considering the Euclidean distance) in neighboring rows. For urban environments, we use: $\alpha^{-1} = 750$, $\beta^{-1} = 900$, $\gamma^{-1} = 2700$ and $\zeta = 1/2$. It should be clarified that the values of the penalty coefficients, when considered individually, allow to set upper bounds for the different factors affecting the computation of the EP (see Section 2.2.1); for example, the value $\alpha^{-1} = 750$ along with a relevance threshold of 75% implies that a parking space located further than 250 meters when the vehicle is expected to be closest to the parking space would not be considered relevant for the driver. For more details about how to fine-tune the penalty coefficients, see [12].
- One for which the relevance is evaluated using the relevance function proposed in [49], called *Time and Distance*. Specifically, the authors use the following function to characterize the relevance of a parking space s:

$$F(s) = -\alpha \times t - \beta \times d \quad (\alpha, \beta \geqslant 0)$$

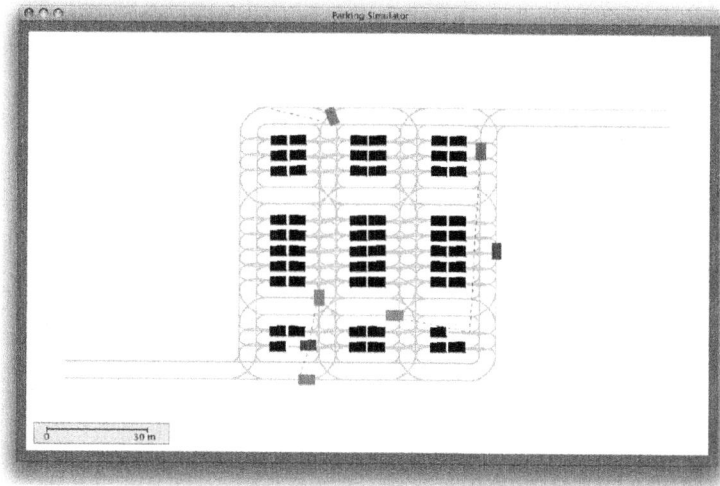

Fig. 4. Scenario for evaluation in a parking lot.

where t is the age of s, d is the distance from the location of s, and α and β are non-negative constants that represent the relative importance of time and distance. We consider $\alpha^{-1} = 30000$ and $\beta^{-1} = 50$ for the coefficients of the function $F(R)$, in order to have the same relative weights than the strategy *Vespa Only*, as we will focus on highly-competitive scenarios where the distance will be the key factor to consider.

The three previous strategies, considered alone, imply that no reservation protocol is used. Besides, different allocation strategies are considered:

- An available space is allocated by the coordinator to the vehicle with the highest EP for the considered space, called *Reservation EP*.
- The relevance function F(s), used by the *Time and Distance* strategy, is considered to allocate available spaces to vehicles, called *Reservation Time and Distance*.

In the rest of this section, we present the experimental results obtained in parking lots and urban environments.

5.3. Experimental results in a parking lot

In the following, we present some of the results obtained in the scenario of a parking lot. A snapshot of the GUI of our simulator with a sample parking lot is shown in Fig. 4.

The results reported consider a parking lot with 60 parking spaces; of course, parking lots of other sizes can be equally managed with our approach. Each vehicle entering the parking lot was considered as searching for an available space. During our experiments, some vehicles leave their parking spaces while others (driving at 10 kmph) are searching for an available slot. We evaluated different configurations with more or less free parking spaces and searching vehicles. For the configurations where the number of available parking spaces is always greater than the number of vehicles searching, the strategies using V2V communications (i.e., all those we evaluated except *View Only*) perform better but no significant difference between them can be observed. So, we present in the following the results for a "heavy" parking configuration where we always consider more searching vehicles than available parking spaces. In this "heavy" configuration, we considered a number of vehicles searching for a parking space ranging between 1 and 10% of the total capacity of the parking facility.

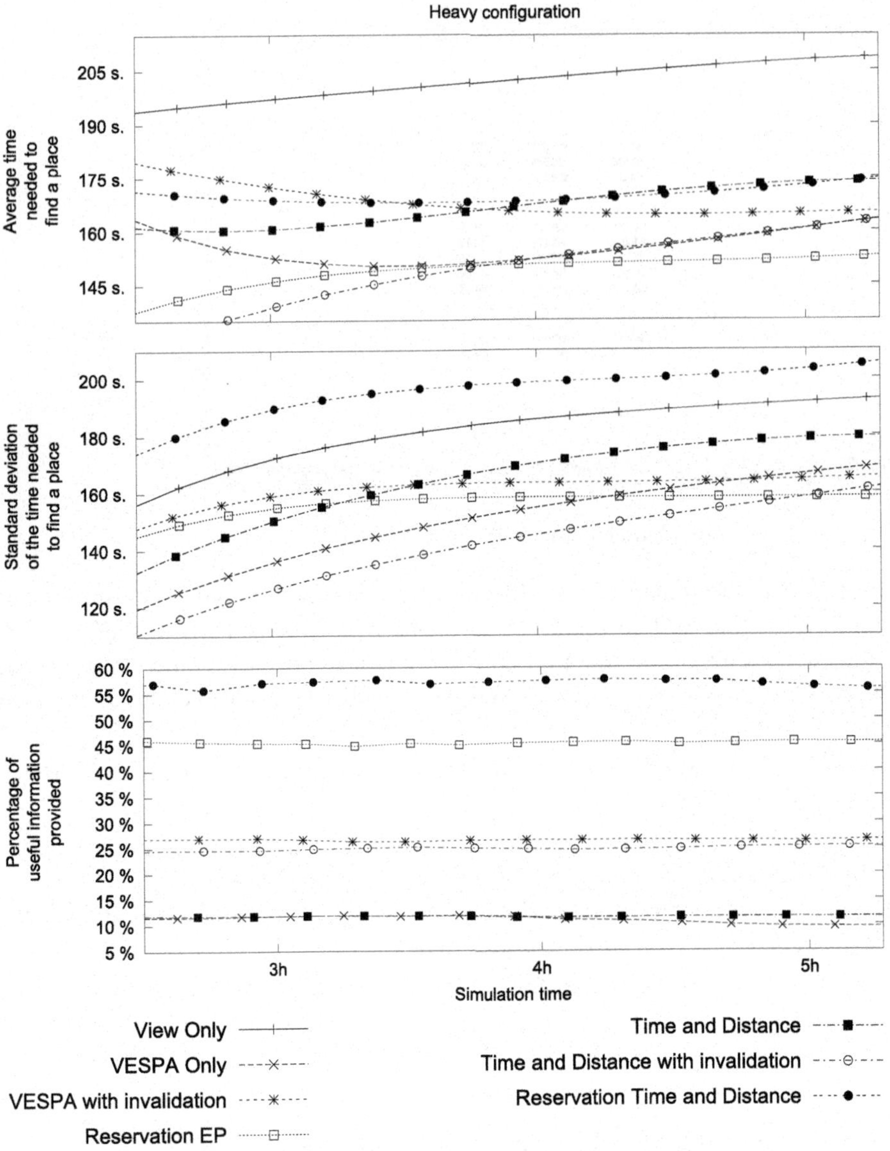

Fig. 5. Experimental results in a parking lot.

Figure 5 shows the average results obtained after 20 simulations for three evaluated criteria: the evolution of the average time needed by the vehicles to park, the standard deviation of the amount of time needed to find a parking space, and finally the percentage of useful information provided to the driver (i.e., parking spaces allocated to a driver that are really obtained by that driver).

As concerns the results of the strategies without reservation protocol, we can first observe, thanks to the average times, the interest of V2V-based solutions. Besides, the results for the *Distance and time only* approach are not very good in terms of average search time, as that relevance function was basically proposed to monitor a set of available parking spaces close to the vehicle. Compared with *VESPA only*, we can notice the importance of the vehicles' direction to determine the best parking space to allocate, since this may avoid u-turns. Moreover, it also maximizes the probability that a vehicle arrives in a parking space earlier than other vehicles trying to park. So, an available parking space in the same row should be preferred even if it is not the closest (i.e., one available parking space may be closer in the next row but the time to reach it will be higher).

To provide vehicles more relevant information, we introduced invalidation messages. These messages are generated by vehicles, once parked, in order to inform the other ones that the slot is not available anymore. The results for the *Time and Distance with invalidation* and *VESPA with invalidation* strategy are presented in Fig. 5. These invalidation messages really improve the result for the *Time and Distance* strategy, since they avoid wasting time to reach a no more free parking space. However, these messages also have an impact on the network load. They indeed require the generation of one additional message per parking space. Furthermore, this message has to be broadcasted using multi-hop techniques in order to (try to) reach all the vehicles previously informed of the available parking space. Even if the results obtained with invalidation messages in this scenario are good, this is also due to the fact that it is not so difficult to quickly communicate the invalidations to all the vehicles in a parking lot. However, ensuring this in a general environment would be really challenging.

Regarding the reservation protocol, the average times are rather good, compared with the other solutions, especially when the Encounter Probability is used to allocate parking spaces. Indeed, since we considered that a driver who sees a parking space will park on that space even if he/she had another parking space allocated, it is better to allocate a parking space close to the vehicle (and if possible in its driving direction). Otherwise, the probability that the parking space is going to be occupied by another driver increases. In that case, the vehicle which had the parking space allocated would have to ask for a new allocation when it receives a new event about an available parking space. This explains the average performance of the reservation protocol deployed on top of the *Time and Distance* strategy due to the "row effect". Indeed, the target parking space is computed here with the Euclidean distance and may be located in the next row. The driver may so need more time to reach it, which increases the probability that the space is no more available when he/she arrives. This also justifies the importance of the standard deviation for this strategy. Besides, the reservation protocol allows maximizing the percentage of useful information provided (i.e., the ratio between the number of resources effectively acquired and the total number of resources communicated to the driver), whatever the allocation strategy used. This means that the reservation protocol allocates a lot of useful parking spaces (i.e., where the driver receiving the position of the available parking space can effectively park his/her vehicle) even if the misses may be very penalizing using the *Time and Distance* allocation strategy (i.e., the driver needs a lot of time to note that the parking space is no more free). Notice that this percentage of useful information is unsurprisingly bad for the strategies where no reservation protocol is used. This means that the drivers received more non-relevant information because the positions of free parking spaces are not communicated to only one driver but to several ones at the same time.

We can conclude that the *Reservation EP* strategy is the best choice with the heavy parking lot configuration. This strategy helps to reduce the average search time for a parking space, especially when the competition is high. By reducing that competition, it also improves the percentage of useful information communicated to the vehicles. Indeed, in spite of the congested situation (in terms of vehicles searching for an available parking space), about 50% of the vehicles received a correct information as opposed to the strategies without reservation protocol (about 15%). We have performed other experiments with different numbers of vehicles and parking spaces, leading to similar conclusions.

5.4. Experimental results in an urban environment

In the previous section we considered parking lots and different data sharing strategies to highlight that our reservation protocol can be used on top of systems other than VESPA. In this section, we present

Fig. 6. Scenario for evaluation in an urban environment: city of Lille in France.

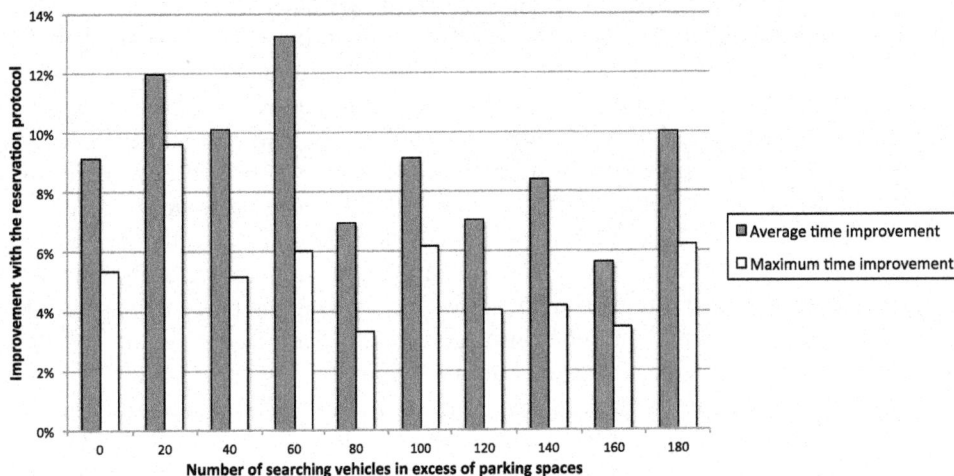

Fig. 7. Experimental results in an urban environment.

the experimental results obtained when evaluating the proposed approach in an urban environment. As in previous papers [12,16,20] we have already shown that sharing data in a vehicular ad hoc network following the VESPA approach (based on the computation of the Encounter Probability) leads to interesting benefits in both urban environments and highway scenarios, in this section we will focus on evaluating the improvement that we can obtain when we add the reservation protocol for the case of parking spaces. It should be noted that strategies such as the use of invalidation messages can be used in small scenarios such as parking lots but are not suitable in urban environments (due to the difficulty of reaching all the vehicles previously informed about the event that is invalidated); besides, the geographic-based EP could behave like the *Time and Distance* relevance function by appropriately setting the values of the penalty coefficients.

We performed several experiments using real road maps corresponding to the city of Lille in France (see Fig. 6). In these experiments, we compared a solution based on sharing data about available parking spaces using VESPA with a solution that also uses the reservation protocol proposed in this paper.

Specifically, we measured the improvement in the average search time and the maximum search time when the reservation protocol is also used. Several situations were considered by varying the amount of competition with respect to the existing number of parking spaces in a scenario (to evaluate a challenging scenario, we assumed that 75% of the vehicles in the scenario search for a parking space).

The results are shown in Fig. 7, where we vary the number of vehicles in excess of the number of parking spaces available (averages of 100 tests for each competition scenario are reported). According to these results, the reservation protocol implies an improvement in terms of the cost of drivers to find an available parking space. Intuitively, it helps to avoid some competitive situations where several vehicles try to reach the same parking space.

6. Related work

In this section, we present several related proposals, that illustrate the interest and originality of the work presented in this paper. First, we focus on research contributions in the academia. Then, we present some existing systems.

6.1. Research on parking spaces

In this section, we present the state of the art on sharing information about available parking spaces. The main features of the most relevant proposals considered are summarized in Fig. 8. In the figure, for each approach we provide the following information: its main *focus*, which can be parking lots or garages (*PL*), on-street parking (*SP*), and/or parking spaces controlled by parking meters (*PM*); the *interaction* mechanism used, which may be *V2V* (vehicle-to-vehicle), *V2I* (vehicle-to-infrastructure), *I2I* (infrastructure-to-infrastructure), and/or client/server (*CS*); the data dissemination *mode*, which can be *push* or *pull* [46]; the communication mechanism (*comm.*), which can assume a wide-area communication system (*WAN*) and/or be *ad hoc* (*AH*); whether it requires (*req.*) a support infrastructure (*I*) and/or monitoring sensors (*S*); whether the approach supports reserving parking spaces (*res.*); and whether it takes competition somehow into account (*comp.*). Usually the use of V2V and V2I implies ad hoc communications and the existence of a central server implies a system based on a support infrastructure; however, arguably it would also be possible to exchange information between vehicles using 3G, so we have decided to separate the criteria for the interaction and the communication features. Similarly, even if it may be unrealistic, it would be possible to communicate with a central server using multi-hop ad-hoc communications. We also consider that V2I implies communication between vehicles and the infrastructure independently of the direction of the communication (i.e., from the vehicles to the infrastructure and/or viceversa). We would also like to clarify that we are considering parking lots and other parking facilities such as garages as equivalent, as even if there may be some differences between them they are not relevant here. Finally, we understand that if a work includes a reservation process, then this is also a mechanism to decrease the competition between vehicles.

Some key aspects about the different proposals discussed in this section are collected in Fig. 9. In the following, we first present general proposals about disseminating parking information. Then, we consider proposals that acknowledge that competition problems may arise when the vehicles receive information about available parking spaces. After that, we describe proposals that focus specifically on parking lots or garages. Finally, we overview other interesting proposals that are less related to the approach in this paper.

Work	focus			interaction				mode		comm.		req.		res.	comp.
	PL	SP	PM	V2V	V2I	I2I	CS	push	pull	WAN	AH	I	S		
ParkNet		•					•		•	•		•	•		
SmartPark		•			•	•			•		•	•	•		
Agent-based		•		•				•			•		-		•
Time-Varying TSP approach		•		•				•			•		-		•
GPA		•		-	-	-	-	-	-	-	-	-	-		•
PMNET			•		•	•		•	•	•	•	•	•	•	•
Caliskan et al. (parking automats)			•	•	•			•			•	•	•		
SPARK	•				•	•		•			•	•	•		
OAPS		•		•				•			•		•		
CAPS		•				•		•	•	•		•	•	•	•
SmartParking	•				•	•		•			•	•	•	•	•
VESPA + reservation	•	•	•	•				•			•		•	•	•

Fig. 8. Main research proposals: Summary of features.

6.1.1. Proposals on disseminating parking information

Some proposals focus on the problem of collecting information about available parking spaces. Thus, for example, *ParkNet* [34] is a mobile sensing system for road-side parking spaces. The vehicles use ultrasonic sensors to collect information about the availability of parking spaces while driving by and they communicate these data to a central server, which aggregates them and builds a real-time map of parking space availability. Interested vehicles can then query the central server to obtain information about available parking spaces. The authors show that this monitoring approach is very effective and much cheaper and convenient than deploying fixed sensors on the parking spaces.

Another related approach is *SmartPark* [39] (see http://smartpark.epfl.ch). In this case, however, there is no centralized server and fixed sensors are used to detect if a parking space is occupied or not. A vehicle searching for parking announces its need to the nearby sensors and the request for parking is re-broadcasted periodically if needed.

A dissemination approach based on the use of fixed hotspots and opportunistic exchanges between mobile peers when they encounter each other is presented in [48]. To reduce the amount of data exchanged, a relevance function is applied, that takes into account the age of the report and the distance to the resource. However, according to [1] (a more recent contribution by some of the authors of [48]), that approach makes the assumption that a driver knows (approximatively): 1) for how long the parking space will remain available, and 2) how much time it will need to reach it.

6.1.2. Proposals that acknowledge the competition among vehicles

In the proposal presented in [7] (*agent-based* community), vehicles exchange information about both available and occupied parking spaces in cities. Based on the preferences of the driver, a decision module selects an appropriate parking space (in the experimental evaluation, parking spaces closer to the current location of the vehicle are preferred) and stops diffusing information about that available parking space in order to maximize the chance to find it free.

Another proposal to maximize the probability to find an available parking space in a city is [45], which computes a route that goes through all the parking spaces considered available. For this purpose,

Work	Main aspects
ParkNet	Mobile sensing, ultrasonic rangefinders. Environmental fingerprinting. Space count, occupancy map.
SmartPark	Parking spot sensors. Request for parking.
Agent-based	Assistant agent. FS (Free Spot) list, OS (Occupied Spot) list. Communication module, Itinerary module and Decision module.
Time-Varying TSP approach	Time-Varying TSP. Exact approach. Clustering-based approach. Live approach.
GPA	Gravity-based Parking Algorithm. Parking places attract vehicles. Parking Slot Assignment Games (PSAG).
PMNET	Parking meters, pm-nodes. Cluster-Based status updates.
Caliskan et al. (parking automats)	Parking automats. Atomic information. Aggregated information. Quadtree-based aggregation.
SPARK	Road-Side Units (RSUs). Parking info dissemination. Parking navigation service. Anti-theft protection.
OAPS	Opportunistically assisted parking search. Mobile storage nodes (msnOAPS).
CAPS	Centrally assisted parking search. First-Come-First-Serve allocation policy.
SmartParking	Based on NOTICE, secure and privacy-aware. Sensor belts, parking belts. Base station, publisher station, order station.
VESPA + reservation	Reservation protocol. Coordinator for parking space allocation. Encounter Probability (EP). Pure ad hoc network.

Fig. 9. Main research proposals: Key aspects.

the *Time-Varying TSP (Travelling Salesman Problem)* is considered. Several algorithms are proposed (the *exact* approach, the *clustering*-based approach, and the *live* approach), including an approach where the path is automatically readjusted when new information from neighboring vehicles is received (the *live* approach). However, according to [1] this type of solution based on the Time-Varying TSP is not appropriate because the availability of the parking spaces can change at any time.

In [1–3], the authors view the competition for parking spaces as an assignment game where the players are the vehicles and the parking spaces are resources with different costs associated. In this context of *Parking Slot Assignment Games (PSAG)*, they assume that the vehicles can receive information about available parking spaces (by using existing proposals) and propose algorithms for vehicles to choose the "ideal" parking space:

- First, they study a *centralized model* that optimizes the total system cost (the "social welfare"). Although an optimal solution can be found in polynomial time, the assignments obtained could imply that some drivers incur in higher costs for the benefit of others; so, they argue that this is not appropriate in distributed scenarios where the drivers make their own choices.
- Then, they study a *distributed PSAG model with complete information* and establish the relation between this parking model and the *stable marriage problem*. In this model, it is assumed that each vehicle has access to the location information of other vehicles, which arises privacy concerns and technical difficulties to perform the real-time tracking.
- Therefore, a *distributed PSAG model with incomplete information* is proposed. The problem is that solving it for an arbitrary number of parking spaces and vehicles is difficult.
- So, finally, a heuristic *gravitational model* (the *Gravity-based Parking Algorithm*, or *GPA*) is proposed. The idea is that parking spaces attract searching vehicles towards them in a way that the vehicles will move towards areas with higher density of parking spaces (even if there are closer parking spaces but with less gravitational pull).

In [2] the authors specifically focus on the GPA approach and compare it with Nash equilibrium strategies with complete information. In [3] the authors consider that vehicles are constrained to move within road networks and propose three variants of GPA adapted to road networks: *Deterministic Angular GPA (DA-GPA)*, *Randomized Magnitude GPA (RM-GPA)*, and *Deterministic Magnitude GPA (DM-GPA)*. These variants were evaluated through simulations.

In [4,5] (*PMNET*), the authors describe a multi-hop wireless network composed of parking meters that exchange information about the availability of their parking spaces based on a hierarchical geographic clustering. The network of parking meters (*pm-nodes*) is considered a distributed database that can be queried by vehicles through a nearby (within communication range) pm-node. They also consider the use of unicast (position-based) routing for drivers to reserve a spot controlled by a certain parking meter. Finally, they consider that the competition between drivers can be managed by enhancing the information provided about available parking spaces with information about potential competitors; this extra information includes data such as the location of the competitors, which may be considered privacy-sensitive information. In this approach, a vehicle must be within range of some parking meter to receive this information. Moreover, no experimental evaluation is presented.

Parking payment terminals (*parking automats*) are also used in [10] to disseminate information about available parking spaces. Both atomic information (information about the availability of specific parking places managed by a single parking automat) and aggregated information (summarized information about the availability in an area covered by several parking automats) are disseminated using ad hoc communications. Atomic information is disseminated in the local proximity of the available parking spaces, whereas aggregated information (which is more stable) is transmitted over wider areas to provide a higher-level perspective of parking availability to vehicles located in more distant locations. The vehicles themselves also periodically re-broadcast the information they have about available parking spaces. The information is prioritized based on the age of the resource and the distance to the resource. The authors acknowledge that several vehicles could try to get the same parking space at the same time, but leave the problem of fair resource sharing as future work.

Considering reservation

To avoid the competition problem, some proposals explicitly consider the importance of booking a parking space for a vehicle. For example, in [22] an agent-based parking reservation facility is presented. However, it is adapted to the specific scenario of a campus and relies on the existence of a support infrastructure (composed of *InfoStations*) and a centralized computer (an *InfoStation Center*) where the information about the parking spaces is managed.

Another interesting example is *SmartParking* [50,51], which is a secure and privacy-aware architecture for the reservation of parking spaces in parking facilities. A support infrastructure based on the use of sensor belts is used to advertise information about parking spaces to nearby vehicles. This infrastructure allows the driver to choose a parking space and reserve it.

In [29,30], the authors assume that each parking space has a sensor providing information about its occupancy status, and compare (through simulations and analysis) three different approaches to discover parking spaces in a city:

- *Non-assisted parking search (NAPS)* or "blind" sequential search, where the driver tries to find a parking space near his/her destination without the help of any software system or external information.
- *Opportunistically assisted parking search (OAPS)*, where the vehicles collect information about the parking spaces that they encounter and exchange it when they meet with other vehicles, filter out this information based on time and space criteria, and then try to reach the available space that is closest to their destination.

– *Centrally assisted parking search* (*CAPS*), where a central server (that has a global knowledge about the parking space availability) processes requests for parking spaces in a First-Come-First-Served (FCFS) manner and guarantees (through reservation) a parking space close to the driver's destination.

The previous approaches are compared by considering two different types of scenarios: one where the destinations of the drivers are uniformly distributed and one where the destinations are mainly focused within a certain road (hotspot). Overall, we could say that in the first scenario the opportunistic scheme performs the best and the centralized solution exhibits the worst behavior (especially for high traffic volumes), but in the second one the centralized solution outperforms the others. That study shows that opportunistically sharing information about parking spaces can indeed increase the competition among vehicles (by synchronizing the movement patterns of the vehicles), and that the cost of setting up and maintaining a centralized infrastructure does not necessarily provide the best results. So, it proves the interest of alternative approaches like the one proposed in this paper.

Summing up, approaches that support the reservation of parking spaces to deal with competition are either restricted to specific scenarios such as parking facilities (as in [22] or in [50,51]) or based on a centralized solution (as in *CAPS* [29,30]). However, up to the authors' knowledge, a general reservation-based solution to avoid the competition in vehicular ad hoc networks is a novelty in the work described in this paper.

Considering probabilities

In the following, we describe some other proposals that take competition into account by considering the availability probabilities at the time of arrival, instead of the current status information about parking spaces. Thus, several proposals acknowledge that the probability to find an available parking space in a certain area, rather than the current occupancy status of specific parking spaces, is an interesting factor to manage. For example, in [28], parking lots (modeled by a continuous-time Markov chain) periodically disseminate in the vehicular ad hoc network some status parameters (specifically, their capacity, the number of occupied spaces, the arrival rate, and the parking rate), thus providing to the vehicles with information that they can use to estimate the probability to find there an available parking space at the time of arrival. A similar approach is described in [9]. As another related example, the availability of parking spaces is predicted in [8] within the context of a parking facility reservation system.

Whereas the previous proposals focus on estimating probabilities for parking lots, other proposals consider parking spaces in general. For example, [49] emphasizes the importance of considering aggregate information and guiding the driver towards areas where finding an available parking space is very likely (instead of guiding the user towards a specific parking space that could be taken by another vehicle in the meanwhile). Within the VESPA project, in [53] the idea is to aggregate information about available parking spaces to extract general knowledge about their overall availability in certain areas and time periods. As commented before, the use of aggregated information about parking spaces is also proposed in [10].

6.1.3. Proposals focused on parking facilities

Several proposals focus on parking lots or garages. Many existing parking management systems just keep track of the number of vehicles within a facility (using a simple barrier system to monitor/count the vehicles entering and leaving the site), whereas others monitor the individual available parking spaces. In this last case, wireless sensor networks are usually used (e.g., see [6,13,38,42,52]). A review of smart parking systems can also be found in [25].

An example is *SPARK* [33] (*Smart PARKing*), which proposes the use of RSUs (Road-Side Units) in parking lots to disseminate information about available parking spaces and to protect the vehicles parked there from theft.

The approach in [37] also considers parking lots and divides the geographic space in *zones* managed by different RSUs (the radius of each zone is smaller than the communication range of vehicles and RSUs). A Voronoi region is assigned to each RSU, such that each RSU keeps the occupancy state of the parking spaces in its region. The main focus of that work is on security aspects. However, no many details are provided, as it is a short paper.

6.1.4. Other proposals

There are other proposals that are less related to the work presented in this paper but that also concern parking spaces. So, in [14] the authors argue that not only drivers compete for parking spaces but car park operators also compete for drivers; with this motivation, and assuming that parking prices are negotiable, they develop an agent-based negotiation platform. In [43], the authors focus on real-time parking reservation systems and propose a fuzzy-based model to decide whether a certain reservation request can be accepted or not (based on the current status of the system regarding space availability, cancellations, etc.) with the goal of maximizing revenues. As a third example, [15] presents an agent-based model to study the interrelation of different factors, such as the parking pricing strategies and the behavior of drivers. These proposals do not concern about communication issues (a wide-area communication infrastructure seems to be assumed).

Finally, it is interesting to mention that technologies such as inductive loops, optical sensors, ultrasonic sensors and magnetic sensors (among others) have been proposed to detect the occupancy status of a parking spot [13,47,50]. Video sensors (specifically, webcams) were also proposed in the *IrisNet* project [36], and [34] proposed the use of ultrasonic range finders along with environmental fingerprinting for mobile sensing from the vehicles. These are complementary technologies to the approach presented in this paper, as the release of a parking space could be detected and disseminated by a sensor located there or by the driver leaving the parking space.

6.2. Working systems and applications

In this section, we overview some existing systems. The main features of the most relevant systems considered are summarized in Fig. 10. In the figure, for each system we indicate its *name* and *URL*, its *availability* regarding the platforms where it is currently supported (as an application for *iOS* and/or *Android* devices, or *Web* when it has a website accessible from web browsers), whether it requires sensors to monitor the available parking spaces (*req. sensors*), whether it supports the reservation of parking spaces (*prov. reserv.*), and the main *types of parking spaces* targeted (the ones under the control of parking *meters*, parking *lots* or garages, private parking spaces *owned* by some person, and/or *public* parking spaces). It should be noted that the features considered in the comparison are different from the ones used for research prototypes (see Section 6.1), as it makes no sense to consider the same elements. Thus, for example, no existing system relies on V2V communication but on wide-range communications (e.g., 3G) and centralized servers, which is a current limitation.

In the following, we describe these systems in three categories: those that are based on the use of sensors, those that rely on the cooperation of people to input the interesting information, and other working systems less related to the work presented in this paper because they focus on very specific scenarios.

name	URL	availability			req. sensors	prov. reserv.	types of parking spaces			
		iOS	Android	Web			meters	lots	owned	public
SFPark	http://sfpark.org	•	•	•	•		•	•		
ParkingCarma	http://www.parkingcarma.com		•		•	•		•	•	
ParkSense	http://www.smartgrains.com	•			•			•		•
Parker	http://www.streetline.com	•	•		•	•	•	•		
SpotScout	http://www.spotscout.com		•			•		•	•	
Apila	http://www.apila.fr	•				•				•
Placelib	http://www.placelib.com	•	•			•				•
ParkShark	http://www.parkshark.mobi	•				•				•
Roadify	http://www.roadify.com	•				•				•

Fig. 10. Examples of existing systems: Summary of features.

6.2.1. Working systems based on sensors

Among the existing systems, one of the frequent examples mentioned in the literature is *SFPark* (http://sfpark.org), that allows checking online the availability of parking spaces in San Francisco, that are shown on a map. The information is obtained thanks to wireless parking sensors embedded in the pavement, that monitor the occupancy status of the different slots. Besides, SFPark adjusts meter and garage prices according to the existing demand, in order to optimize the parking resources. This is a very interesting system. However, according to [34], deploying the required infrastructure is quite expensive.

ParkingCarma (http://www.parkingcarma.com) allows to search and reserve a parking space in several cities in the United States through the Internet. Owners of parking lots or other parking assets can sign up with ParkingCarma and share information about the availability of their parking spaces. Then, ParkingCarma will be able to match the available parking spaces with the needs of potential customers. It can also make price adjustments in real-time.

ParkSense (http://www.smartgrains.com) provides solutions for on-street parking, public car parks, shopping centers and airports. It is interesting to mention that this system considers not only parking spaces but also charging stations for electric vehicles. Parking sensors are organized into a mesh network and communicate status information with their neighbors (by radio) until it is finally collected by a gateway node. According to the information provided in the website, several case studies have been considered for shopping centers and there is also an experience for on-street parking in Issy-les-Moulineaux (with 100 parking spots equipped).

Streetline (http://www.streetline.com) offers several products: ParkEdge (an information platform for parking providers), ParkSight (for managing parking facilities in cities), and Parker (for drivers). In the context of this paper we focused our attention on *Parker*, which allows drivers to find and reserve spots controlled by parking meters. A wireless sensor mesh network communicates the information about parking spaces to mirrored data sites with the help of relays located on streetlights and telephone poles.

6.2.2. Working systems based on information shared by people

Several systems rely on information provided by people (drivers) instead of using sensors to monitor parking spaces:

– *SpotScout* (http://www.spotscout.com) supports "SpotScouting" (searching for parking spots) and "SpotCasting" (broadcasting information about a parking spot owned so that others can find it and use it) in cities in the United States. The spots can be reserved and paid through the Internet. No sensors are used because the availability of the parking spaces is determined based on the information provided by the "SpotCasters" (who can define the periods when the space is available) and the actual reservations made. A rating system similar to the one used in eBay is used to provide information to users before the agree to trade a parking space. The system can also show information about parking spaces with an electric outlet available to charge electric vehicles (EV).

- *Apila* (http://www.apila.fr) relies in the cooperation among drivers to facilitate searching and finding a public parking space in France. The idea is to actually perform an exchange of the parking space. First, the vehicle that is going to release a parking space announces it. Then, if there is any vehicle interested in that parking space, it requests the spot. Finally, the vehicle releasing the parking space will wait for a while (an average of three minutes, according to the description in the website) until the other interested vehicle comes and then it will complete the transaction by liberating the space for the arriving vehicle. In the exchange of the parking space, the vehicle taking the spot pays a "ticket" to the vehicle releasing it. This is an interesting idea, but reserving a public space for a specific driver may lead to disputes and even be illegal in some cities.
- *Placelib* (http://www.placelib.com), formerly ShareMySpot, proposes a similar idea to Apila (also for France). A driver announces that he/she is going to release a parking space in a few minutes and the system decides which vehicle searching for a parking space is the ideal candidate for that spot. Before releasing the parking space, the driver will wait for the arrival of that vehicle. The transaction is paid with "virtual nuts". The main difference with Apila is that the system automatically matches parking spaces and searching vehicles.
- *ParkShark* (http://www.parkshark.mobi) is a similar application that allows drivers to share information about the parking space that they are going to release. It uses a rating system to encourage cooperation (good "sharers" and "reservers" are expected to receive positive reviews), such that users with a high rating will more likely receive information about available parking spaces earlier than others.
- *Roadify* (http://www.roadify.com) also supports sharing information about available or soon-to-be-available parking spots in the United States. However, we did not find much information about it on the Web.

Finally, it is interesting to mention *Waze* (http://www.waze.com), that allows the exchange of information about traffic, events on the roads, and maps. However, it does not focus on parking spaces, even if parking lots can be represented on the maps.

6.2.3. Other working systems

In many cities, information about available parking spaces is usually posted in electronic panels located near major parking sites. Such *Parking Guidance and Information* (*PGI*) systems usually provide information about the location, direction and status of parking sites (mainly, regarding the availability of parking spaces within). Some systems try to provide very precise information by monitoring the individual parking spaces (e.g., *Intelligent Parking*, http://www.intelligentparking.com). There are also products focused on helping managing parking facilities, such as *myParkfolio* (http://www.parkeon.com).

A number of services are available nowadays that allow a driver to book a parking space in a certain parking facility by submitting a reservation request through the Internet [26]. One of these services is *E-Z Park* (http://www.ezparkinc.com) in the city of Philadelphia, but the examples are numerous. Some of these services focus on specific parking facilities, such as *eparking* (http://www.eparking.uk.com), about airport parkings in the United Kingdom, or *AboutAirportParking* (http://www.aboutairportparking.com), with information about many international airports. Other web sites provide information about the existence of parking spaces and their associated fees (e.g., http://www.chicagometers.com in the city of Chicago), but do not allow reservations or provide the occupancy status in real-time. It is also interesting to mention *Parkopedia* (http://en.parkopedia.com), which provides information about parking facilities in 28 countries and sometimes even supports booking; private parking spaces can be listed at http://www.parkatmyhouse.com and then be provided by Parkopedia.

Finally, a number of similar applications are available for smartphones; for example, for iOS devices we could mention *BestParking* (to find garages and compare parking rates), *EasyPark* (interface between parking operators and drivers to support parking payments), *ParkMe* (that provides parking recommendations based on their location and price), *Parking Mate* (to help keep track of timers and other aspects related to pay parking spots), *SocialParking* (that allows drivers to share information about the parking spaces that they are releasing), etc. Existing systems show the interest of applications to help with parking-related issues.

7. Conclusions and future work

In this paper, we have presented a solution to disseminate and allocate available parking spaces to drivers. The originality of our contribution resides in our reservation protocol, which is to the best of our knowledge the first solution to allocate parking spaces to vehicles in VANETs. We have evaluated our approach in different scenarios (including parking lots and urban environments), obtaining positive results: our system reduces both the time needed to find a parking space and the competition among the vehicles. Besides, an extensive study of related work has been performed to properly situate our approach within the state of the art and highlight its novel contributions.

We could envision an adaptation of this work to enable its use with other types of resources. For example, in hybrid networks with vehicles and mobile users we could use a similar approach to help users to find available taxi cabs. Similarly, in car sharing applications we can see a shareable car as a limited resource with some capacity and availability. Finally, charging stations for electric vehicles could become a scarce resource in the future. Each of these scenarios would probably benefit from a different resource allocation policy, but the basics of the general schema proposed in this paper could be used. Besides, it could be interesting to analyze the benefits of a pull-based query processing approach about parking spaces based on the work presented in [21].

Acknowledgments

We thank the support of the French ANR agency in the scope of the MURPHY project and the CICYT project TIN2010-21387-C02-02. We also gladly acknowledge the funding provided by the Ministry of Education (Ministerio de Educación) in Spain through the program "Programa Nacional de Movilidad de Recursos Humanos del Plan Nacional de I-D+i 2008-2011" ("José Castillejo", ref. JC2011-0196). Finally, we thank Dorsaf Zekri for her help with some aspects related to the simulations.

References

[1] D. Ayala, O. Wolfson, B. Xu, B. Dasgupta and J. Lin, Parking slot assignment games, in: *19th ACM SIGSPATIAL International Conference on Advances in Geographic Information Systems (GIS 2011)*, pp. 299–308, ACM Press, 2011.

[2] D. Ayala, O. Wolfson, B. Xu, B. Dasgupta and J. Lin, Parking in competitive settings: A gravitational approach, in: *13th International Conference on Mobile Data Management (MDM 2012)*, IEEE Computer Society, 2012. To appear.

[3] D. Ayala, O. Wolfson, B. Xu, B. DasGupta and J. Lin, Stability of marriage and vehicular parking, in: *Second International Workshop on Matching Under Preferences (MATCH-UP 2012)*, 2012. To appear.

[4] P. Basu and T.D.C. Little, Networked parking spaces: Architecture and applications, in: *IEEE Vehicular Technology Conference (VTC'02)*, pp. 1153–1157. IEEE Computer Society, 2002.

[5] P. Basu and T.D.C. Little, Wireless ad hoc discovery of parking meters, in: *MobiSys Workshop on Applications of Mobile Embedded Systems (WAMES'04)*, 2004.

[6]   J.P. Benson, T. O'Donovan, P. O'Sullivan, U. Roedig and C. Sreenan, Car-park management using wireless sensor networks, in: *31st IEEE Conference on Local Computer Networks (LCN'06)*, pp. 588–595. IEEE Computer Society, 2006.

[7]   N. Bessghaier, M. Zargayouna and F. Balbo, An agent-based community to manage urban parking, in: *Advances on Practical Applications of Agents and Multi-Agent Systems*, volume 155 of *Advances in Intelligent and Soft Computing*, pp. 17–22. Springer, 2012.

[8]   F. Caicedo, C. Blazquez and P. Miranda. Prediction of parking space availability in real time, *Expert Systems with Applications* **39**(8) (2012), 7281–7290.

[9]   M. Caliskan, A. Barthels, B. Scheuermann and M. Mauve. Predicting parking lot occupancy in vehicular ad hoc networks, in: *Vehicular Technology Conference (VTC'07)*, pp. 277–281. IEEE Computer Society, 2007.

[10] M. Caliskan, D. Graupner and M. Mauve, Decentralized discovery of free parking places, in: *Third ACM International Workshop on Vehicular Ad Hoc Networks (VANET'06)*, pp. 30–39. ACM Press, 2006.

[11] C. Campolo, A. Molinaro, A. Vinel and Y. Zhang, Modeling prioritized broadcasting in multichannel vehicular networks, *IEEE Transactions on Vehicular Technology* **61**(2) (2012), 687–701.

[12] N. Cenerario, T. Delot and S. Ilarri, A content-based dissemination protocol for VANETs: Exploiting the encounter probability, *IEEE Transactions on Intelligent Transportation Systems* **12**(3) (2011), 771–782. Special issue on exploiting wireless communication technologies in vehicular transportation networks.

[13] J. Chinrungrueng, U. Sunantachaikul and S. Triamlumlerd, Smart parking: An application of optical wireless sensor network, in: *International Symposium on Applications and the Internet Workshops (SAINTW'07)*, IEEE Computer Society, 2007.

[14] S.-Y. Chou, S.-W. Lin and C.-C. Li, Dynamic parking negotiation and guidance using an agent-based platform, *Expert Systems with Applications* **35**(3) (2008), 805–817.

[15] M. Dell'Orco and D. Teodorović, *Applied Research in Uncertainty Modeling and Analysis*, volume 20, chapter "Multi agent systems approach to parking facilities management", pp. 321–339. Springer, 2005.

[16] T. Delot, N. Cenerario and S. Ilarri, Vehicular event sharing with a mobile peer-to-peer architecture, *Transportation Research Part C: Emerging Technologies* **18**(4) (2010), 584–598.

[17] T. Delot, N. Cenerario, S. Ilarri and S. Lecomte, A cooperative reservation protocol for parking spaces in vehicular ad hoc networks, in: *Sixth International Conference on Mobile Technology, Applications and Systems (Mobility Conference 2009)*, pp, 1–8. ACM Press, 2009. Best paper award.

[18] T. Delot and S. Ilarri, Data gathering in vehicular networks: The VESPA experience (invited paper), in: *Fifth IEEE Workshop On User MObility and VEhicular Networks (LCN ON-MOVE 2011)*, pp. 801–808. IEEE Computer Society, 2011.

[19] T. Delot and S. Ilarri, Introduction to the special issue on data management in vehicular networks, *Transportation Research Part C: Emerging Technologies* **23** (2012), 1–2.

[20] T. Delot, S. Ilarri, N. Cenerario and T. Hien, Event sharing in vehicular networks using geographic vectors and maps, *Mobile Information Systems* **7**(1) (2011), 21–44.

[21] T. Delot, N. Mitton, S. Ilarri and T. Hien, GeoVanet: A routing protocol for query processing in vehicular networks, *Mobile Information Systems* **7**(4) (2011), 329–359.

[22] I. Ganchev, M. O'Droma and D. Meere, Intelligent car parking locator service, *International Journal on Information Technologies and Knowledge* **2**(2) (2008), 166–173.

[23] E. Gantelet and A. Lefauconnier, The time looking for a parking space: strategies, associated nuisances and stakes of parking management in France, in: *European Transport Conference*, 2006.

[24] A.M. Hanashi, I. Awan and M. Woodward, Performance evaluation with different mobility models for dynamic probabilistic flooding in MANETs, *Mobile Information Systems* **5**(1) (2009), 65–80.

[25] M. Idris, Y. Leng, E. Tamil, N. Noor and Z. Razak, Car park system: A review of smart parking system and its technology, *Information Technology Journal* **8** (2009), 101–113.

[26] K. Inaba, M. Shibui, T. Naganawa, M. Ogiwara and N. Yoshikai, Intelligent parking reservation service on the Internet, in: *2001 Symposium on Applications and the Internet – Workshops (SAINT'01 Workshops)*, pp. 159–164. IEEE Computer Society, 2001.

[27] G. Karagiannis, O. Altintas, E. Ekici, G.J. Heijenk, B. Jarupan, K. Lin and T. Weil, Vehicular networking: A survey and tutorial on requirements, architectures, challenges, standards and solutions, *IEEE Communications Surveys and Tutorials* **13**(4) (2011), 584–616.

[28] A. Klappenecker, H. Lee and J.L. Welch, Finding available parking spaces made easy, *Ad Hoc Networks*, Available online: 17 March, 2012.

[29] E. Kokolaki, M. Karaliopoulos and I. Stavrakakis, Value of information exposed: wireless networking solutions to the parking search problem, in: *Eighth International Conference on Wireless On-Demand Network Systems and Services (WONS 2011)*, pp. 187–194. IEEE Computer Society, 2011.

[30] E. Kokolaki, M. Karaliopoulos and I. Stavrakakis, Opportunistically assisted parking service discovery: Now it helps, now it does not, *Pervasive and Mobile Computing* **8**(2) (2012), 210–227.

[31] E. Kulla, M. Hiyama, M. Ikeda, L. Barolli, V. Kolici and R. Miho, MANET performance for source and destination moving scenarios considering OLSR and AODV protocols, *Mobile Information Systems* **6**(4) (2010), 325–339.

[32] T.-H. Lin, H.-C. Chao and I. Woungang, An enhanced MPR-based solution for flooding of broadcast messages in OLSR wireless ad hoc networks. *Mobile Information Systems* **6**(3) (2010), 249–257.

[33] R. Lu, X. Lin, H. Zhu and X. Shen, SPARK: A new VANET-based smart parking scheme for large parking lots, in: *28th IEEE International Conference on Computer Communications (INFOCOM'09)*, pp. 1413–1421. IEEE Computer Society, 2009.

[34] S. Mathur, T. Jin, N. Kasturirangan, J. Chandrashekharan, W. Xue, M. Gruteser and W. Trappe, ParkNet: Drive-by sensing of road-side parking statistics, in: *Eighth International Conference on Mobile Systems, Applications, and Services (MobiSys 2010)*, pp. 123–136. ACM Press, 2010.

[35] H. Mousannif, I. Khalil and S. Olariu, Cooperation as a service in VANET: Implementation and simulation results, *Mobile Information Systems* **8**(2) (2012), 153–172.

[36] S. Nath, A. Deshpande, Y. Ke, P. B. Gibbons, B. Karp and S. Seshan, IrisNet: An architecture for internet-scale sensing services, in: *29th International Conference on Very Large Data Bases (VLDB'03)*, **29**, pp. 1137–1140. VLDB Endowment, 2003.

[37] R. Panayappan, J. M. Trivedi, A. Studer and A. Perrig, VANET-based approach for parking space availability, in: *Fourth ACM International Workshop on Vehicular Ad Hoc Networks (VANET'07)*, pp. 75–76. ACM Press, 2007.

[38] B. Panja, B. Schneider and P. Meharia, Wirelessly sensing open parking spaces: Accounting and management of parking facility, in: *Americas Conference on Information Systems (AMCIS 2011) – All Submissions*, 2011. Paper 270.

[39] M. Piórkowski, Collaborative transportation systems, in: *Wireless Communications and Networking Conference (WCNC 2010)*, IEEE Computer Society, 2010.

[40] R. Saqour, M. Shanuldin and M. Ismail, Prediction schemes to enhance the routing process in geographical GPSR ad hoc protocol, *Mobile Information Systems* **3**(3–4) (2007), 203–220.

[41] D. Shoup, Cruising for parking, *Access* (30) (2007), 16–22.

[42] V.W. Tang, Y. Zheng and J. Cao, An intelligent car park management system based on wireless sensor networks, in: *First International Symposium on Pervasive Computing and Applications*, pp. 65–70. IEEE Computer Society, 2006.

[43] D. Teodorović and P. Lučić, Intelligent parking systems, *European Journal of Operational Research* **175**(3) (2006), 1666–1681.

[44] S. Ukkusuri, Y. Wang and T. Chigan, Introduction to the special issue on exploiting wireless communication technologies in vehicular transportation networks, *IEEE Transactions on Intelligent Transportation Systems*, **12**(3) (2011), 633–634.

[45] V. Verroios, V. Efstathiou and A. Delis, Reaching available public parking spaces in urban environments using ad-hoc networking, in: *12th International Conference on Mobile Data Management (MDM 2011)*, pp. 141–151. IEEE Computer Society, 2011.

[46] A.B. Waluyo, B. Srinivasan and D. Taniar, Research in mobile database query optimization and processing, *Mobile Information Systems* **1**(4) (2005), 225–252.

[47] J. Wolff, T. Heuer, H. Gao, M. Weinmann, S. Voit and U. Hartmann, Parking monitor system based on magnetic field sensors, in: *2006 IEEE Intelligent Transportation Systems Conference (ITSC'06)*, pp. 1275–1279. IEEE Computer Society, 2006.

[48] O. Wolfson, B. Xu and H. Yin, Dissemination of spatio-temporal information in mobile networks with hotspots, in: *Second International Workshop on Databases, Information Systems, and Peer-to-Peer Computing (DBISP2P'04)*, volume 3367 of *Lecture Notes in Computer Science*, pp. 185–199. Springer, 2004.

[49] B. Xu, A. M. Ouksel and O. Wolfson, Opportunistic resource exchange in inter-vehicle ad-hoc networks, in: *Fifth International Conference on Mobile Data Management (MDM'04)*, pp. 4–12. IEEE Computer Society, 2004.

[50] G. Yan, S. Olariu, M.C. Weigle and M. Abuelela, SmartParking: A secure and intelligent parking system using NOTICE, in: *11th International Conference on Intelligent Transportation Systems*, pp. 569–574. IEEE Computer Society, 2008.

[51] G. Yan, W. Yang, D.B. Rawat and S. Olariu, SmartParking: A secure and intelligent parking system, *IEEE Intelligent Transportation Systems Magazine* **3**(1) (2011), 18–30.

[52] Y.-Z. Bi, L.-M. Sun, H.-S. Zhu, T.-X. Yan and Z.-J. Luo, A parking management system based on wireless sensor network, *ACTA AUTOMATICA SINICA*, **32**(6) (2006), 968–977.

[53] D. Zekri, B. Defude and T. Delot, Summarizing sensors data in vehicular ad hoc networks, *RAIRO – Operations Research* **44**(4) (2010), 345–364.

Accurate mobility modeling and location prediction based on pattern analysis of handover series in mobile networks

Péter Fülöp, Sándor Imre, Sándor Szabó and Tamás Szálka
Budapest University of Technology and Economics, Department of Telecommunication, 2, Magyar tudósok körútja, Budapest 1117, Hungary
E-mail: {fulopp,imre,szabos}@hit.bme.hu; szalkat@mul.hu

Abstract. The efficient dimensioning of cellular wireless access networks depends highly on the accuracy of the underlying mathematical models of user distribution and traffic estimations. Mobility prediction also considered as an effective method contributing to the accuracy of IP multicast based multimedia transmissions, and ad hoc routing algorithms. In this paper we focus on the tradeoff between the accuracy and the complexity of the mathematical models used to describe user movements in the network. We propose mobility model extension, in order to utilize user's movement history thus providing more accurate results than other widely used models in the literature. The new models are applicable in real-life scenarios, because these rely on additional information effectively available in cellular networks (e.g. handover history), too. The complexity of the proposed models is analyzed, and the accuracy is justified by means of simulation.

Keywords: Location prediction, mobility model, Markovian approach, random walk, complexity, accuracy, handover, user movements

1. Introduction

Random Walk Mobility model is often used in network planning and in analyzing network algorithms, because of its simplicity [1]. On the other hand this very simple model presumes unrealistic conditions, like uniform user distribution in the mobile network. However, in real-life networks, geographical characteristics, such as streets and parks influence the cell residence time (dwell time) and movement directions of users in the network, and result in a non-uniform user density. While these models are appropriate for mathematical analysis, easy to use in simulations and for trace-generation, they fail to capture important characteristics of mobility patterns in specific environments, e.g. time variance, location dependence, unique speed and dwell-time distributions, etc. [3,6,12,24].

User movements in a cellular network can be described as a time-series of radio cells the user visited. The handover event of active connections (e.g. cell boundary crossing) is recorded in the network management system's logs, thus the information can be extracted from the management system of cellular mobile networks, such as GSM/GPRS/UMTS networks. The users movements are described by the dwell-time and outgoing probabilities (the probability of a user leaving for each neighboring cell). These parameters can be calculated for each cell based on the time-series of visited cells of the users. However, in some cases, these two parameters – dwell-time and outgoing probabilities – are not enough to capture all the information in the time-series of user movements. In many situations, the outgoing probabilities are correlated with the incoming direction of the users, so the movement contains memory.

This paper is organized as follows. In Chapter 2 we give a brief on the mobility models, and in Chapter 3 we summarize their applications in mobility prediction, Call Admission Control, etc. [22].

In Chapter 4 we introduce extensions to widely used mobility models in order to enhance the accuracy of the models, meanwhile aiming at keeping the models as simple, as possible. In Chapter 5 we propose a novel mobility model, called Path Based Mobility model, especially designed to predict movements in cellular networks in urban areas. We calculate the theoretical accuracy of the models, and compare them by simulation to other models can be found in the literature in Chapter 6. Chapter 7 concludes the paper.

The results of this paper are applicable in engineering tasks of location based services, network dimensioning, can provide input for more effective Call Admission Control algorithms in order to ensure user's satisfaction and optimal resource usage in cellular wireless mobile networks [19].

2. Mobility models in the literature

Different mobility models have been proposed in the literature to cope with user mobility in different wireless and mobile networks (e.g. cellular networks, ad hoc networks etc.) In this chapter we give a short overview on the most widely used mobility models.

In the *Random Walk* mobility model [15], the node moves from its current location to a new location by randomly choosing a direction and a speed. The Random Walk model defines user movement from one position to the next with memoryless, randomly selected speed and direction. Each movement in the Random Walk Mobility Model occurs in either a constant time interval t or a constant distance traveled d. When a mobile node reaches a simulation boundary, it simply bounces off the simulation border with an angle determined by the incoming direction, then the node continues along this new path. Many derivatives of the Random Walk Mobility Model have been developed including the *one, two, three –* and d-dimensional walks.

Chiang's mobility model [21] defines the *Probabilistic Version of Random Walk* model, which utilizes a probability matrix to determine the position of a particular mobile network in the next time step. The position of the node is represented by three different states for position x and three different states for position y. In this mobility model state 0 represents the current (x or y) position of a given node, state 1 represents the node's previous (x or y) position, and state 2 represents the node's next position if the node continues to move in the same direction. The probability matrix used in the probabilistic random walk model is

$$P = \begin{bmatrix} P(0,0) & P(0,1) & P(0,2) \\ P(1,0) & P(1,1) & P(1,2) \\ P(2,0) & P(2,1) & P(2,2) \end{bmatrix},$$

where each entry $P(a,b)$ represents the probability that a node will go from state a to state b. The values within this matrix are used for updates to both the node's x and y position.

In the *Random Waypoint* model [7] the user stays at a particular location for a specified time period before moving on to the next in a randomly chosen direction with speed uniformly distributed between zero and maximum speed. The model derived from the Random Walk model breaks down the entire movement of the user into a series of pause and motion periods.

A density wave is the clustering of nodes in one part of the simulation area. The *Random Direction Mobility Model* [9] was created to overcome *density waves* in the average number of neighbors produced by the Random Waypoint Mobility Model. In case of the Random Waypoint Mobility Model, this clustering occurs near the center of the simulation area.

The *Modified Random Direction Mobility Model* is a slight modification to the Random Direction Mobility Model [9]. In this modified version, the node continues to choose random directions, but they are no longer forced to travel to the simulation boundary before stopping to change direction. Instead, a node chooses a random direction and selects a destination anywhere along that direction of travel. The node then pauses at this destination before choosing a new random direction.

In the *Boundless Simulation Area Mobility Model*, a velocity vector $\vec{v} = (v, \theta)$ is used to describe a node's velocity v as well as its direction θ. In this model a relationship between the previous direction of travel and velocity of a node with its current direction of travel and velocity exists [27]. Both the velocity vector and the position are updated at every Δt time steps according to the following formulas:

$$v(t + \Delta t) = \min[\max(v(t) + \Delta v, 0), v_{\max}],$$

$$\theta(t + \Delta t) = \theta(t) + \Delta \theta$$

$$x(t + \Delta t) = x(t) + v(t) + \cos \theta(t),$$

$$y(t + \Delta t) = y(t) + v(t) + \sin \theta(t).$$

Here, $V_{\max}$ is the maximum velocity, Δv is the change in velocity which is uniformly distributed between $[-A_{\max}\Delta t, A_{\max}At]$, and $A_{\max}$ is the maximum acceleration of a given node. The parameter $\Delta \theta$ means the change in direction which is uniformly distributed between $[-\alpha\Delta t, \alpha\Delta t]]$, and α is the maximum angular change in the direction.

To adapt different levels of randomness, one can use the *Gauss-Markov Mobility Model*. In this model initially each node is assigned a current speed and direction. In this mobility model at fixed intervals of time, n movement occurs by updating the speed and direction of each node. The value of speed and direction at the *n-th* instance are calculated based upon the value of speed and direction at the $(n+1)$-*st* instance and a random variable using the following equations:

$$s_n = \alpha s_{n-1} + (1 - \alpha)\,\vec{s} + \sqrt{(1 - \alpha^2)}s_{x_{n-1}},$$

$$d_n = \alpha d_{n-1} + (1 - \alpha)\,\vec{d} + \sqrt{(1 - \alpha^2)}d_{x_{n-1}},$$

where $\vec{s}$ and $\vec{d}$ are the new speed and direction of the node at time interval n, and $\alpha = 0$ is the tuning parameter used to vary the randomness, while s and d are constants representing the mean value of speed and direction. The parameters $s_{x_{n-1}}$ and $d_{x_{n-1}}$ are random variables from a Gaussian distribution. Totally random values are obtained by setting $\alpha = 0$ and linear motion is obtained by setting $\alpha = 1$ [5], while intermediate levels of randomness are obtained by varying the value of α between 0 and 1.

In the *City Section Mobility Model* we represent a section of a city where the ad hoc network exists [21]. The streets and speed limits on the streets are based on the type of city being simulated. The streets might form a grid in the downtown area of the city with a high-speed highway near the border of the simulation area to represent a loop around the city.

If a flexible mobility framework for hybrid motion patterns is needed, one can rely on the *Mobility Vector model* [26]. A mobility vector expresses the mobility of a node as the sum of two sub vectors: the Base Vector $\vec{B} = (bx_v, by_v)$ and the Deviation vector $\vec{V} = (vx_v, vy_v)$. The base vector defines the major direction and speed of the node while the deviation vector stores the mobility deviation from the base vector. The mobility vector $\vec{M}$ is expressed as $\vec{M} = B + \alpha V$ where α is an acceleration factor.

In mobile networks there are many situations where it is necessary to model the behavior of nodes as they move together as a group. These models are called group mobility models.

The most general of these group models is the *Reference Point Group Mobility (RPGM)* model. In the Reference Point Group Mobility model [25], the motion of the group center completely characterizes the movement of its corresponding group of nodes. The RPGM model represents the random motion of a group of nodes as well as the random motion of each individual node within the group. The logical center for the group is used to calculate group motion via a group motion vector $\vec{GM}$. In this mobility model the individual nodes randomly move about their own pre-defined reference points, whose movements depend on the group movement. When the updated reference points $RP(t+1)$ are calculated, they are combined with a random motion vector $\vec{RM}$, to represent the random motion of each node about its individual reference point. Specifically, three group mobility models (Column, Nomadic, and Pursue) can be implemented as special cases of the RPGM model.

In the *Exponential Correlated Random* Mobility Model [25] a motion function is used to create node movements. Given a node or group position at time t, and $\vec{b}(t)$ is used to define the next position at time $(t+1)$:

$$\vec{b}(t+1) = \vec{b}(t)\,e^{-\frac{1}{\tau}} + \left(\sigma\sqrt{1 - \left(e^{-\frac{1}{\tau}}\right)^2}\right)r,$$

where τ adjusts the rate of change from the node's previous location to its new location, thus small τ equates to large change and r is a random Gaussian variable with variance σ. We notice, that it is not easy to create a given motion pattern by selecting appropriate values for (r, σ) in the Exponential Correlated Random Mobility Model [25].

This *Column Mobility Model* [20] represents a set of nodes that move around a given line or column, which is moving in a forward direction. The model is useful for scanning or searching purposes and a slight modification of the Column Mobility Model allows the individual nodes to follow one another. Each node is placed in relation to its reference point in the reference grid; the node is then allowed to move randomly around its reference point via an entity mobility model.

In the *Nomadic Community Mobility Model*, each node uses an entity mobility model like the Random Walk Mobility Model, to roam around a given reference point. The model represents groups of nodes that collectively move from one point to another [14]. However, within each community or groups, the individuals move in random ways. When the reference point changes, all nodes in the group travel to the new area defined by the reference point and then begin roaming around the new reference point.

The *Pursue Mobility Model* [14] represents nodes tracking a particular target. The model consists of a single update equation for the new position of each node. The *random vector* value is obtained via an entity mobility model like the Random Walk Mobility Model. The amount of randomness is limited in order to maintain effective tracking of the node being pursued. The model combines the current position of a node, a random vector, and an acceleration function to calculate the next position of the node.

3. Related work

Mobility prediction provides useful input for dimensioning and planning of wireless mobile networks, ad hoc routing algorithms, efficient multicast transmission and call admission control [19,23].

In this chapter we present a brief summary on mobility prediction algorithms.

The *shadow cluster* scheme [8] estimates future resource requirements in a collection of cells in which a mobile is likely to visit in the future. The shadow cluster model makes its prediction based on the mobile's previous routes. In this model, the highway traffic with various constant speeds is simulated and users travel in forward and backward directions. The shadow cluster model improves estimation of resources and decision of call admission. In the study by Chao and Chen (1997), user mobility is estimated based on the aggregate history of handoff observed in each cell Shadow Cluster Concept takes its prediction, based on the mobile's previous routes.

The *proximity model* [2] minimizes the requirement for precise mobility information and computes the initial baseline link availability assuming random independent mobility. This model aims to quantify the future proximity of adjacent nodes and reflects the future stability of a given link. The model adapts future computations depending on the expected time-to-failure of the link. The total link availability between two nodes m and n is expressed as:

$$A_{m,n}^T(t) = A_{m,n}^i(t) P_i + A_{m,n}^c(t)(1 - P_i),$$

where $A_{m,n}^T(t)$ is total link availability, $A_{m,n}^i(t)$ is the availability when mobility is independent and $A_{m,n}^c(t)$ is the availability when mobility is correlated. The metric reflects independent or correlated behavior as given by the value of P_i. A value of $P_i = 1$ reflects independent movement with respect to the total link availability.

The *Sectorized ad hoc mobility prediction* scheme [18] achieves maximum accuracy in movement prediction. In this model the prediction process should be restricted to areas of high cluster change probability. The sectorized ad hoc mobility prediction scheme makes use of the cluster-sector numbering scheme to predict user movements in an ad hoc network.

In the *cluster based* [18] model the cluster head has complete knowledge of each of its member nodes. In this model the location of the user is defined with respect to its position with that of the cluster head. Assuming a circular cluster structure there is a region of the cluster in which all the nodes belonging to the cluster are in closest proximity to each other. All nodes in this region of the cluster are within communication range of each other, and the nodes in this region of the cluster will not satisfy the requirements for membership to any of the neighbouring clusters.

A prediction mechanism for *link expiration time* between any two ad hoc nodes has been observed [11] to enhance various unicast and multicast ad hoc routing protocols. In this model, if node i and node j at positions (x_i, y_i) and (x_j, y_j) are travelling at speeds v_i and v_j with moving directions θ_i and θ_j respectively with a transmission range r, then the time period D_t during which they would stay connected is predicted as:

$$D_t = \frac{-(ab + cd) + \sqrt{(a^2 + c^2) r^2 - (ad + bc)^2}}{a^2 + c^2},$$

where $a = v_i \cos\theta_i - v_j \cos\theta_j, b = x_i - x_j, c = v_i \sin\theta_i - v_j \sin\theta_j$, and $d = y_i - y_j$.

In wireless ad hoc networks, *network partitioning* occurs when mobile nodes moving with diverse mobility patterns. If the network consists of two mobility groups C_j and C_k, and each moving at velocities W_j, and W_k, the relative mobility is obtained by fixing one group stationary. The effective velocity W_{jk} at which C_k is moving away from C_j is given as $W_{jk} = W_k + (-W_j)$. Assuming that all groups have a circular coverage region of diameter D wherein the nodes are uniformly distributed and are in overlap, C_k must move past a distance of the diameter D of C_j's coverage area. The time taken for the two groups to change from total overlap to complete separation is given as:

$$T_{jk} = \frac{D}{\sqrt{w_{jk,x}^2 + w_{jk,y}^2}},$$

where $W_{jk} = \left(w_{jk,x}^+ \, w_{jk,y} \right)$. The partition prediction method employed in a clustering algorithm exhibits perfect accuracy of node classification [13].

4. Methods to enhance the accuracy of mobility models

The basic idea of our work is to utilize the additional mutual information available in the time-series of mobile users' movement patterns in cellular mobile networks. In our work we assume that given traces can be extracted from the mobile service provider's network history. The network management dataset consists of network management signals that were transferred in the network during the examined time interval. Beside many other network parameters and properties, two main information sets can be recovered:

– the cell-path that each user visited before
– the time intervals users have spent in each cell

The series of visited cells is crucial to analyze the similarities in the users' motion. Based on the motion patterns of the terminals amongst the cells, we can describe some drifts of the users' motion in a given cell or point. A drift may be caused by geographical or infrastructural objects (like highways, etc.) or some time-dependent circumstances (like mass events, concert, football matches, etc.).

From a mathematical point of view, the drift of the motion can be interpreted as different transition probabilities from one cell to another. A probability-vector can be defined for each cell that describes the probabilities of moving from the source cell to an adjacent one. This vector is called Handover Vector (HOV) [10]. The HOV is a discrete distribution, it contains transition probabilities to neighboring cells on the condition that the user moves out from the given cell [4]. The HOV might have time dependent components (e.g. the morning or the afternoon rush hours). We describe the usage of the HOV in Section 4.2.

The cell dwell time is an important parameter that helps to estimate the velocity of the motion. Velocity can vary in a wide range since the users frequently change their speed or motion direction.

Motion dynamics depends on the speed and the drift. From the network operator's point of view the accurate prediction of these components are useful to estimate the users' position in the near future, especially in case of handoffs. The analytically precise model that is able to predict both, helps the network operator in dynamic resource planning for short intervals based on actual network state. A long-term resource planning scheme (i.e. day-night scheme) can be supported by a short-term prediction that can follow dynamic escalations of call-initiations or degradation of the quality of the radio channel.

In this section we propose extensions to well known mobility models, in order to model motion drift and velocity with direction and cell dwell time prediction.

4.1. Random walk

Although the Random Walk model is easy to be used, it has disadvantages in user motion modeling:

– Without any improvements, the RW model is not capable of simulating the mobility in an environment where geographical or infrastructural objects determine the motion behavior. Beside this, the RW model simulates the user distribution in the network in a uniform way which is clearly not applicable in real-life situations.

– The model steps in each time slot. The previously visited states or the origin state are recurrent in one and two dimensions, but the model does not allow the same state in two following time slots. Thus the time spent in each state is not taken into account, each time tick means a transition to another cell that can be interpreted as a constant velocity user-motion.

– The model does not include the user's motion history, the states that were visited in the past. The uniformly distributed successor directions in a given state are not precise enough in a real-life application. It can be seen that the possibility of moving forward in a user drift is higher than stepping backward. The sophisticated weighted possibilities of successor directions can be constructed by knowing the previously visited cells [7] or by mapping the geographical or infrastructural circumstances of the area covered by the given cell.

4.2. Random walk model extensions

We have decided to use the RW model since it can be applied in case of a wide range of motion patterns (from pedestrian walk to highways) and can be specialized for uncommon scenarios also. The focus of our work is the elimination of the drawbacks outlined in Section 4.1. With a few extensions the RW model becomes able to more accurately estimate future user distribution in a network or cell cluster.

Since the RW model assumes a motion with constant absolute value of velocity it cannot simulate different movement speeds. The best method to implement this feature is ensuring the possibility of staying in the same RW state for arbitrary amount of time. This can be achieved by two different approaches that mostly give similar results.

According to the basic RW model the user remains in the actual cell until the end of the actual time slot. The lengths of slots are equal, since it is a discrete time system. We propose two methods to replace the constant time slot-lengths with a distribution that converges better to real-life motion patterns.

The handoff trace can be used to calculate several dwell times for each cell. Based on these derived data-series, we modeled each cell by a phase-type (PH) system, which produces a distribution of the cell dwell time with the appropriate absorption time. The phase-type system is defined in Eq. (1) with a transient matrix ($\underline{\mathbf{A}}$), a vector that contains the transition intensities to the absorbing state (**a**) and the initializing vector (α).

$$Q = \begin{bmatrix} \mathbf{A} \ \mathbf{a} \\ 0 \ 0 \end{bmatrix} \quad pdf_{PH} = f(A, \alpha) \tag{1}$$

The Cell Phase Type System is defined based on the trace, a unique distribution can be created for each cell. It is started when a user registers in the given cell, and the model unregisters the user when the PH system absorbs. Immediately after the absorption the user registers into one of the adjacent cells.

In Fig. 1 the simplest PH system can be seen, combined with the uniform successor distribution. The PH contains one transient state which means that the absorption time is exponentially distributed.

Based on the handoff trace another dwell time simulation method can be derived. We provided the possibility of staying in the same cell for the next fixed length time slot as an extension of the RW model. There is an additional transition, the loopback direction. The simple RW model does not allow the user to stay in the same state, so our proposition enables the model to simulate different cell dwell times.

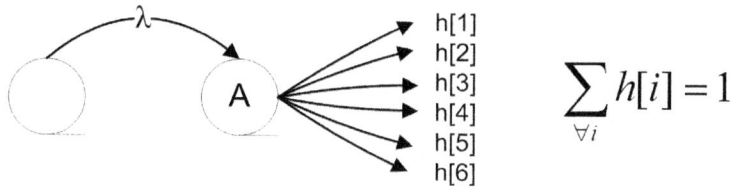

Fig. 1. Exponentially distributed (exp(λ)) dwell time simulator with HOV.

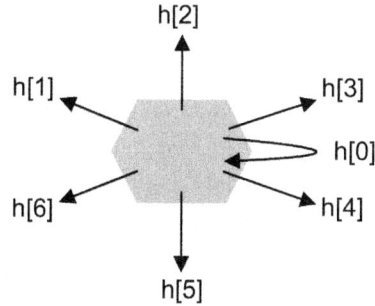

Fig. 2. Looping transition in the modified RW structure.

Formally that means the probability of stepping into the same cell at the end of the slot, such as a looping step shown in Fig. 2.

With the correct tuning of the loopback value, arbitrary distributions of cell dwell times can be simulated if the time slot length is appropriately chosen.

$$\mathbf{h'} = \big[\mathbf{h'}[0], \mathbf{h'}[1], \mathbf{h'}[2], \mathbf{h'}[3], \mathbf{h'}[4], \mathbf{h'}[5], \mathbf{h'}[6]\big]$$
$$\mathbf{h'}[0] = pstay \tag{2}$$
$$\mathbf{h'}[i] = (1 - pstay) \cdot \mathbf{h}[i]$$

Equation (2) shows that the outgoing transition probabilities have to be weighted in order to keep them as a distribution. This model is the Extended Random Walk (ExtRW).

The remaining drawback of the Extended Random Walk model is the lack of handling different motion directions. Our extension substitutes the uniformly distributed random successor direction in ExtRW to a special distribution that is valid only in the given cell. This distribution is represented by the Handover Vector (HOV). The HOV can be determined from the trace of handoffs made in an arbitrary network. The given data of the trace provides measures of the relative frequency of handoffs between each adjacent cell-pairs. Based on the relative frequency, the probabilities of the HOV can be easily calculated. The HOV is cell-specific, each cell has its own geographical properties that affects the users' motion drift and the transition probabilities into the adjacent cells.

The Extended Random Walk model combined with the cell-dependent transition probabilities gives the best estimation of future user distribution. In following sections the simulation results prove that the Handover Vector can simulate the constant or time-dependent changing density of the users distributed in the different cells of the network.

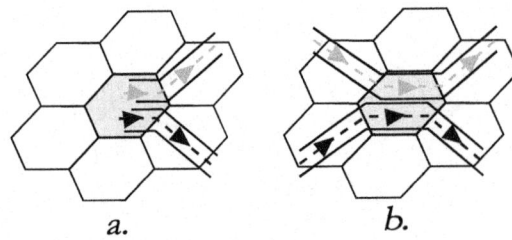

a.   b.

Fig. 3. User prediction methods; a. model without memory, unknown where is the users come from; b. model with memory, the previous steps of the users taken in account.

4.3. The effect of memory in the model

The Random Walk model, as it was analyzed in the previous section, is the most commonly used mobility model to describe individual movement behavior. The extended RW and HOV eliminate the disadvantages of the RW model with variable dwell time and weighted transition probabilities calculated from known parameters of the system in the past.

The limit of accuracy even in the most sophisticated HOV model is caused by the lack of memory, since the RW-like models do not possess the knowledge of the actual distribution of recent moments. Beside the trends of user movement flows of a longer period in the past, the application of the recent user locations have a capital importance in variable, directional user motion. Neglecting the actual transition series of a user in the cluster, i.e. the model is memoryless, the estimation works with a significantly higher error rate. Figure 3 shows a simple example which represents the effect of memory in the model. If we consider two roads shown is Fig. 3b, the accuracy of transition probability estimations are better when the model knows where the users come from than the RW-like estimation which cannot differentiate the users on the two roads (Fig. 3a).

To clarify the error rate of a memoryless HOV model compared to an algorithm that possesses memory, we use the example cells shown in Fig. 3. Let us define the following parameters:

- the number of incoming users on the upper road at timeslot t is $in1_t$,
- the number of incoming users on the lower road at timeslot t is $in2_t$,
- similarly the number of users leaving on the upper road at timeslot t is $out1_t$,
- the number of users leaving on the lower road at timeslot t is $out2_t$,
- and the user movement directions with a simple transition matrix

$$\mathbf{P} = \begin{bmatrix} p_{11} & p_{12} \\ p_{21} & p_{22} \end{bmatrix}$$

The HOV model in Fig. 3a calculates the number of leaving users ($out1$, $out2$) with a historical estimation of the $\mathbf{P}$ matrix, and with the sum of $in1$ and $in2$ (the total number of users in the observed cell), but without the knowledge of $in1$, $in2$.

In our proposed algorithm, the use of $in1$, $in2$ means that the model calculates based on the information of the incoming users. The incoming users can be considered uniform or different (marked users). Since the latter is more accurate, in this comparison we use marked users.

Assume that the historical estimation of the HOV model's $\mathbf{P}$ matrix is based on the previous timeslot. That is the P($out1_{t+1}$) and P($out2_{t+1}$) probabilities are $out1_t/(out1_t+out2_t)$ and $out2_t/(out1_t+out2_t)$, respectively. Applying the same assumption on the algorithm with memory, the number of leaving users can be calculated with the $\mathbf{P}$ matrix itself, that is $out1_{t+1} = in1_t*p_{11} + in2_t*p_{21}$ and $out2_{t+1} = in1_t*p_{12} +$

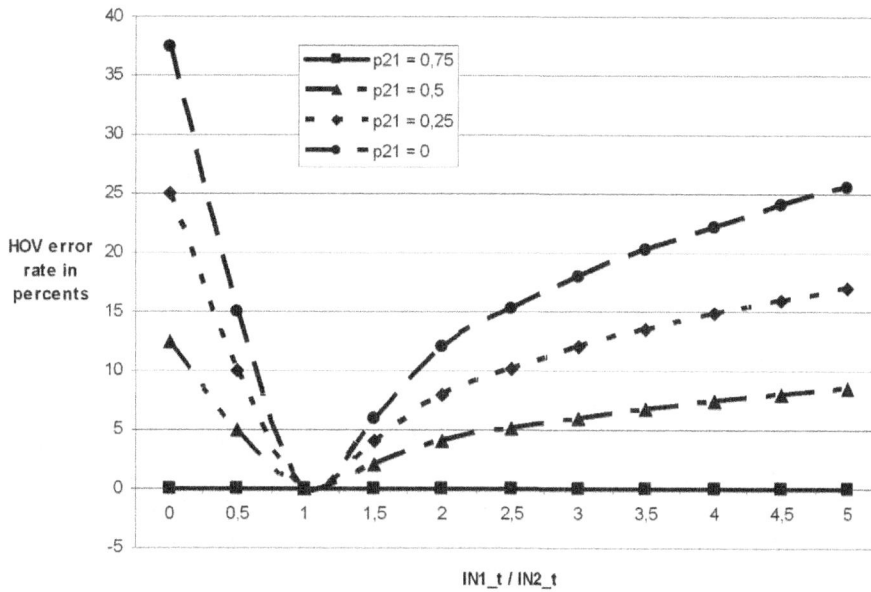

Fig. 4. HOV prediction error in percents – Given $in1_{t-1}/in2_{t-1} = 1$ and P $= \{\{0.75, 0.25\}.\{p_{21}, 1\text{-}p_{21}\}\}$, p_{21} plotted with four different values.

$in2_t{}^*p_{22}$. At a given and constant **P** matrix let us assume that the incoming user distribution varies, that is the $in1_t/in2_t$ ratio (Incoming Distribution – ID) changes. Figure 4. shows the error of HOV compared the estimation using memory.

The HOV model works with error if ID_{t-1} is different from ID_t which is caused by the fact that the HOV historical **P**-estimation in this special case equals the number of leaving users of the previous timeslot. That is it does not include the actual ID_t value. Contrarily the memory-model calculates with the actual number of incoming users and the **P** matrix itself, which results the exact probabilities of the leaving users distribution. The error rate caused by the lack of memory increases as the variance of ID increases, that is $in1_t/in2_t$ ratio changes.

Using memory cannot enhance furthermore the accuracy of the estimation if $p_{21} = p_{11}$ and Fig. 4. shows a constant zero error rate ($p_{11} = p_{21} = 0.75$). In this case the leaving direction of each user is independent of the incoming direction and the memory is useless, since users arriving from each direction are leaving towards a given direction with the same probabilities.

The results show that our proposition of using memory in a mobility model siginifically increases the accuracy of the model in cases where the ID distribution in an arbitrary cell has high variance, or has periodicities without stationary distribution.

To increase the accuracy with the use of memory in the mobility model, we introduce our Discrete Time Markov model with memory (M_n, where n denotes the number of Markov states). The states of the model show and store where the users came from. The memory can be interpreted in two meanings. Time dimension memory shows the number of timeslots in the past that the model considers. Thus a model with m time dimensions in time t can calculate the next transtion based on the user position in $(t - m, \ldots t - 2, t - 1)$. Direction dimensions memory shows the number of directions that the model can differentiate. In general a cell cluster consists of hexagonal cells. The direction dimension of a model on this cluster is maximum 6. If transition from the central cell to two adjacent cells are not differentiated then the direction dimension is decreased by 1.

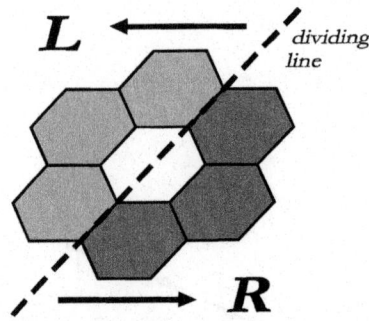

Fig. 5. Neighbour cells separated into two groups.

Obviously with increasing the dimensions of time and direction memory exponentially increases the complexity of the model, contrarily the logarithmical increase of accuracy. Our work was motivated to find an optimal size of estimation parameter space with the highest accuracy beside tolerable complexity. In the following subsections we will give example models with different time and direction dimension. We will show the coherence of the complexity and accuracy in a model with a given statespace.

4.4. Markov model extensions

In our method the direction of a user is identically distributed between 0 and 2π. The user's speed is between 0 and $V_{\max}$. After moving in a direction with a randomly chosen speed for a given Δt time, the user changes its direction and speed. N_{cell} denotes the number of cells in the network.

In a simple Markov-chain based model a user can be located in three different states during each time slot, the stay state (S) and the left-move state (L) and the right-move state (R). This classification of cells can be seen in Fig. 5.

Let us define $X(t)$ random variable, which represents the movement state of a given terminal during time slot t. We assume that $\{X(t), t = 0, 1, 2,...\}$ is a Markov chain with transition probabilities p, q, v.

We assume that a user is in cell i at the beginning of a timeslot. If the user is in state S, similarly to the previous model, it remains in the given cell in the next slot. If the user is in state R, it moves to one of the cells on the right-hand side, if in state L, it moves to the left-hand side of the dividing line (direction dimension is 3).

Since the transition propensities are not symmetric, the left-move state and the right-move state have different probabilities. Figure 6 depicts the Markov chain and transition (II) matrix.

Transition probabilities p, q and v can be determined based on the network parameters.

The balance equations for this Markov chain are given in Eq. (3).

$$P_S \cdot (p_1 + p_2) = P_L \cdot (1 - q_1 - v_2) + P_R \cdot (1 - q_2 - v_1)$$
$$P_L \cdot (1 - q_1) = P_S \cdot p_1 + P_R \cdot v_1 \tag{3}$$
$$P_R \cdot (1 - q_2) = P_S \cdot p_2 + P_L \cdot v_2$$

We also know that $P_S + P_L + P_R = 1$ thus the steady state probabilities can be calculated.

With knowledge of the result we can predict the number of mobile terminals for time slot $t + 1$ for each cell, using Eq. (4), where $S^i_{adj(r)}$ is the set of right neighbours of cell I.

$$N_i(t + 1) = N_i(t) \cdot P_S(i) + + \frac{1}{3} \sum_{l, C_l \in S^i_{adj(l)}} N_l(t) \cdot P_R(l) + \frac{1}{3} \sum_{r, C_r \in S^i_{adj(r)}} N_r(t) \cdot P_M(r) \tag{4}$$

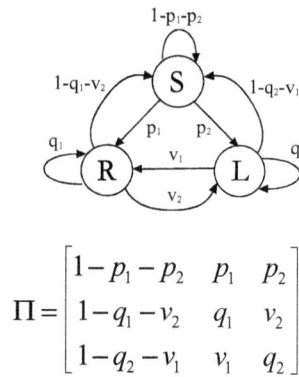

$$\Pi = \begin{bmatrix} 1-p_1-p_2 & p_1 & p_2 \\ 1-q_1-v_2 & q_1 & v_2 \\ 1-q_2-v_1 & v_1 & q_2 \end{bmatrix}$$

Fig. 6. State diagram and Ï matrix of 3-state M-model (M_3).

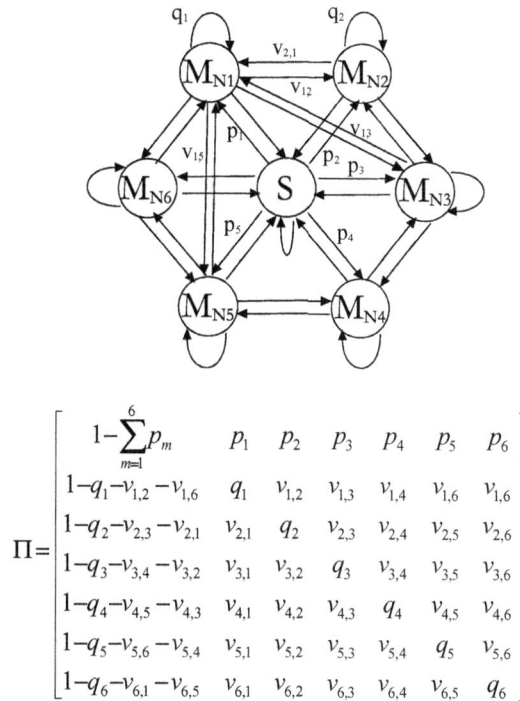

$$\Pi = \begin{bmatrix} 1-\sum_{m=1}^{6}p_m & p_1 & p_2 & p_3 & p_4 & p_5 & p_6 \\ 1-q_1-v_{1,2}-v_{1,6} & q_1 & v_{1,2} & v_{1,3} & v_{1,4} & v_{1,6} & v_{1,6} \\ 1-q_2-v_{2,3}-v_{2,1} & v_{2,1} & q_2 & v_{2,3} & v_{2,4} & v_{2,5} & v_{2,6} \\ 1-q_3-v_{3,4}-v_{3,2} & v_{3,1} & v_{3,2} & q_3 & v_{3,4} & v_{3,5} & v_{3,6} \\ 1-q_4-v_{4,5}-v_{4,3} & v_{4,1} & v_{4,2} & v_{4,3} & q_4 & v_{4,5} & v_{4,6} \\ 1-q_5-v_{5,6}-v_{5,4} & v_{5,1} & v_{5,2} & v_{5,3} & v_{5,4} & q_5 & v_{5,6} \\ 1-q_6-v_{6,1}-v_{6,5} & v_{6,1} & v_{6,2} & v_{6,3} & v_{6,4} & v_{6,5} & q_6 \end{bmatrix}$$

Fig. 7. State diagram and Π matrix of seven-state Markov model.

As we mentioned earlier this model performs well when the user's movement has only one typical direction, because in this case the handover intensities of the right-move cells does not differ significantly. It is the same in the aspect of the left-move cells.

If we try to predict the user's distribution in a city having irregular, dense road system, or in a big park where people are able to move around then the handover intensities could differ thus the calculations above could produce errors. From this point of view the best way is if we represent all of the neighbour cells as a separated Markov state, so we create 7 states, because we assume 7 elements in a cluster (direction dimension is 7) :

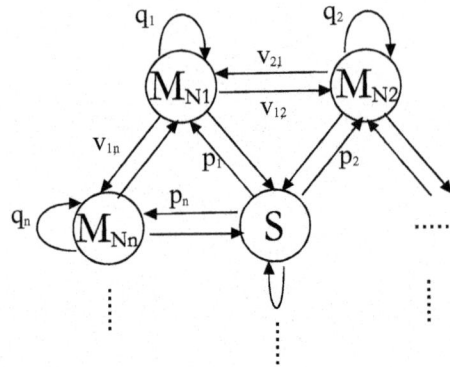

$$\Pi = \begin{bmatrix} 1-\displaystyle\sum_{m=1}^{n} p_m & p_1 & p_2 & \cdots & \cdots & \cdots & p_n \\[2mm] 1-q_1-v_{1,2}-v_{1,n} & q_1 & v_{1,2} & \cdots & \cdots & \cdots & v_{1,n} \\[1mm] \cdots & \cdots & \cdots & \cdots & \cdots & \cdots & \cdots \\ \cdots & \cdots & \cdots & \cdots & \cdots & \cdots & \cdots \\ 1-q_i-v_{i,i+1}-v_{i,i-1} & \cdots & \cdots & v_{i,i-1} & q_i & v_{i,i+1} & \cdots \\ \cdots & \cdots & \cdots & \cdots & \cdots & \cdots & \cdots \\ 1-q_n-v_{n,1}-v_{n,n-1} & v_{n,1} & \cdots & \cdots & \cdots & v_{n,n-1} & q_n \end{bmatrix}$$

Fig. 8. State diagram and Π matrix of n+1-state Markov model (M_{n+1}).

– stationary state (S)
– move to neighbour 1..6 state $(M_{N1}..M_{N6})$

The Markov chain and transition matrix are more complex as it can be seen in Fig. 7. As in the previous cases the steady state probabilities can be evaluated.

To calculate the number of users in C_i for time slot t, then Eq. (5) is to be used, where the neighbour cells of cell i are indexed from 1 to 6.

$$N_i(t+1) = N_i(t) \cdot P_S(i) + \sum_{j, C_j \in S^i_{adj}} N_j(t) \cdot P_{M_{Nj+3\mathrm{mod}6}}(l) \tag{5}$$

It is to be taken into account that in the real networks a cell does not always have six neighbours depending on the coverage. This model has to be generalized for a common case when a cell has n neighbouring cells. We expanded our previous mentioned model to n-neighbour case (Fig. 8), when all the n neighbours are represented with a Markov state:

– stationary state (S)
– move to neighbour 1..n state $(M_{N1}..M_{Nn})$

The steady state probabilities can be calculated as in the previous cases Eq. (6).

$$P_S \cdot \sum_{i=1}^{n} p_i = \sum_{j=1}^{n} (1 - q_j - \sum_{l \neq k}^{n} v_{k,l}) P_{Ni}$$

$$
\begin{array}{c|cccc}
 & U_1 & U_2 & U_3 & ... \\
\hline
t_1 & C_i & C_j & C_j & ... \\
t_2 & C_k & C_k & C_j & ... \\
t_3 & C_k & C_l & C_j & ... \\
t_4 & C_j & ... & ... & ... \\
\end{array}
$$

Fig. 9. The format of the result trace (U_x means the user x).

.... (6)

$$
P_{Nk}(1 - q_k) = P_S \cdot p_n + \sum_{i \neq k} P_{Ni} \cdot v_{i,k} \quad 1 \leqslant k \leqslant n
$$

...

Using the result the predicted number of users in the next time slot is given in Eq. (7), where $P_{M_i}(j)$ denotes the steady state probabilities of moving to the cell i from cell j.

$$
N_i(t + 1) = N_i(t) \cdot P_S(i) + \sum_{j, C_j \in S^i_{adj}} N_j(t) \cdot P_{M_i}(j)
$$

(7)

4.5. Determining the parameters and calculating the error of the model

In this chapter we present the transition probabilities and its determination of our Markov model. We introduce a method to process the network traces. In this section we show the error of the module based on the parameter determination.

The logs from the network provider contain the handover information and current time. Namely the handover information consist of the user ID and a source and destination cell. We split the examined interval into Δt parts. In every timeslot a cell is appended to every user. This results the input for the determination algorithm (Fig. 9).

First of all, the transition probabilities, used in previous section, are defined as:

$p_n =$ the probability of moving from stationary state (S) to M_n
$q_n =$ the probability of staying in a move state (M_n)
$v_{n,m} =$ the probability of moving from a move state (M_n) to another one (M_m)

These values are calculated based on above mentioned result trace. Let us introduce the following terminologies to process the traces and to determine the probabilities mentioned above:
Pa(X,Y)

– A two-step move pattern (Fig. 10) from C_i to C_j, in other way a user, who is in C_i at time t, and in C_j at time t+1. This trace could be used for ExtRW and HOV parameter determination also, in fact this is the "memory less trace"; we don't know, where the user came from, we only know is where it is heading to : $Pa(C_i, C_j)$.

Pa(X,Y,Z)

– A three-step move pattern (Fig. 10) from C_h to C_j, across C_i, in another way a user, who is in C_h at time t-1, in C_i at time at t, and in C_j at time t+1. This trace could be called a "memory trace", because during we are in C_i, we know, that the user came from C_h): $Pa(C_h, C_i, C_j)$.

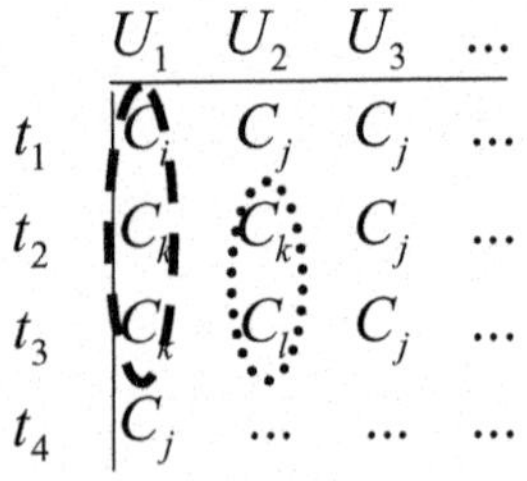

$$
\begin{array}{c c c c}
 & U_1 & U_2 & U_3 & \cdots \\
t_1 & C_i & C_j & C_j & \cdots \\
t_2 & C_i & C_k & C_j & \cdots \\
t_3 & C_i & C_l & C_j & \cdots \\
t_4 & C_j & \cdots & \cdots & \cdots
\end{array}
$$

Fig. 10. Example for the move patterns. Dashed line: three-step move pattern, Dotted line: two-step move pattern.

- For all users and for all timeslots we search the $Pa(C_a, C_b)$ pattern in the trace . The number of the found patterns:
  $Num(Pa(C_a, C_b))$.
- The index of a cell, which is in z direction from the current C_j: $Ind(z,i)$.

Using these determinations the probabilities belonging to C_j can be given by the following way from real network traces:

$$
\hat{p}_{(j)\,i} = \frac{Num(Pa(C_j, C_j, C_{Ind(j,i)}))}{\displaystyle\sum_{l=0}^{n}\sum_{k=0}^{n} Num(Pa(C_{ind(j,l)}, C_j, C_{ind(j,k)})) = Sum[j]} \tag{8}
$$

$$
\hat{q}_{(j)\,i} = \frac{Num(Pa(C_{ind(j,i)}, C_j, C_{Ind(j,i)}))}{Sum[j]} \tag{9}
$$

$$
\hat{v}_{(j)\,i} = \frac{Num(Pa(C_{ind(j,i)}, C_j, C_{Ind(j,i)}))}{Sum[j]}, \tag{10}
$$

Based on these equations a prediction can be given for the users distribution. The question is the accuracy of the predicting model.

It can be seen that these $\hat{p}_i, \hat{q}_i, \hat{v}_{i,k}$ values were defined based on relative frequency in the network traces. Accuracy of this calculation depends mainly on the number of samples. That is a finite set of samples could not be sufficient, there is always a minimal error which is the difference between the real, model $(p_i, q_i, v_{i,k})$ and the calculated $(\hat{p}_i, \hat{q}_i, \hat{v}_{i,k})$ values. Let us define the $\varepsilon_p, \varepsilon q, \varepsilon_v$ as the error of $p_i, q_i, v_{i,k}$ determination of the parameters of the Markov model (As ε_p in Eq. (11), $\varepsilon_q, \varepsilon_v$ can be determined similarly).

The mean value of calculated parameters (for example $E(\hat{p}_i)$) is the real value because of the law of averages.

$$
\varepsilon_p = |\hat{p}_i - p_i| = |\hat{p}_i - E(\hat{p}_i)| \rightarrow p_i = \hat{p}_i + \varepsilon_p \tag{11}
$$

As we defined earlier the n-state Markov model (see Chapter 2.3.1) satisfies the general matrix equation $P = P\Pi$, which can be solved. But we are not able to determine exactly the Π matrix. Calculating with the relative frequencies we get a $\hat{\Pi}$ matrix, which estimates the Π matrix with calculation errors. When we solve the matrix equation $P = P\Pi$ using $\hat{\Pi}$ instead of Π, we get a $\hat{P}$ equals P with the addition of the model error. Summarizing these coherences:

- theoretical solution: $P = P\Pi$

– practical solution: $\hat{P} = \hat{P}\hat{\Pi}$

$$\hat{P} = P + H \tag{12}$$

$$\hat{P} = \hat{P}\hat{\Pi} \rightarrow P + H = (P + H)(\Pi + E) \tag{13}$$

where H denotes the error of the steady state probability comeing from the parameter calculation error, and E means the matrix derived from the $\varepsilon_p, \varepsilon_q, \varepsilon_v$ values. Our aim by the following is to calculate the dependency of the model error on the network parameters.

Let us start from the Eq. (6). described with the calculated parameters instead of the real parameters Eq. (14).

$$\hat{P}_S \cdot \sum_{i=1}^{n} \hat{p}_i = \sum_{j=1}^{n} \left(1 - \hat{q}_i - \sum_{l \neq j}^{n} \hat{v}_{j,l} \right) \hat{P}_{Ni}$$

$$\ldots \tag{14}$$

$$\hat{P}_{Nk}(1 - \hat{q}_k) = \hat{P}_S \cdot \hat{p}_n + \sum_{i \neq k} \hat{P}_{Ni} \cdot \hat{v}_{i,k} \quad 1 \leqslant k \leqslant n$$

$$\ldots$$

Replace the variables by Eq. (12) and 11 to get the following equation Eq. (15).

$$(P_S + H_S) \cdot \sum_{i=1}^{n} (p_i + \varepsilon) = \sum_{j=1}^{n} (1 - (q_i + \varepsilon) - \sum_{l \neq j}^{n} (v_{j,l} + \varepsilon))(P_{Ni} + H_{Ni})$$

$$\ldots$$

$$(P_{Nk} + H_{Nk})(1 - (q_k + \varepsilon)) = (P_S + H_S) \cdot (p_n + \varepsilon) + \sum_{i \neq k} (P_{Ni} + H_{Ni}) \cdot (v_{i,k} + \varepsilon) \tag{15}$$

$$\ldots$$

$$1 \leqslant k \leqslant n$$

As next step we denote H_S, H_{Nk} using the terms from Eq. (6). and the coherence $\varepsilon = \max(\varepsilon_p, \varepsilon_q, \varepsilon_v)$:

$$H_S = \sum_{j=1}^{n} (1 - q_i - \sum_{l \neq j}^{n} v_{j,l})H_{Ni} + (1 - \sum_{i=1}^{n} p_i)H_S - n\varepsilon(nH_{Ni} + 1 + H_S)$$

$$\ldots$$

$$H_{Nk} = H_S p_m + \sum_{i \neq k} H_{Ni} \cdot v_{i,k} + H_{Nk}q_k + H_S\varepsilon + (nH_{Nk} + 1)\varepsilon \tag{16}$$

$$\ldots$$

$$1 \leqslant k \leqslant n$$

We have an equation system with $n + 1$ equation Eq. (16). Let us do some simplifying by calculating average of some parameters in the following way:

$$\begin{aligned} p &= avg(p_i) \ \forall i \\ q &= avg(q_i) \ \forall i \\ v &= avg(v_{i,j}) \ \forall i, \forall j \end{aligned} \tag{17}$$

These three general probability parameters can describe typical user movements in the current cell.

Parameter p means the probability of the moving after stop, q is for describing how the user can hold its moving direction, and at last v, the opposite of q, namely how often the moving direction of the users changed.

Using these general parameters yields an upper estimation with an equation system in 2 variables:

$$H_S = n(1 - q - (n-1)v)H_N + (1 - np)H_S - n\varepsilon(nH_N + 1 + H_S)$$
$$H_N = H_S p_m + (n-1)H_N v + H_N q_+ H_S \varepsilon + (nH_N + 1)\varepsilon$$

(18)

If we examine the ε error rate, we find that an upper estimation can be applied for it as well using the Chebyshev inequality. The variance of relative frequency is well known:

$$\sigma(\hat{p}) = \sqrt{\frac{p(1-p)}{m}}$$

(19)

where $p = E(\hat{p}_i)$, and m are the number of samples in the trace.

Let us make an upper estimation on Eq. (19) to eliminate variable p:

$$\sqrt{\frac{p(1-p)}{m}} \leqslant \sqrt{\frac{1}{4m}} = \frac{1}{2\sqrt{m}}$$

(20)

Using this result in the Chebyshev inequality:

$$P(|\hat{p} - E(\hat{p})| \geqslant \varepsilon) \leqslant \frac{1}{2\sqrt{m}\varepsilon^2}$$

(21)

If we assume that $|\hat{p} - E(\hat{p})|$ is not greater than ε with 99% probability, and that $m = 10000$, we reach an upper estimation for ε:

$$0.01 \leqslant \frac{1}{2\sqrt{m}\varepsilon^2}$$
$$\dots$$
$$-\frac{1}{\sqrt{2}} \leqslant \varepsilon \leqslant \frac{1}{\sqrt{2}} \to \varepsilon = \frac{1}{\sqrt{2}}$$

(22)

In order to get an equation system that depends only on p, q, v, the ε is replaced in Eq. (18).

$$H_S = n(1 - q - (n-1)v)H_N + (1 - np)H_S - n\frac{1}{\sqrt{2}}(nH_N + 1 + H_S)$$
$$H_N = H_S p_m + (n-1)H_N v + H_N q + H_S \frac{1}{\sqrt{2}} + (nH_N + 1)\frac{1}{\sqrt{2}}$$

(23)

H_S, H_N, the error of the steady state probability can be calculated from the equation system, but due to the limitation of this paper we do not present it in detail.

As the last step we define H_{Model}, the model error from H_S, H_N: $H_{Model} = H_S + nH_N$

Using the plotting abilities of *Mathematica* [16] we examined this model error depending on p, q, v, n. The next figures depict the results. The x axis shows the number of states (n), the y axis represents the error of the module.

We analyzed four different scenarios considered as typical user movements. The result of the first combination is depicted on Fig. 11. In this case the all of the general transition parameters (p, q, v) take low values, meaning that the users in the current cell move slowly, stop several times. The function of model error increases linearly, as it is depicted.

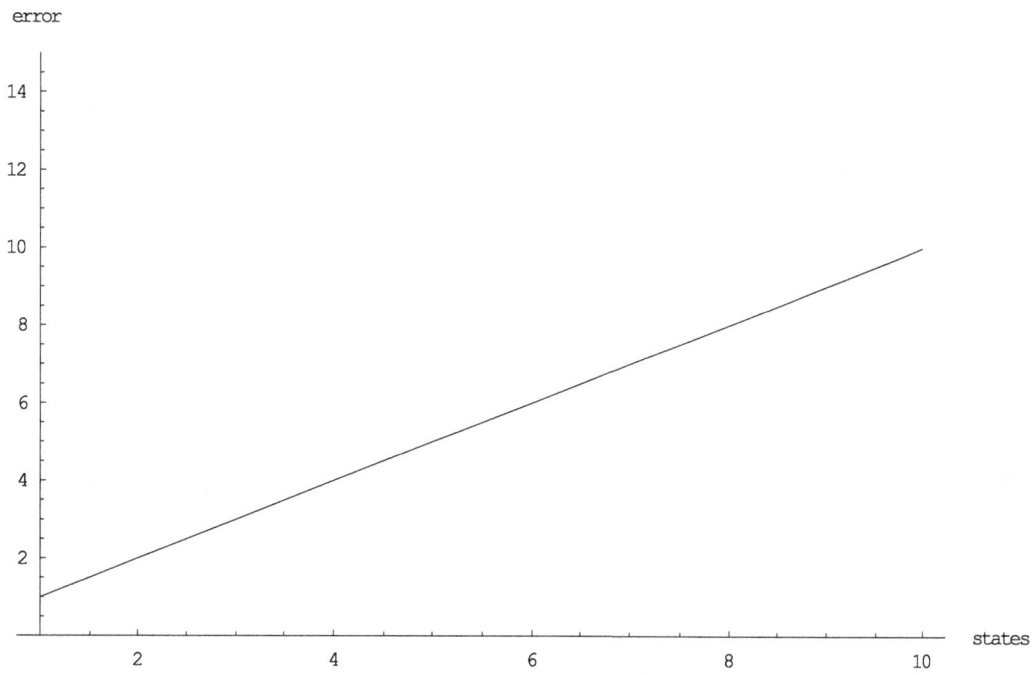

Fig. 11. Model error, $p = 0.1$, $q = 0.1$, $v = 0.1$; Slow motion in current cell.

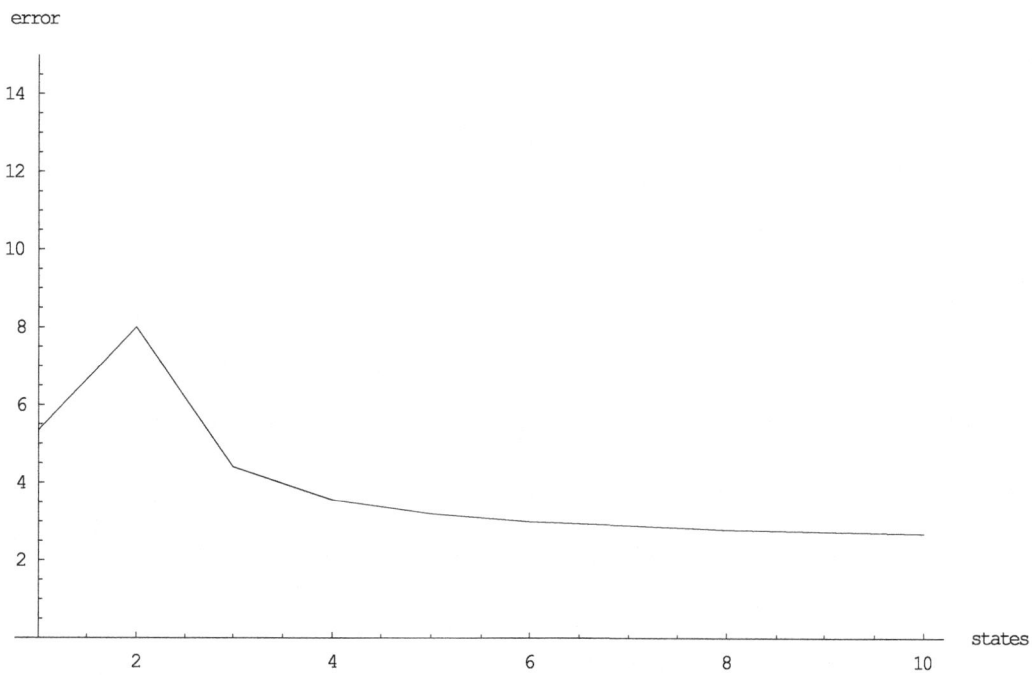

Fig. 12. Model error, $p = 0.2$, $q = 0.9$, $v = 0.9$; Fast motion with direction changes and direction holding as well.

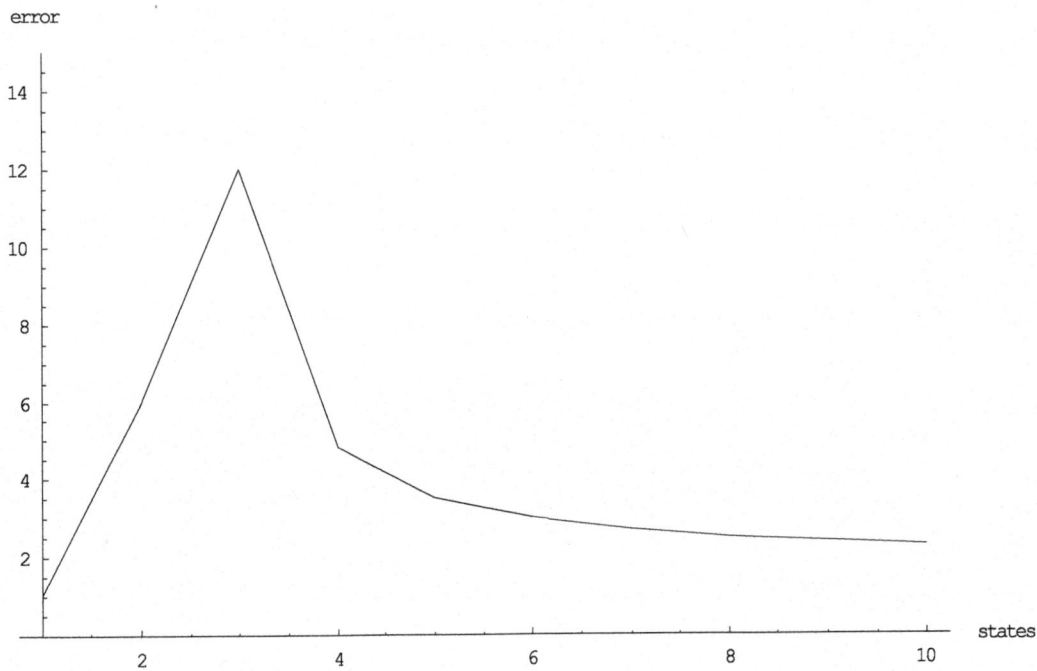

Fig. 13. Model error, $p = 0.2$, $q = 0.1$, $v = 0.9$; Fast motion with direction changes.

The second scenario shows the opposite of the previous case, namely fast movement in all direction, none of the users stops (Fig. 12).

Third case simulates a fast, direction-changing movement environment. The function on Fig. 13 has a local minimum in $x = 3$, after that it is decreasing logarithmical.

In the last situation the users move fast, and hold their direction. The model error increasing slowly (Fig. 14).

In this section we determined the previously introduced Markov model parameters from the network traces, defined that the model error derived from the parameter calculations, and examined it dependently on the user movement behaviors.

4.6. Complexity and accuracy of the extended markov model

In the previous section we introduced a Markov model generator method, an optional model can be derived depending on the resource requirements (complexity) and the demanded precision (accuracy).

The accuracy of the model is increasing in function of the number of states. The number of states is increasing when the memory (time dimension) is increased, or when the number of direction (direction dimension) is increased. Time dimension increases the states exponentially, direction dimension increases it linearly. With the state-space increasing, the computational complexity of the Markov steady state calculations are also following a rising curve. The question is the characteristic of these functions and the existence of a theoretical or practical optimum point.

It is assumed that each cell has N neighbors and the 3-state (stay, left and right-move state) model is used to determine the user movement. It is also assumed that N/2 cells belong to both left and right Markov-states, and the users are uniformly distributed between cells. A theoretical error can be derived

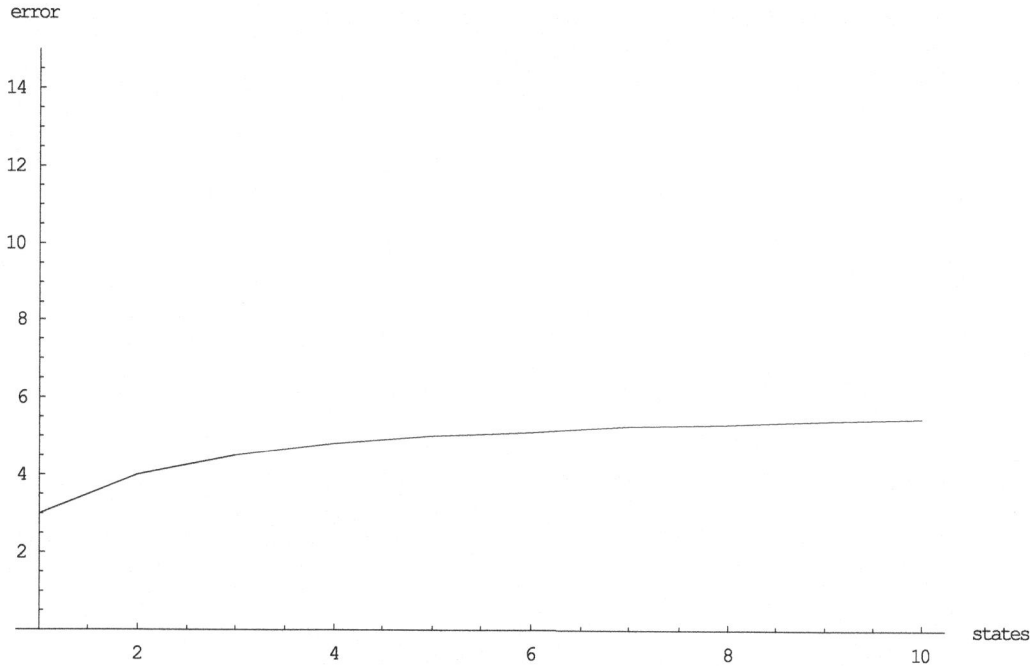

Fig. 14. Model error, $p = 0.3$, $q = 0.9$, $v = 0.1$ Fast motion with direction holding.

from this assumption since in most cases the user motion pattern does not result in a uniform distribution in the N/2 cells. In the worst case the users move with 1 probability into one of the neighbouring cells. In our estimation calculations the error can be measured with the difference between the uniform distribution and the worst case. This difference in mathematical equation is the following:

$$1 - \frac{1}{N/M} + \frac{1}{N/M} * ((N/M) - 1), \tag{24}$$

where N means the neighbour numbers, and M means the direction numbers in the model.

We measured the computation complexity also as the function of state number. This enables us to compare complexity and prediction error in an easy way. Based on the 1..M-state model the prediction computation is calculated with the costs of Markov steady state mathematical operations and other procedures necessary for transition probabilities. The complexity can be estimated with o($M_{\text{states}}^3 + M_{\text{states}} + 1/ M_{\text{states}}$).

Figure 15 shows the complexity and error characteristic. In the given model calculations the optimal point of operation is around 5 states where error is minimal at this level of complexity.

In the previous comparison the direction dimension sizing is used only. If we want to look back for the estimation, than memory has to be introduced into the model. In fact this means that every state in the Markov chain has to be changed with M states. This causes exponential state number explosion that can be seen on the Fig. 16.

As we have extended our model step by step in time and direction dimension, its precision increased as well as the complexity of the model. In order to decrease it, we developed a simple algorithm for direction dimension, which is able to minimize the number of states based on merging adjacent cells. The input parameters are the following:

Theoretical worst case error vs. Mathematical complexity

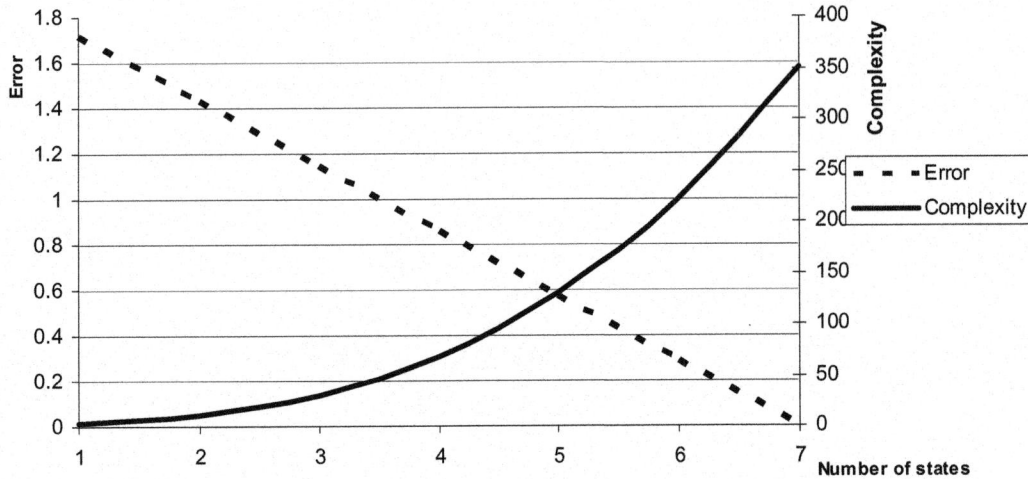

Fig. 15. Complexity (solid) and accuracy of the calculation (dashed).

State number explosion in case of time level

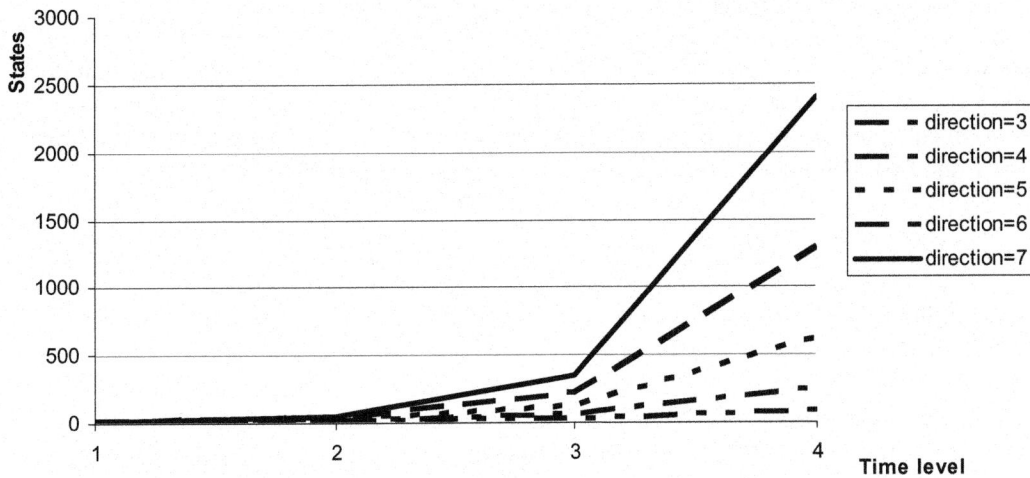

Fig. 16. The increase of number of states in case of different direction dimensions (3,4,5,6,7 states), and different time dimensions (1,2,3,4).

– an acceptable *Err* rate of error caused by the merging
– h_{kl} elements of the handover matrix (where h_{kl} is the handover probability from C_l to $C_k, \forall C_k, \forall C_l \in S^i_{adj}, k, l = 1 \ldots N_{cell}$). $Q_{Hi}(n_i + 1 \times n_i + 1)$ matrix is defined (where n_i is the number of neighbour cells of cell i). The elements in the first row and column are the handover probabilities of the examined cell i.

Based on the Q_{Hi} matrix (elements denoted by h^i_{kl}) and the error rate a group of cells is to be found, where the absolute difference between h^i_{ki} and h^i_{li}, and between h^i_{ik} and h^i_{il} ($\forall C_l, \forall C_k$ in a group) is less than *Err*.

The algorithm is given with its pseudo code:

original_Q = QHi;
minimal_grouping_Q = QHi;
minimal_grouping_num = ni;
For x = 2 to ni +1 {
  grouping.clear();
  nullstations.clear();
  QHi = original_Q;
  For j = x − 1 to ni + x − 1 {
    k = (j mod 7) + 1;
    If ((k! = x) && (k! = i) && (!grouping.member(k))) {
    *group_row = hik_*
    group_column = trans(hi_k);
    temp_g_row = 0;
    temp_g_column = trans(0);
    group[k].create();
    group[k].add(k);
    For l = 2 to ni {
      if (!grouping.member(l)) {
      greater_than_Err = false;
      For m = 1 to ni{
        if((!nullstation.member(m)) && (m! = l)){
        if((|hikm- hilm | >Err) || (|himk- himl | >Err)){
        greater_than_Err = true;
        break;
        } else{
        temp_g_row[m]=(|group_row[m]+ hilm |/2);
        temp_g_column[m]=(|group_column[m]+ himl |/2);
  }}} if (!greater_than_Err){

  nullstations.add(m);
  group[k].add(l);
  group_row = temp_g_row;
  group_column = temp_g_column;
  }}}
  QHi[k,_] = group_row;
  QHi[_,k] = trans(group_column);
  }}
  if(grouping.length<minimal_grouping_num) {
  minimal_grouping_Q = QHi;
  minimal_grouping_num = grouping.length;
}}

As a result of the algorithm the simplified equation is Eq. (25), where K denotes a group of cells, m_K^j the number of elements in group K, and $P_{M_K}(j)$ is the steady state probability of moving to group K

from cell j.

$$N_i(t+1) = N_i(t) \cdot P_S(i) + \sum_{\substack{j, C_j \in S^i_{adj}}} N_j(t) \cdot \frac{1}{m^j_K} P_{M_K}(j) \tag{25}$$

$$C_i \in K$$

In this section we examined complexity and theoretical errors. We exhibited the state number explosion at time dimension increasing. Chapter 3. contains measurements with simulations, which confirms our mathematical results.

5. Path based mobility model

At the definition of the prediction the importance of the time dimension, besides the direction dimension, was mentioned at the introduction and detailed presentation of the Markov model. At the Markov model we treated the m depth time dimension of the users arriving from different directions uniformly, time dimension being the number of steps that was taken into account from the user's past. As we could see with the inclusion of each new step, the number of states increased exponentially. However treating this factor generally might not be optimal. Let us check if it is useful to look back in a certain direction, if yes how much, examine how much the prediction is influenced by taking a certain direction's different depths into account during the calculations. Also let us analyze how the past directions can be combined, similarly to the previously introduced state combination algorithm. In this chapter the previously introduced mobility modeling will be put into a slightly different aspect, a second model, the *path based mobility* model, including the analysis of the time dimension's each past direction and multi-level predictions will be introduced besides the Markov solution.

5.1. The basic idea of the path based mobility model

The importance of the time dimension was introduced on Fig. 3 and in the corresponding mathematical calculations. The future movement of the users is highly influenced by the path they have taken in the past until the investigated point. Leaving this out of consideration would introduce large errors into the mobility model. However it is not always useful to look back into each direction, or to look back in equal depth into each direction. One of the path based mobility model's – which gives a prediction for a given cell similarly to the Markov model – main idea is to analyze the importance of each direction traveled by the users, and decide its importance for consideration. The importance is decided based on two basic criteria:

– the rate of the users arriving from a certain direction into the cell.
– the fluctuation of the number of users arriving from a certain direction based on previous data.

Taking these and the previous steps of the users into consideration we get the different time level paths pointing into the actual cell (Fig. 17a).

In the second step we analyze which outgoing predictions correspond to the different length, incoming paths. The outgoing predictions can be examined for more future steps, contrary to the Markov model, also depending on the rate of the users leaving in a certain direction. If outgoing predictions of two incoming path matches with a certain limit, the two incoming path should not be treated separately but combined (Fig. 17c). Therefore the resulting paths consist of more slices; each slice might contain multiple cells (Fig. 17e).

As result we get reduced incoming paths and corresponding, also multi-level, different length predictions, which lead to a more accurate and less complex model.

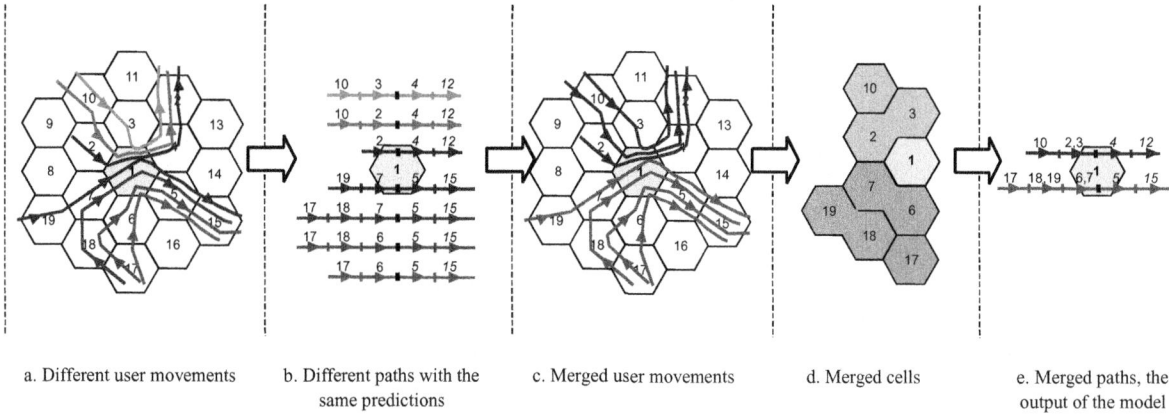

a. Different user movements   b. Different paths with the same predictions   c. Merged user movements   d. Merged cells   e. Merged paths, the output of the model

Fig. 17. Path based mobility model.

5.2. Determine the different paths

In this chapter we present the mathematical repository and the parameters of the path based mobility model and the algorithm determining the optimal and minimal numbers of paths.

As earlier mentioned two criteria have to be examined to decide if it is useful to take, if yes in what depth, a certain direction into consideration. The first is the main value of the users in the certain cell:

$$E^{i,j}_{user\_rate} = E\left(\frac{N_{i,j}}{N_i}\right) \tag{26}$$

where i is the identifier of the cell, j is the analyzed direction, $N_{i,j}$ is the number of users from direction j into cell i, and N_i is the number of users in cell i. The second value is the variance of the number of users:

$$V^{i,j}_{user\_rate} = E\left(\left|E\left(\frac{N_{i,j}}{N_i}\right) - \left(\frac{N_{i,j}}{N_i}\right)\right|\right) = \sigma\left(\frac{N_{i,j}}{N_i}\right). \tag{27}$$

After this we got incoming paths for the actual cell. We use the simple handover model (HOV) to determine the predictions for the paths, which result in a vector p. Determination of p_i element of vector p is done by analog method with pattern matching described in Section 2.3.2, but in this case we calculate forward move patterns (Fig. 18) to determine the relative probabilities.

In the next step, not to analyze paths corresponding to different arriving directions if their predictions match, let's introduce the mean difference between predictions for two paths:

$$D^{a,b}_{pred} = \frac{\sum\limits_{k \forall n_{dir}} \left|p^a_k - p^b_k\right|}{n_{dir}} \tag{28}$$

where a, b are the two incoming paths, p^a_k is the element of HOV vector with direction k, for path a, and n_{dir} is all the possible direction. If $D^{a,b}_{pred}$ is smaller than a given parameter, that is adequately small, then the paths a and b should be combined and treated together. Let us introduce the model's parameters before the detailed description of the algorithm.

- $\varepsilon_{userr\ variance}$ – limit of variance, if $V^{i,j}_{user\_rate}$ is higher than this, than direction j corresponding to cell i has to be further analyzed.

$$
\begin{array}{c|cccc}
 & U_1 & U_2 & U_3 & \dots \\
\hline
t_1 & C_i & C_j & C_j & \dots \\
t_2 & C_k & C_k & C_j & \dots \\
t_3 & C_k & C_l & C_j & \dots \\
t_4 & C_j & \dots & \dots & \dots
\end{array}
$$

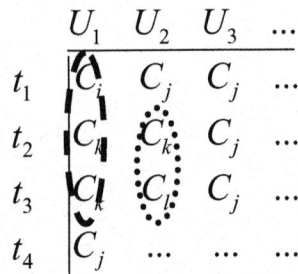

Fig. 18. Example for the move patterns for HOV determination (C is theactual cell). Dashed line: three-step forward move pattern, Dotted line: two-step forward move pattern.

- ε_{pred} – if for any k direction corresponding to path a, p_k is less than this, then the predicted path can be disregarded; taken into account when determining vector p.
- $\varepsilon_{diff\ pred}$ – limit of the mean difference between predictions for two paths; if $D_{pred}^{a,b}$ is smaller than this, paths a and b can be combined and further treated together.
- n_{past} – the maximum number of past steps
- n_{pred} – the maximum number of future steps, taken into account when determining vector p.

Beside the introduced parameters other variable types and functions belong to the algorithm:

- *group* – list of cells, container for a slice of the path.
- *path* – list of groups, container for paths.
- *get_neighbours(ListofCells)* – the common neigh-bours of the list of cell.
- *path.get_sector(depth)* – returns the cells from the depth th slice of a path.
- *get_variance_recursive(cell, depth)* – it analyzes the $V_{user\_rate}^{i,j}$ deviation of the users heading to cell, starting from a given cell, recursively within maximum depth steps.

Based on these the algorithm run for Cell C_a is the following:

```
group temp_group;
group temp_group_set[];
root path;
path path[];
path closed_path[];
root.add(C_a, 0);
path.add(root);
for i = 0 to n_past{
  foreach j in path_cont.get_paths(){
    neighbours = get_neighbours(j.get_sector(i));
    foreach k in neighbours {
      if !temp_group_set.is_in(k) {
      temp_group.clear();
      temp_group.add(k);
      foreach l in neighbours {
        if (k! = l) && !temp_group_set.is_in(l) {
        fitable_to_group = true;
        foreach m in temp_group {
```

$if\,(D_{pred}^{a,b}\!>\!\varepsilon_{diffpred})\,\{$
　$fitable\_to\_group = false;$
　　$break;\}\}$
　$if\,(fitable\_to\_group)\,temp\_group.add(l);\}\}$
　$temp\_group\_set.add(temp\_group);\}\}$
$temp\_group.clear();$
$foreach\ k\ in\ temp\_group\_set\ \{$
$foreach\ l\ in\ k.elements()\ \{$

　$if\,(V_{user\_rate}^{a,neighbours\ ,j}\!<\!\varepsilon_{userrvariance})\,\&\&$

　$(get\_variance\_recursive(j,\ \left\lceil E_{user\_rate}^{i,j}\cdot(n_{past}-i)\right\rceil)\!<\ \ \varepsilon_{userrvariance})$

　$k.remove(l);$

　$temp\_group\_set.update(k);$

　$temp\_group.add(l);$

　$\}\}\}$
$temp\_path.add(temp\_group,i);$
$closed\_path.add(temp\_path);$
$foreach\ k\ in\ temp\_group\_set\ \{\ path\_cont.add(k,i);\}$
$\}\}$
$foreach\ i\ in\ path\ \{closed\_path.add(i);\}$

In the algorithm $V_{user\_rate}^{i,S,j}$ is a reduced version of Eq. (25), when we only take the users arriving from S direction entering into cell i into account at determination of the deviation.

The algorithm with the given parameter results in minimal path numbers.

Proof: the algorithm basically performs a reduction from all possible paths, leaving out any of the steps results in more paths or the given boundary conditions are not fulfilled.

5.3. Complexity of the model

The complexity of the model is determined by the number of executed steps. Let us examine how the number of steps changes as the function of the parameters. Consider the processing of the trace file, accessing the read data and the simpler mathematical operating as one unit. The complexity of the algorithm can be determined as follows:

$$2n_{user} + n_{past}\cdot n_{nb}\cdot(n_{cell}^{path}\cdot n_{cell}^{path}(n_{pred}^{path}+2)+n_{nb}\cdot n_{cell}^{path}+n_{nb}) \tag{29}$$

where n_{user} is the number of users in the used network trace, n_{nb} is the average number of neighbor cells from the point of the cells in one slice of the path, n_{cell}^{path} is the average number of cells in one slice of the path, n_{pred}^{path} is the average number of predicted paths for one path. The parameters defined depend on the size of the trace, the time of sampling and the mobility speed of the users. Perform the simplification, take the largest from the parameters and denote it with n.

$$n = \max(n_{past}, n_{user}n_{pred}^{path}, n_{cell}^{path}, n_{nb}) \tag{30}$$

In this case the complexity of the algorithm can be expressed the following simple way (Fig. 19):

$$o(n^5 + n^4 + n^3) \tag{31}$$

Complexity of the Path Based Mobility Model

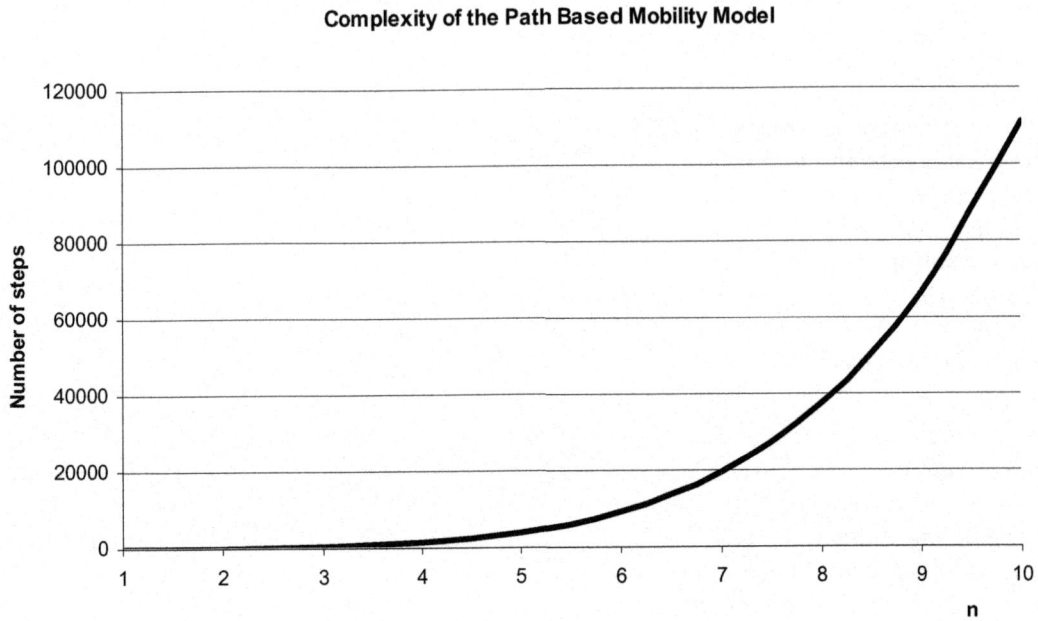

Fig. 19. Complexity of the algorithm.

Path based mobility model predictions based on different networks

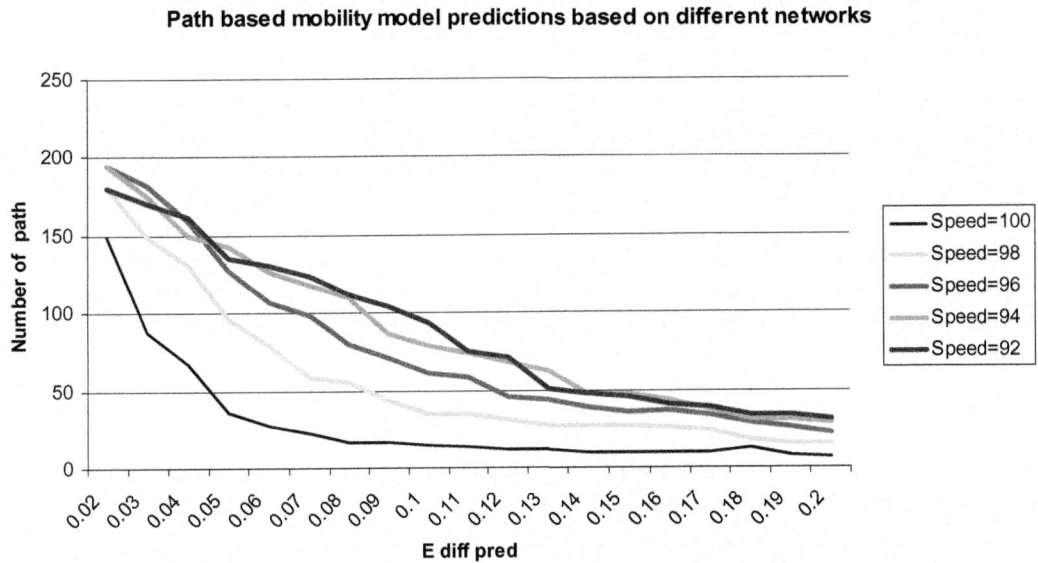

Fig. 20. Effect of the $\varepsilon_{diffpred}$ parameter with high speed.

5.4. Measures

The Path Base Mobility solution's one basic output that can be analyzed besides the actually resulting paths is the number of paths. The more paths determined by the algorithm, the more difficult the realization of the prediction will be. Taking this into consideration, it is really important that the input parameters get appropriate values as function of a defined goal.

Path based mobility model predictions based on different networks

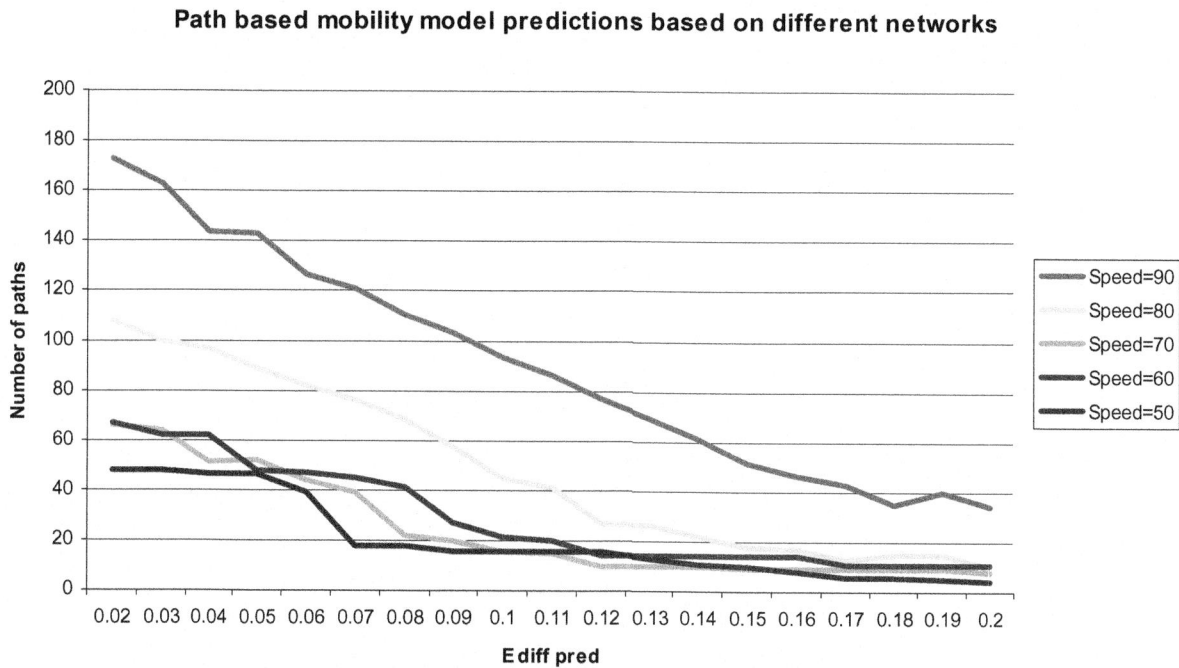

Fig. 21. Effect of the $\varepsilon_{diffpred}$ parameter with lower speed.

An erlang simulation was prepared to justify the algorithm and to monitor the effects of the parameters. The simulation was prepared based on mobility traces taken for different characteristic user speeds. The mobility traces were provided by OMNet simulation described in Chapter 6. The value of the speed was a relative number between 0 and 100, 100 meaning fast users, moving in each timeslot, 0 meaning slow, nearly steady users.

The change in the two parameters $\varepsilon_{diffpred}$ and n_{past} is shown in relation to the number of paths. Changing the other parameters does not result in significant change in the number of calculated paths.

Two separate measurements were executed to analyze the change of $\varepsilon_{diffpred}$. In the first one (Fig. 20) the algorithm was run on trace data resulting from user movements with rather higher, 100 and 92 speeds. It can be seen that curve corresponding to speed 100 strongly breaks down by increasing the parameter and always results in far less calculated paths as compared to lower speeds. By increasing the speed, the resulting curve approaches a linear curve with larger path numbers on average.

In the second measurement for $\varepsilon_{diffpred}$ (Fig. 21) the change in the number of resulting paths was examined with user movement speeds 90 and 50. The curves' characteristics are linear, but changed in relation to the number of paths. The algorithm calculated more paths for higher speeds than for lower speeds.

The curve representing the change in the number of paths in function of the speed has a local extreme at speed 90; this clearly shows on Fig. 22.

The number of paths is increased rapidly, by increasing the limit of looking back into the past. This is shown on Fig. 23.

The path based mobility model, the corresponding algorithms were introduced, its complexity and a few simulation results were discussed in this chapter.

In the next chapter the accuracy of the created models is proved by simulation results in contrary to the other models in the bibliography. The Markov model can be considered as a certain specialization

Path based mobility model predictions based on different networks

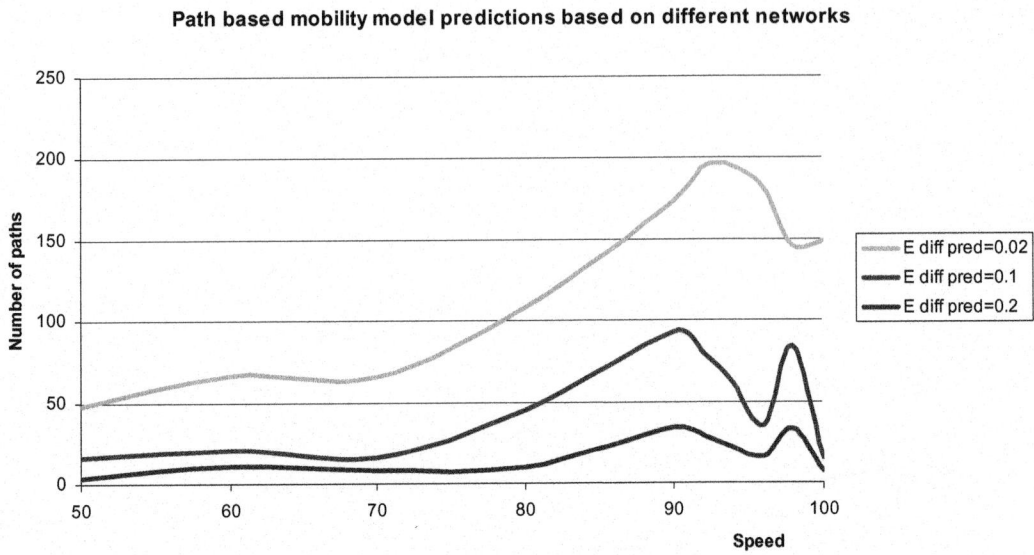

Fig. 22. Relationship between speed and number of path with different $\varepsilon_{diffpred}$ parameters.

Path based mobility model predictions based on different networks

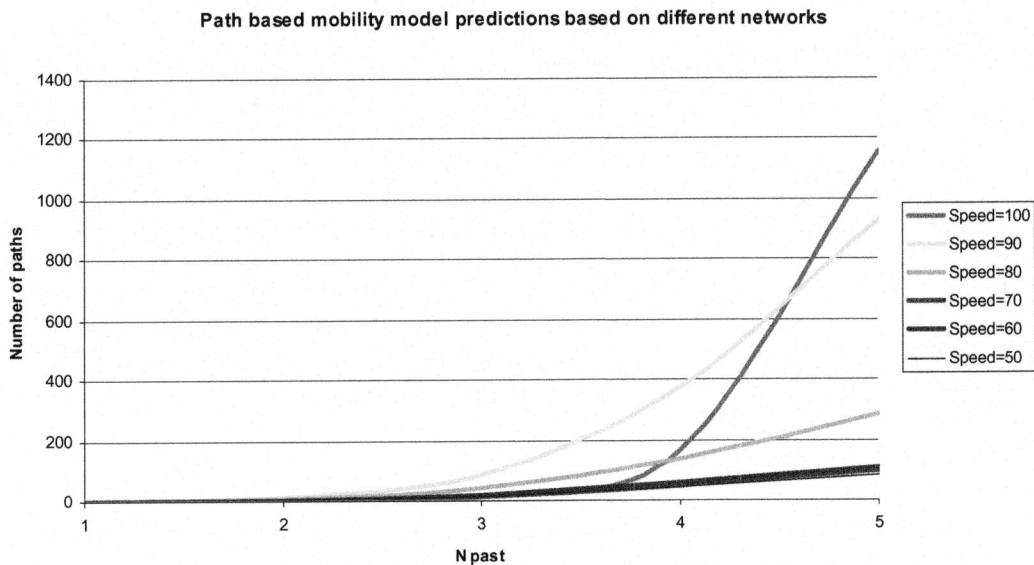

Fig. 23. Effect of the n_{past} parameter.

of the path based mobility model. Due to this, the results of the Markov model in the simulation chapter also prove the path based mobility model as well.

6. Accuracy measurements by simulation

In this chapter we compare the accuracy of the proposed extensions and the Path Based mobility solution to other models can be found in the literature. The Random Walk-based mobility models

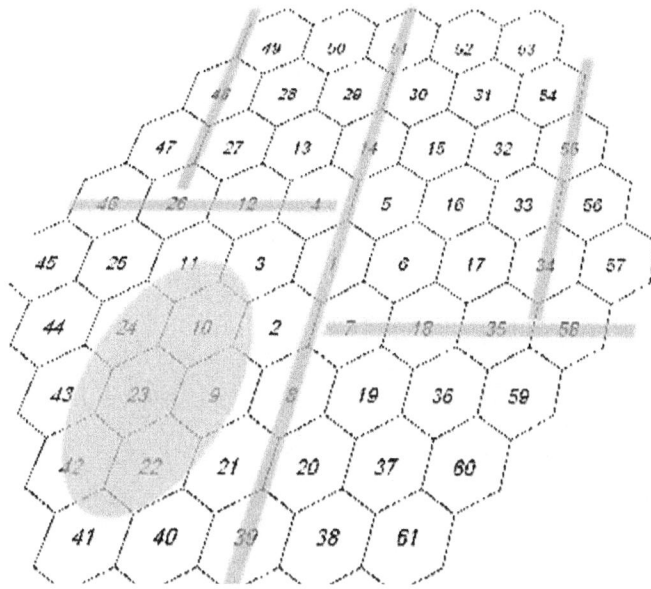

Fig. 24. Cell cluster with streets and park.

depend on the transition probabilities. The RW model is only capable of accurate prediction of user movements in case of uniform movement distributions (e.g. all elements in the transition probability vector are equal to 1/6).

According to the calculations and simulations, the Markovian approach of user movement produces better estimation of the users' distribution in a cell cluster.

The estimation procedure was validated by a simulation environment of a cell cluster shown on Fig. 24.

The simulation was written in the open source OMNet++ (Fig. 25) using C++ language. The cluster consisted of 61 named cells, the simulation environment included geographical data that is interpreted as streets on the cluster area. The drift of the movement is heading to the streets from neutral areas.

The simulation used 610 mobile terminal (10 for each cell), in the initial state uniformly distributed in the cluster. The average motion velocity of the users is parameterized with a simple PH cell dwell time simulator (reciprocal of exponentially distributed values).

The simulation consists of two parts. The trace simulation is the series of cell-transitions that the mobiles have initiated. It produces a time-trace that contains the actual location data for each mobile terminal in the network. We have used this trace simulation as if it was a provider's real network trace.

We described estimation procedures based on the different mobility model discussed in this paper. The first reference model is the simple Random Walk (RW) with constant speed and uniformly distributed direction-probabilities. The RW estimation draws a direction out of the possible six in every timeslot for every simulated user, and forwards the given user to the adjacent cell in the drawn direction.

A widely used modified Random Walk estimation was used in the simulation as a secondary reference. In this case, every user has a preferred direction. The distribution of the directions to draw are following normal distribution. The expected value is the preferred direction, variance equals 1, that is the the difference between to adjacent directions. The draw is expected to result a direction that is near the preferred direction, thus the user moves with a drift the points to the preferred direction.

We used these models as a comparison basis to the 7-state Markovian model simulations.

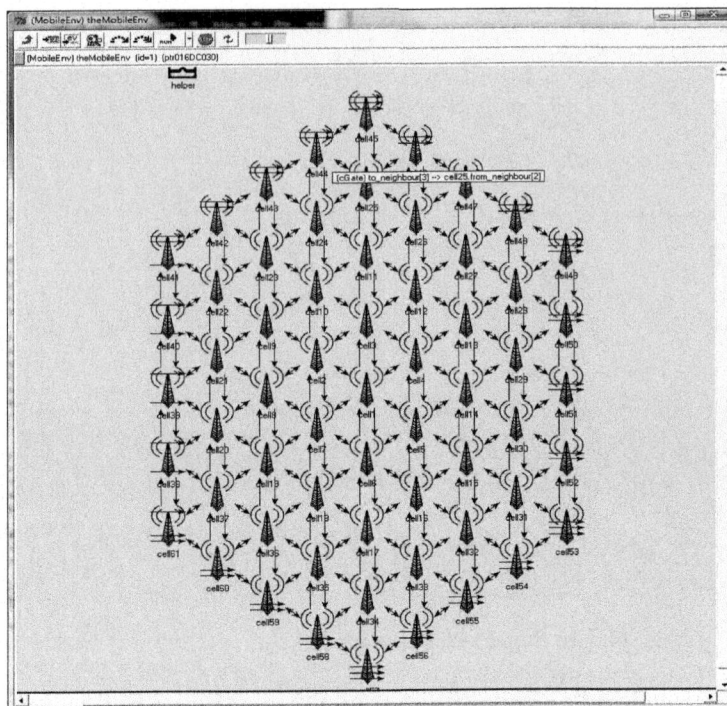

Fig. 25. The mobility simulation environment in OMNet++.

The simulation uses the past and the current trace simulation results to estimate future number of users in each cell and future transition direction of each user. The estimation produces an alternative user distribution in the network. Th estimation error is defined as the difference of the user-distribution in the trace and the model. The quantity of estimation error is interpreted as the measure of accuracy of each mobility model in this paper.

The trace simulation starts earlier than the estimations, during the initial steps the trace reaches its operating stability without cold-start transients. During the warm-up process the trace simulation produces sample data for the estimation procedures which uses the previous trace results as an input to estimate the future user distribution. Each user-transition in the warm-up period is used to derive transition probabilities, motion speed and patterns in the simulation cell-space. These patterns serve as an input for the estimation threads of each mobility model. The models have the same input throughout the simulation process thus the results are comparable.

In the following we discuss the results of two basically different simulation environments. In the first environment the trace simulation follows a strict pattern of forward-backward motion. That is the users are on a self-closing straight path and there is no practical chance of stepping down from this path. In this case the trace simulation works with two directions, forward and backward.

The other environment represents a crowded urbanized area, where each direction has a non-zero probability. Although, the motion is not uniform, each cell has at least one preferred neighbor.

The following plots show the summarized error of the estimations simulated. In the plot-pair, the first shows the forward-backward scenario, while the second plots the result of the urbanized environment.

Figure 26(a,b) shows the sum of the absolute value of error for every cell during the simulation.

Figure 26a) shows that the two Random Walk based models cannot follow the patterns in user fluctuation, the estimation works with a significantly higher error rate than the 7-state Markovian model.

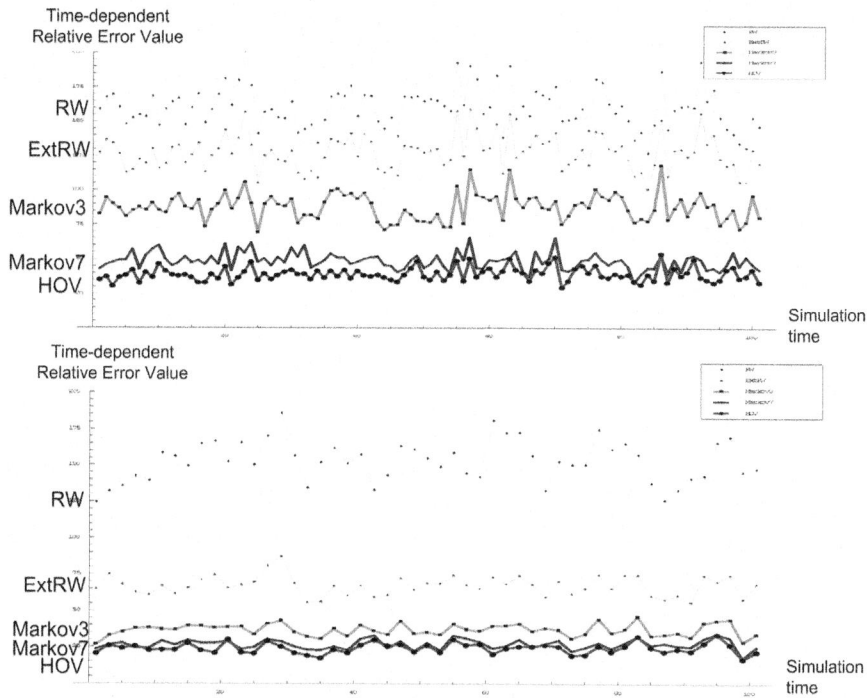

Fig. 26. a. Forward-backward environment error summary; b. Urbanized environment error summary.

CPU usage of the models

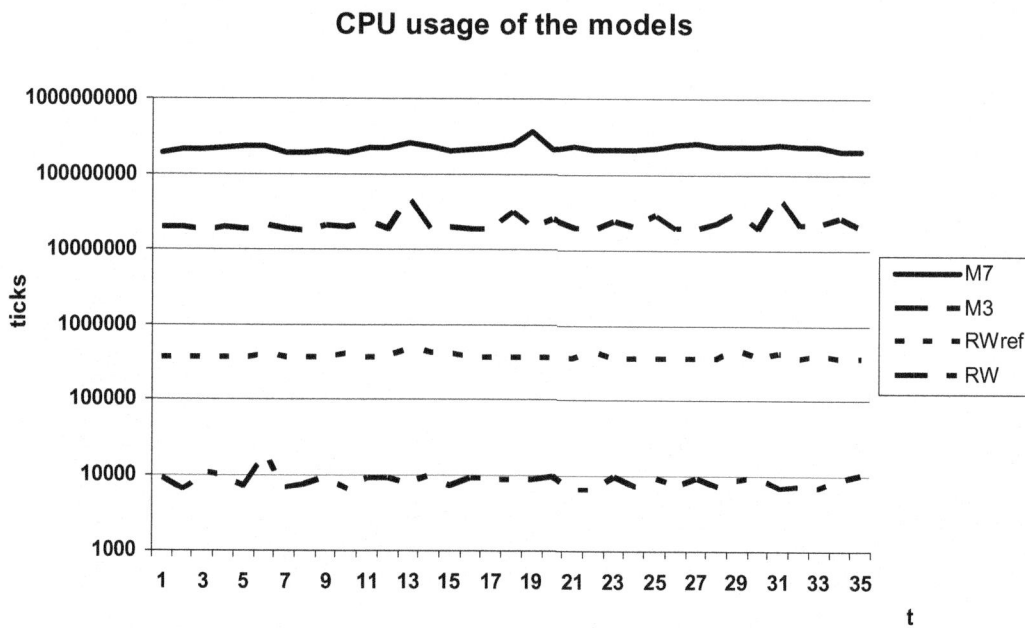

Fig. 27. CPU usage of the models on logarithmical scale.

The 7-state Markovian model has a decreasing error rate ongoing the simulation, along with the Markov chain learning the motion patterns.

Figure 26b) shows a less sharp difference between RW-based and our model, the 7-state Markovian model produces better results. The patterns of the urbanized environment are closer to the simple uniform motion pattern that is used by the Random Walk. This confirms that the RW-like estimations are capable of modeling cases where users are fluctuating randomly without significant or strict patterns.

The relative performance of the models can be seen on the Fig. 27. The execution time of the methods of model are plotted in each timeslot on logarithmical scale.

7. Conclusion

In this paper we proposed mobility models that are capable of simulating real-life user movement accurately. We proposed an alternative Path Based and a Markov-chain based method. The simulation results proved the analytical properties of the proposed mobility models.

The widely used and simple RW model works with a significant error rate at all times in mobility modeling. Since the user movement patterns in the simulation are not completely random due to the streets and geographical circumstances, the uniformly distributed Random Walk pattern cannot model it.

The algorithms that work with memory can estimate with a significantly higher accuracy. That is the 7-state Markovian model produces better results in the prediction of each user position.

We proposed mobility accurate models that are producing better estimations of user distribution in the network, based on present and past network operator data. A network operator may use the 7-state Markov model to make predictions on the future distribution and location of users among radio cells to justify CAC or other QoS decisions.

Acknowledgements

This work was supported by the Mobile Innovation Center, Hungary.

References

[1] A. Jardosh, E.M. Belding-Royer, K.C. Almeroth and S. Suri, Towards Realistic Mobility Models for Mobile Ad hoc Networks, In proceedings of *The Annual International Conference on Mobile Computing and Networking, MobiCom* (September 2003), 217–229.

[2] A. McDonald and T. Znati, Predicting Node Proximity in Ad Hoc Networks: A Least Overhead Adaptive Model for Selecting Stable Routes. In proceedings of *First Annual Workshop on Mobile Ad Hoc Networking and Computing,(MobiHoc),* (2000) Boston, pp. 29–33.

[3] A. Fongen, C. Larsen, G. Ghinea, S.J.E. Taylor and T. Serif, Location based mobile computing - A tuplespace perspective, *Mobile Information Systems* 2(2–3) (2006), 135–149.

[4] A.U. Jayasuriya, *Improved Handover Performance Through Mobility Prediction*, PhD thesis, University of South Australia, 2001.

[5] B. Liang and Z. Haas, Predictive distance-based mobility management for PCS networks, In Proceedings of the *Joint Conference of the IEEE Computer and Communications Societies (INFOCOM)* (March 1999), 1377–1384.

[6] C. Abondo and S. Pierre, Dynamic location and forwarding pointers for mobility management, *Mobile Information Systems* 1(1) (January 2005), 3–24.

[7] D.B. Johnson and D.A. Maltz, Dynamic Source Routing in Ad-Hoc Wireless Networks, *Mobile Computing* 353 (1996), 153–181.

[8] D. Levine et al., A Resource Estimation and Call Admission Algorithm for Wireless Multimedia Networks Using the Shadow Cluster Concept, *IEEE/ACM Transaction On Networking* **5**(1) (February 1997), 1–12.

[9] E. Royer, P.M. Melliar-Smith and L. Moser, An analysis of the optimum node density for ad hoc mobile networks, *In Proceedings of the IEEE International Conference on Communications (ICC)* **3** (2001), 857–861.

[10] H. Liao, L. Tie and Z. Du, A Vertical Handover Decision Algorithm Based on Fuzzy Control Theory, *Computer and Computational Sciences, 2006. IMSCCS '06. First International Multi-Symposiums* **2** (24 April 2006), 309–313.

[11] J.J. Garcia-Luna-Aceves and M. Spohn, Source-tree routing in wireless networks, In Proceedings of *The 7th International Conference on Network Protocols* (ICNP), (31 Oct–3 Nov 1999), 273–282.

[12] J. Yoon, B.D. Noble, M. Liu and M. Kim, Building realistic mobility model from coarse-grained wireless trace. In proceedings of the *Fourth International Conference on Mobile Systems, Applications, and Services (MobiSys)*, Uppsala, Sweden, pp. 177–190, June 2006.

[13] K. Wang and B. Li, Group Mobility and Partition Prediction in Wireless Ad-Hoc Networks, *IEEE International Conference on Communications, (ICC 2002)* New York, **2** (April 2002) 1017–1021.

[14] M. Sanchez, Mobilitymodels, from http://www.disca.upv.es/misan/mobmodel.htm (2002, May 30).

[15] M.M. Zonoozi and P. Dassanayake, User mobility modelling and characterization of mobility patterns, *IEEE Journal on Selected Areas in Communications* **15**(7), (1997 September).

[16] *Mathematica Documentation.* Wolfram Research Inc. from http://www.wolfram.com/.

[17] S. Michaelis and C. Wietfeld, Evaluation and comparison of prediction stability for user movement pattern detection algorithms, *European Wireless Conference (EW2006)*, Athens April, 2006.

[18] R. Chellappa, A. Jennings and N. Shenoy, A comparative study of mobility prediction in fixed wireless networks and mobile ad hoc networks. In proceedings of the *IEEE International Conference on Communications 2003 (ICC2003)*, Alaska, USA, **2** (May 2003), 891–895.

[19] S. Michaelis and C. Wietfeld, Comparison of User Mobility Pattern Prediction Algorithms to increase Handover Trigger Accuracy, *IEEE Vehicular Technology Conference*, Melbourne, **2** (May 2006), 952–956.

[20] T. Camp, J. Boleng and V. Davies, A Survey of Mobility Models for Ad Hoc Network Research, Wireless Communications and Mobile Computing (WCMC), *Special Issue on Mobile Ad Hoc Networking: Research, Trends and Applications* **2**(5) (2002), 483–502.

[21] V. Davies, Evaluating mobility models within an ad hoc network, Master's thesis, Colorado School of Mines, 2000.

[22] V. Simon and S. Imre, A simulated annealing based location area optimization in next generation mobile networks, *Mobile Information Systems* **3**(3–4) (2007), 221–232.

[23] Vincent W.-S. Wong and V.C.M. Leung, Location Management for Next-Generation Personal Communication Network, *IEEE Network* (September/October 2000).

[24] Yuguang Fang Chlamtac and I. Yi-Bing Lin, Portable movement modeling for PCS networks, *IEEE Vehicular Technology* **49** (2000), 1356–1363.

[25] X. Hong, M. Gerla, G. Pei and C. Chiang, A group mobility model for ad hoc wireless networks, In Proceedings of *The ACM International Workshop on Modeling and Simulation of Wireless and Mobile Systems (MSWiM)*, (1999), pp. 53–60.

[26] X. Hong, T. Kwon, M. Gerla, D. Gu and G. Pei, A Mobility Framework for Ad Hoc Wireless Networks. In proceedings of *ACM 2nd International Conference on Mobile Data Management, (MDM 2001)*, Hong Kong, (January 2001), 185–196.

[27] Z.J. Haas, A new routing protocol for reconfigurable wireless networks, In Proceedings of *The IEEE International Conference on Universal Personal Communications (ICUPC)*, (1997), 562–566.

6

Extending the Internet of Things to the Future Internet through IPv6 support

Antonio J. Jara[a,*], Socrates Varakliotis[b], Antonio F. Skarmeta[a] and Peter Kirstein[b]

[a]*Department of Information and Communications Engineering, Computer Science Faculty at the University of Murcia, Murcia, Spain*
[b]*Department of Computer Science, University College London, London, UK*

Abstract. Emerging Internet of Things (IoT)/Machine-to-Machine (M2M) systems require a transparent access to information and services through a seamless integration into the Future Internet. This integration exploits infrastructure and services found on the Internet by the IoT. On the one hand, the so-called Web of Things aims for direct Web connectivity by pushing its technology down to devices and smart things. On the other hand, the current and Future Internet offer stable, scalable, extensive, and tested protocols for node and service discovery, mobility, security, and auto-configuration, which are also required for the IoT. In order to integrate the IoT into the Internet, this work adapts, extends, and bridges using IPv6 the existing IoT building blocks (such as solutions from IEEE 802.15.4, BT-LE, RFID) while maintaining backwards compatibility with legacy networked embedded systems from building and industrial automation. Specifically, this work presents an extended Internet stack with a set of adaptation layers from non-IP towards the IPv6-based network layer in order to enable homogeneous access for applications and services.

Keywords: Internet of Things, network communications, internetworking, wireless sensor networks, backwards compatibility, system architecture, IPv6

1. Introduction

The Internet of Things (IoT) [1] is one of the main applications driving the evolution of the Internet towards the Future Internet. Here sensors, actuators and devices (now called things), are connected to the Internet through gateways and Supervisory Control and Data Acquisition platforms (SCADAs). This Intranet of Things [2] is being extended to smart things [3] with a higher scalability, pervasiveness, and integration into the Internet Core.

The ongoing and future work aims to create an extended Internet of Things. This requires both an architecture and products that allow for the extension of the Internet technologies, in order to reach a homogeneous integration of the Future Internet, Services, People, and Things with the Future Internet of Things, Services and People.

This drive to integrate everything into the Internet Core is motivated by the market wish to have all processes remotely accessible – while at the same time understanding that re-engineering an infrastructure to allow this for each application would be prohibitively costly and time-consuming. Moreover, the current evolution from uniform mass markets, to personalized ones, where the customization and

---

*Corresponding author: Antonio J. Jara, Department of Information and Communications Engineering (DIIC), Computer Science Faculty at the University of Murcia, Murcia, ES-3100, Spain. E-mail: jara@ieee.org.

user-specified adaptation is a requirement, makes the sort of uniform infrastructure found in the Internet, imperative. This allows many components to be re-used, and services to be shared, with correspondingly huge economies of scale and shortened implementation times.

The Internet of Things fills the gap between the needs arising from the evolution of the market, information, users, and things, by moving all of them to a common framework, the Internet. This is different from the current approach in such applications, where they are based usually on stand-alone and monolithic solutions designed for a narrow application domain. Users now require more flexibility and freedom. Offering a common framework allows choice among the available manufacturers, suppliers, service providers, delivery options, and payment services. While this obviates the need for standalone or proprietary solutions, it also requires a high level of integration.

This work describes an integrated approach to the Future Internet that supports existing Internet of Thing technologies and extends, and bridges to IPv6, the existing IoT building blocks. The basic network technologies proposed are the use of 6LoWPAN [4]; 6LoWPAN offers IPv6 over IEEE 802.15.4, the EPC codes in RFID [5], and Bluetooth Low Energy (BT-LE). We propose mappings between the three technologies. These involve a novel header compression and adaptation protocol for BT-LE and IEEE 802.15.4 called GLoWBAL IPv6 [6], which presents a better performance for header compression when Global IPv6 addresses are used.

Second, an integration solution has been proposed using a multiprotocol card for maintaining backwards compatibility with legacy, networked, embedded systems. As a specific example, we present the integration of EIB/KNX [7], X10 [8], Control Area Networks Bus (CAN) [9], and digital/analog I/O for buildings and industrial automation. These technologies were integrated into a novel IPv6 addressing structure, in order to achieve integral support of IPv6 throughout the proposed framework, which is complementary to the aforementioned integration of IPv6.

This approach provides a further step for the integration of IPv6 into the IoT; this is part of the co-existence strategy for the management of heterogeneous technologies and architectures, on the way to achieving interoperability across businesses, service providers, and users [10].

2. Related technologies and standards

IoT is the main driver for the Future Internet, where IPv6 is the fundamental technology. It is estimated that the Future Internet will number hundreds of billions of *connected things* by 2020. Unlike IPv4, IPv6 can address this number of objects. The IPv6 address space supports 2128 unique addresses (approximately 3.4×1038); specifically, it can offer 1.7×1017 addresses on an area about the size of the tip of your pen. The advantages of the IPv6 integration in the IoT are not limited to a universal addressing space; its main advantages are to offer stable, scalable, extensive, and tested protocols for global end-to-end communications, node/service discovery, mobility, end-to-end security, and relevant features such as stateless addressing auto-configuration, multicast addressing for group operations and its extensibility for application layers with technologies such as Web Services.

The initial step for the integration of a dual IPv4/IPv6 stack in embedded systems was lwIP/uIP [11]. This approach focused mainly on the reduction of the memory requirements and code size for the communication stack. This stack is commonly used for the integration of IPv6 in embedded systems for their communication interfaces such as Ethernet.

Embedded systems and sensor networks have been extended during the last decade or so with wireless technologies. Therefore, IPv6 support is needed also in technologies for Wireless Sensor Networks such as IEEE 802.15.4. The constraint of these networks is not only the limited memory, but also the reduced payload available in view of the low power consumption features desired. For this reason, a new protocol, 6LoWPAN [4], was designed as an adaptation layer to carry IPv6 datagram over the IEEE 802.15.4 link, taking into account the limited bandwidth, memory and energy resources. This adaptation layer has focused on header compression in order to reduce the processing load.

In addition to 6LoWPAN, the GLoWBAL IPv6 protocol [6] presents a lower overhead than 6LoWPAN header and an optimized approach for global communications, since 6LoWPAN has limitations in the compression of global IPv6 addresses. For this reason, GLoWBAL IPv6 is being considered for new wireless technologies with higher payload limitations such as Bluetooth Low Energy.

An important aspect of the Internet is that there is a uniform interface between applications and network services. Hence Web technologies such as HTTP, REST, SOAP, JSON and XML [12], which provide access to resources and services, are being adapted to the IoT. Thereby, the application layer for smart things is being defined via Web Services, thus becoming the Web of Things [13].

This is considered the most generic and homogeneous route to access services from the IoT. Specifically, it is based on a constrained version of RESTFul, denominated CoAP [14]. DNS-SD [15] also defines a description of CoAP Web Services which follow the semantic and naming conventions that describe how services will be represented in DNS records. In addition, the CoRE IETF working group is defining the Web Linking description, termed the Link format, and other protocols for CoAP such as Observe and Blockwise Transfer [14].

The IETF is also in the process of defining light versions of XML and SOAP (specified as EXI [16], and SOAP lightweight [17], respectively). Regarding security, DTLS is being simplified to DTLS for CoAP [18], JSON offers security with JOSE, and ID/Locator split architectures such as HIP in its "diet" version, i.e. HIP DEX [19].

When this entire infrastructure has been provided, it is necessary to apply it to a specific use case to ensure all hangs together.

3. Integration of things in the IPv6 stack

The previous section has already presented some adaptation techniques defined for the integration of IPv6 in smart things. However, to account for the degree of heterogeneity encountered in the real world one has to consider how legacy technologies would integrate into the IoT using IPv6. While it would be convenient if the legacy systems were immediately abandoned, such an approach is completely unrealistic. Just as it is clear that there will be a long period of overlap between the use of IPv4 and IPv6 in the Internet, there will also be a lengthy period of overlap between use of legacy systems for application domains, and their transition to systems more tailored to the IoT.

While the use of IPv6 and 6LoWPAN are strongly advocated for the Internet side of the IoT, the integration of legacy, non-IPv6-enabled technologies, require additional mechanisms in order to map the different address spaces to the IPv6 one. These legacy technologies have been tailored to, and are heavily used in, areas such as building and industrial automation. For this reason, this work aims to provide a transparent mechanism for users, devices and control systems to map the different address spaces to a common IPv6 one. Using the proposed integrated Internet stack, with legacy-system-specific translation gateways, every device from each technology will get a common framework based on IPv6 and protocols over IPv6 such as Web Services and any other protocol via TCP/UDP sockets.

For these reasons, a key contribution of this paper is the definition of *Half-Gateways (HGWs)* that bridge non-IPv6-enabled technologies with an IPv6-powered environment. This integration operates on

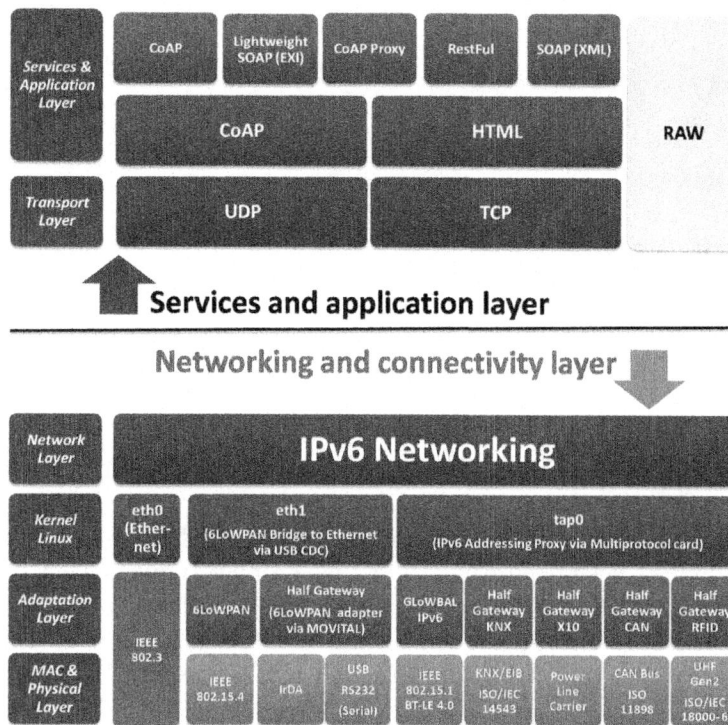

Fig. 1. Integration of Things in IPv6 stack.

top of the Future Internet infrastructure, i.e. IPv6. Thus, users and clients discover and use homogeneous IPv6-based resources.

At the same time, the application domains will continue to use well-known established protocols and already deployed technologies such as KNX/EIB for building automation, and new technologies such as Bluetooth Low Energy, which are not based on IPv6 networks, and other technologies such as RFID and its identifiers, e.g. Electronic Product Code (EPC) [20] and UID [5].

Our principal goal is to focus on IPv6 network mechanisms in order to homogenize the discovery, access and use of resources through the Internet infrastructure, i.e. through the IPv6 network. This makes services reachable via homogeneous and interoperable technologies. For example, Web Services and the discovery of services should be conducted through network-based Information Systems that are already deployed, such as the Domain Name System with Service Discovery (DNS-SD) [15].

Specifically, Fig. 1 presents the technology stack proposed for the full integration of smart things, building automation technologies, RFID tags, and embedded technologies into a homogeneous IPv6 networking layer.

Under the IPv6 layer exists the sub-system adaptation and integration modules. These modules are based on hardware and/or software. The hardware adaptation provides the physical interface between the proprietary or native protocol and our platform. The software module transforms the native functionality into a set of homogeneous Web Services accessible via IPv6.

This architecture provides to the layers above IPv6 an environment for services and applications totally independent of the underlying technology.

Just as in the rest of the Internet, a transport layer is defined; in the case of smart things this is mainly focused on UDP because of its simplicity. For example, 6LoWPAN offers compression for UDP headers;

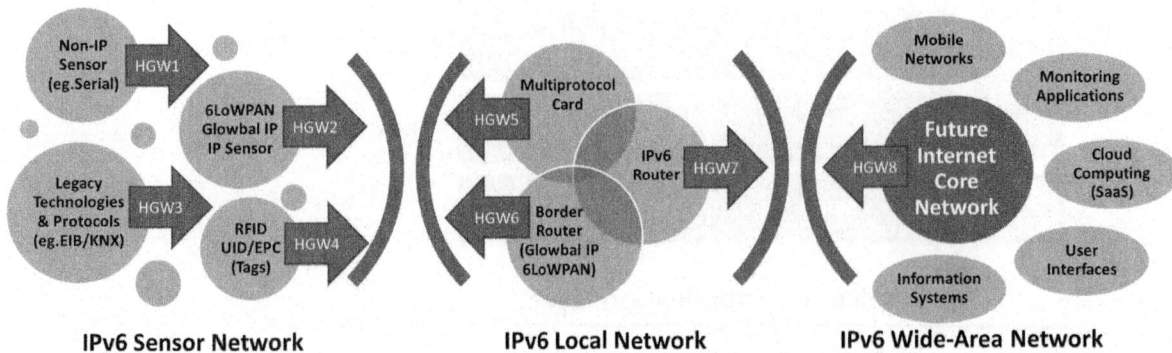

Fig. 2. Schematic of the topology envisaged to integrate in IPv6 through half-gateways from the sensor level to the Future Internet Core Network.

however, implementations exist for TCP in most extended operating systems used for smart things such as Tiny and Contiki OS.

In the application layer, due to the dominance of the aforementioned Web of Things, the main tendency is to offer Web Services based on the Constrained Application Layer for smart things, i.e. CoAP; this is a light version of RESTFul/HTTP, paying attention to the limitations of these devices.

Stacks like those in Fig. 1 require adaptation layers at many different levels; thus a stack is being built for the IoT analogous to the existing one for the current Internet. Specifically, a set of gateways, proxies, and border routers provide the adaptation. Some examples are the 6LoWPAN Border Router [4], the CoAP proxy [21], and now the HGWs presented in this work.

HGWs interface between the Internet and the different proprietary architectures. The properties in each network/architecture must be considered separately; then, the two stacks must be connected together through the HGWs into an IPv6 network.

Figure 2 illustrates our vision of the integration of the things into the Future Internet Core Network. There are usually three regions.

The left region of the Fig. 2 represents the "IPv6 Sensor Network" domain, a term we use to denote a set of technologies including legacy protocols, RFID, non-IP sensors, and 6LoWPAN/GLoWBAL IPv6 sensors. This contains both legacy and IPv6-enabled subsystems that are directly concerned with things. It is here that each subsystem may require a technology-specific HGW in order to adapt/bridge the native protocol with an IPv6-enabled interface. The technologies from the left region (usually called the 'fringe' of the Internet) connect to an "IPv6 Intranet" through different HGWs. Each HGW bridges a specific technology and provides adaptation layers specific to this technology. For that reason, this "IPv6 Intranet" can be seen as an "Application or technology-dependent Internet", since we need to take into account the specific features from the applications and technologies integrated. This Intranet is located at the central region, which is a domain-specific, but IPv6 technology. This contains the domain-specific operations, and has access to the domain-specific resource databases such as is the case of a multi-protocol card, residential gateway or Border Routers for 6LoWPAN. Section IV presents the technologies proposed to support the HGWs functionality used in this work.

This central region is also composed of the network termination broadband adapter such as xDSL modems and IPv6 routers which link the local and domain-specific networks to the Wide Area Network. Thereby, all access to the things resources are via standard Internet procedures.

The right region represents the Wide Area Network accessed through the Future Internet Core Network. This offers the different technologies and services of the Future Internet, such as monitoring applications, Software as a Service (SaaS) solutions, and any information system.

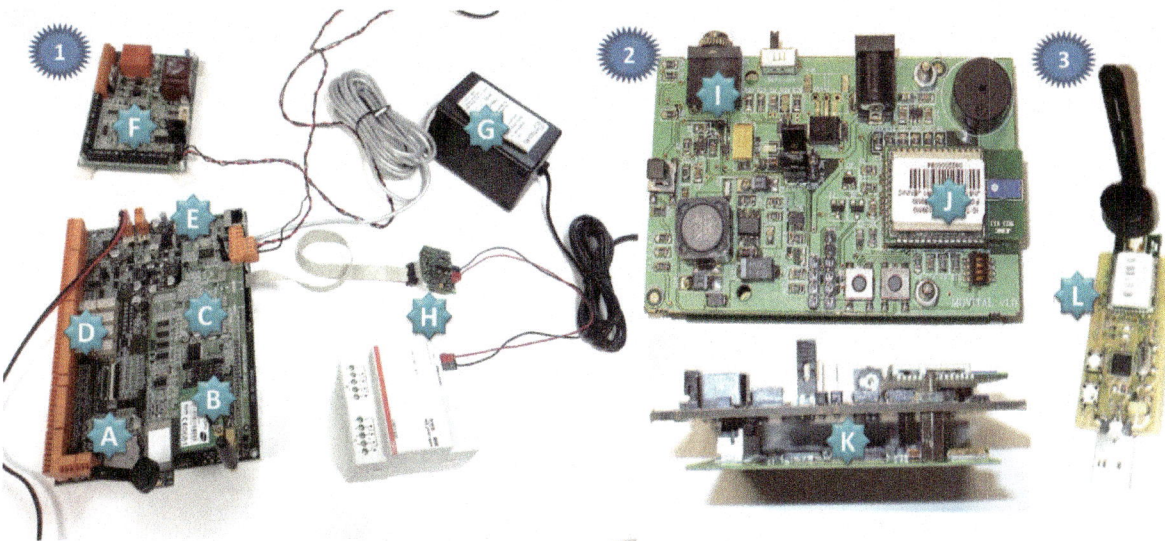

Fig. 3. (1) Multiprotocol card with 6LoWPAN (A), GPRS (B), Bluetooth (C), native interfaces (D), Ethernet (E), CAN (F), X10 (G), and KNX (H). (2) Movital adapter with native interfaces (I), 6LoWPAN (J) and RFID (K). (3) 6LoWPAN Bridge based on USB (L) from 6loWPAN to Ethernet.

In summary, this integration of things with IPv6, through the presented communications stack and architecture, offers the aforementioned advantages for interoperability and homogeneity. In addition, it also enables the related IPv6 standards to be used, offering a wide set of tested, open, and extended technologies. Smart things, and solutions based on them, are able to benefit from the standard implementations.

4. Building Blocks of the proposed IPv6 integrated Stack for Smart Things

The integration of things in the IPv6 stack presents several advantages to use standard, tested, open and useful solutions through open platforms such as Linux OS, which offers all the required IPv6-related technologies and protocols. For this reason, in the proposed stack, Fig. 1, we have defined an adaptation of the technologies to an interface compatible with the Linux Kernel, i.e., Linux Operating System (OS).

Linux OS allows to use the existing implementations for routing ('routed'), security policies ('iptable'), Neighbor Discovery ('radvd'), DNS-SD ('bind') and mDNS ('avahi'). In addition, this makes the framework more secure and scalable; since it can be configured to contain sub-networks, one for each technology, and allocate a virtual network kernel devices built with virtual interfaces such as tun/tap. Thereby, the different access policies may apply for each virtual interface, thus allowing isolation and independent managing of each sub-network. It can allow the definition of different namespaces for DNS-SD/mDNS [15]. This also allows the upgrade of platform components without requiring reconfiguration of the other components; this makes the system easier to expand and allows the installation of installing software that has already been tested and made available for Linux thus – adding a degree of robustness to the proposed IoT integration framework.

In order to implement the integration of the IoT using IPv6, we consider the platforms presented in Fig. 3. Here, one can see various instances of HGWs: some bridging towards legacy and non-IPv6 sensor technologies and some acting as gateways to interconnect to the Future Internet Core Network via the Local Network (both being IPv6-based). The main platform is the multiprotocol card presented in Fig. 3.1. It is based on the Atmel ARM9 processor running at 400 MHz (32-bit) with 256 MB RAM and 256 MB NAND memory, which supports Linux OS. It features 6LoWPAN adaptation (with Contiki OS) (point A). It also offers the following interfaces: GPRS from Wavecom (B), Bluetooth from BlueGiga (C), USB 2.0 ports, I/O digital/analog/relays (D), and Ethernet 100 Mbps (E).

This multiprotocol card supports, through its extension interfaces (serial RS232 and SPI ports), the technologies for industrial and building automation. More specifically, the extension interfaces support Control Area Networks (CANs) (F), X10 (G) and European Industrial Bus (EIB)/Konnex (KNX) (H). To put these into the context of Fig. 2, one can see how the previous technologies represent HGW2 from the IPv6 Sensor network and HGW5 from the IPv6 Intranet points of view.

The following subsection describes the technologies and protocols proposed for the HGWs and some of the adaptations carried out for the different technologies, which define the building blocks of the presented stack.

4.1. 6LoWPAN

6LoWPAN defines an adaptation protocol based on header compression for IPv6 datagrams. More specifically, 6LoWPAN defines one reduced header format for IP (IPv6) and one reduced format for UDP. 6LoWPAN is one of the most important technologies for the integration of IPv6 in smart objects based on Wireless Sensor Networks with low power, limited bandwidth and reduced memory capabilities.

In order to bridge 6LoWPAN subsystems in our framework with the Kernel Linux and IPv6, a USBNet bridge based on CDC-ECM (Ethernet Networking Control Model) has been built, (see Fig. 3.3) along with a USB module (L). This way the 6LoWPAN-bridge defines an Ethernet Interface making access to the 6LoWPAN-connected devices transparent. Specifically, 6LoWPAN bridge carries out the translation from the 6LoWPAN header of the packets received via the WPAN interface to the IPv6/UDP headers for the packets transmitted through the Ethernet Interface, and vice versa, in a transparent way. Thereby, it also allows to use the existing Linux protocols such as 'radvd' for neighbor discovery, in order to announce the prefix of the network assigned to the 6LoWPAN. Comparing this to Fig. 2, this feature corresponds to the functionality of HGW6.

An ancillary component to this 6LoWPAN bridge is the Movital, wireless and personal device, (see Fig. 3.2), which is an adapter used for the integration of specific devices such as clinical sensors in personalized health solutions [22]. This connects legacy technologies through USB, Serial RS232, or IrDA, this is the functionality of HGW1 (I). Movital also offers an interface based on RFID of High Frequency (HF) for the user interaction with other users and objects through contactless identification, with the module from Skyetek, i.e. HGW4 (K), and finally the communication through 6LoWPAN, i.e. HGW3 (J).

The 6LoWPAN modules are based on the Jennic JN5139 module. This is an OpenRISC 32-bit processor, which supports IEEE 802.15.4, ZigBee-Pro, 6LoWPAN, and GLoWBAL IPv6. In addition, a port that supports Contiki OS has been built for them. Finally, this also implements an advanced cryptography stack based on Elliptic Curves in order to offer authentication, integrity and confidentiality. These are highly relevant to the IoT as attested in [18].

4.2. GLoWBAL IPv6

The compression headers originally defined for 6LoWPAN in RFC4944 were insufficient for many practical uses of IPv6 with smart things. This was because, the header compression method proposed in RFC4944 was primarily conceived to serve effectively unicast communications in local and personal communication scopes, where IPv6 addresses carry the link-local prefix and an Interface Identifier (IID) directly derived from IEEE 802.15.4 addresses. In this case, both addresses may be completely elided. When global communications are considered end-to-end, including smart things addressable by IPv6, the existing mechanism proved inefficient. To resolve the issue a new encoding format was standardized in the revised RFC6282, to improve the compression of Unique Local, Global, and multicast IPv6 addresses. This new encoding format is based on shared state within contexts. Although usable, the RFC6282 method yields header overheads of 26 bytes; while it does better than the 41 bytes required by RFC4944, it can still be inefficient considering the 102 bytes available for a LoWPAN frame (127 bytes less the 25 bytes from the MAC layer header).

For this reason, GLoWBAL IPv6 [6] has been proposed to optimize global addressing involving LoW-PANs. This has the further advantage that it provides efficient addressing and integration to both IEEE 802.15.4 sensor devices, which have no native support for 6LoWPAN, and also to other technologies which do not cater for IPv6 communication capability into their stacks.

Take for example a smart device with a Bluetooth Low Energy interface, such as a smart phone. Usually, a Smart Phone would offer Internet connectivity through its GPRS/GSM network interface. GLoW-BAL IPv6 fills the IPv6 addressing requirement for any smart thing connected to the smart phone's Bluetooth Low Energy network (compare Fig. 2) by acting as the mapping protocol between the Local Network and the wide-area network using appropriately constructed IPv6 addresses. Thereby, this smart phone can efficiently enable with IPv6 through GLoWBAL IPv6 to the smart things connected to it through its Bluetooth Low Energy interface.

GLoWBAL IPv6 defines an Access Address/Identifier (AAID), and an AAID-IPv6 address translation mechanism for different technologies, in order to adapt any device to the IoT architecture with IPv6. In this respect, AAID simplifies IPv6 communication parameters (source and destination addresses/ports, originally 36 bytes long) to a single 4-byte communication identifier augmented by one byte for the 'Dispatch' header field, totaling 5 bytes for the GLoWBAL IPv6 header. Thus, the IPv6/UDP headers are significantly reduced. This mechanism achieves an efficient frame format for global communications in networks that do not have native support for IPv6.

The implementation of the GLoWBAL IPv6 mechanism is done in a gateway dedicated to carrying out the translation from AAID to IPv6 and vice versa. In reality the gateway is a software module built over the smart phone or the multiprotocol hardware platform depicted in Fig. 3.1, which utilizes the mentioned virtual network kernel devices such as tun/tap. The device operates at layer 2 and simulates Ethernet frames. As such, it can exchange frames with the Future Internet core network (right side of the Fig. 2).

4.3. IPv6 addressing proxy

The current situation in industrial and building automation is a rather fragmented set of technologies. Each technology comes with a set of fit-for-purpose sensors and their respective application environments with lack of efficient interoperability among them. Some associations of manufactures have been formed to build common technology frameworks, e.g. Konnex (KNX) for building automation. While such de facto standards enjoy widespread adoption to date, this does not discourage use of other relevant protocols such as the emerging ZigBee and the older X10. Due to this fragmentation, various IoT initiatives are considering a shift towards a common access and communication framework based on IPv6. Adoption of the Internet Protocol implies that addressing of devices in each legacy technology needs to

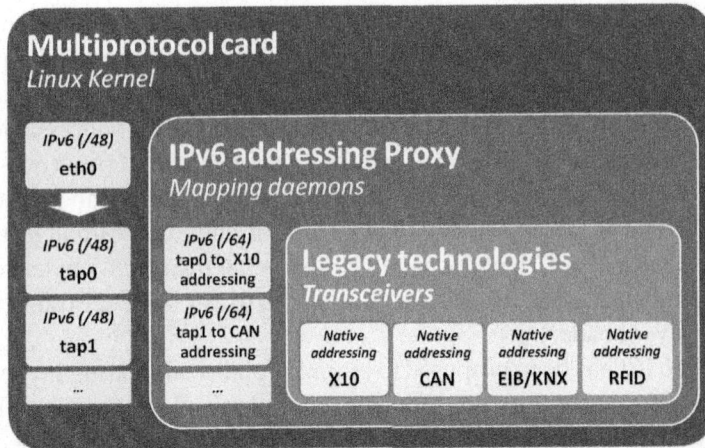

Fig. 4. IPv6 addressing proxy integration in the multiprotocol card.

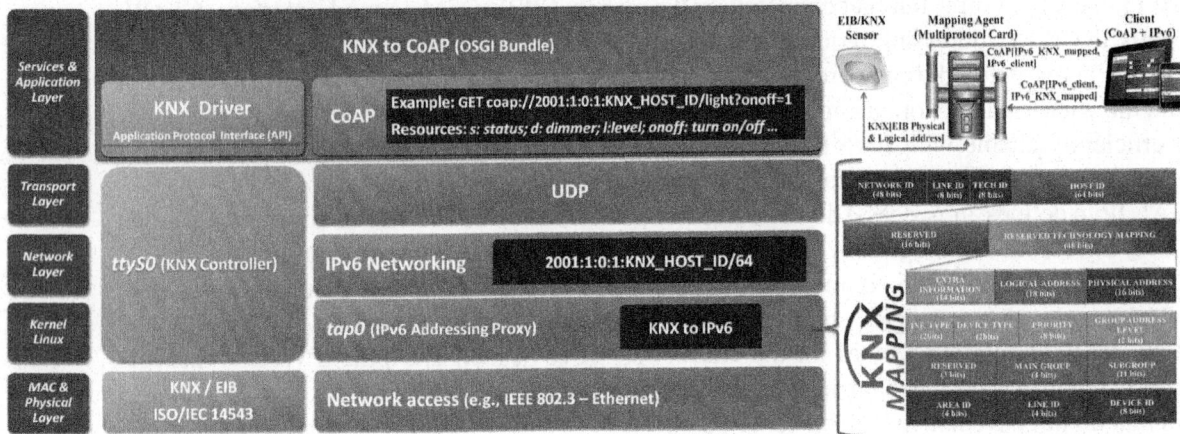

Fig. 5. Left: Internet integration stack instance for KNX integration. Right: KNX mapping from native addressing to IPv6 and translation process.

be redefined. With our work we aim to provide a transparent mechanism for the users and devices to map the different addressing spaces from each legacy technology to a common IPv6 addressing space.

The challenge for IPv6 here is to embrace all existing native addressing schemes, so that the features and functional specifications of existing devices in the legacy building automation networks are maintained. To this respect, we have proposed IPv6 mappings for each native addressing scheme by use of an IPv6 Addressing Proxy which handles the translations between an IPv6 address and its corresponding technology addressing, i.e. the native addressing depending on the technology.

Figure 4 describes the mapping hierarchy from the multiprotocol card presented in the Fig. 3. This mapping is managed by a set of tun/tap virtual network kernel device as in the GLoWBAL IPv6 implementation. The IPv6 Addressing Proxy requires information for each technology, this has been implemented for legacy networks, such as the EIB/Konnex, CAN, X10, and it is currently being adapted for BACNet and DALI networks.

Figure 5 presents an example of the mapping technique for the native addressing from EIB/KNX based on logical and physical addressing. Logical addressing is focused on the organization in groups/families, and physical addressing is more focused on the location of the device. These native addressing fields, presented in the right of the Fig. 4, have been mapped into an IPv6 address structure directly in conjunction with the properties specific to each technology.

Mapping allows that a system can easily locate and identify the device in a multi-protocol card platform such as the one we presented in Section III (Fig. 3.1).

The mapping is carried out in the lowest 48 bits of the "Host ID" half of an IPv6 address, leaving the highest 16 bits for application-level sub-networking. In addition, a network prefix of 48 bits has also been assumed; this arrangement leaves 8 bits for sub-networking of the technology (*LINE ID*) and another 8 bits for sub-networking of the 'group' (*TECH ID*). Thus access policies can be managed and a more scalable management of the different technologies is enabled.

Native addressing for EIB/KNX is based on the concept of physical and logical addresses. Physical addresses are defined by 'lines'. The lines are grouped by 'areas', and finally, areas connect to the backbone of the network. Logical addresses are used to associate a group of devices with similar functionality.

The proposed mapping takes into account the definition of physical and logical addresses, in order to make the mapping easier. Therefore, the various bits of the EIB/KNX addresses maintain their semantics and they are extended with new features.

Figure 5 depicts how the physical address structure is maintained in the lower (least significant) 16 bits. These fields contain information about the device identifier, line and area. In addition, the 'Extra Information' field is defined in the remaining 14 bits. These have been further sub-divided into 4 fields:

- The 'Information Type ID' is an identifier of the information's type that the device is able to manage. It can differentiate between 4 different types of information, which are defined by the following field the "Device type",
- The Device Types for EIB/KNX are sensor, actuator, line coupler, and area coupler.
- The Priority field, which matches with the priority field of the EIB message, for future Quality of Service use
- The Group Address Level to indicate the logical level addressing. In our case this field is always set to Level 2.

4.4. Other technologies

6LoWPAN has been presented as suitable for integrating smart things into IPv6. We now propose GLoWBAL IPv6 and an IPv6 addressing Proxy as an optimization for 6LoWPAN and to enable IPv6 to be used with new programmable technologies such as Bluetooth Low Energy. In addition, the IPv6 addressing Proxy is now proposed also for the integration of technologies which are not programmable; here a proxy is needed in order to translate from the assigned IPv6 address for each end device to the native addressing defined by the legacy technology. This new technology provides a solution that is applicable for any current addressing scheme. Moreover, it can be considered also for the current translations schemes from non-IPv6 addressing to an IPv6 one for the identifiers from RFID technologies, Digital Objects Identifiers (DOIs) and Universal Identifiers (UID). For example, the protocol and identifier deployed most widely is the Electronic Product Code (EPC) from EPCGlobal. EPC codes are 96-bits unique codes. An architecture has been proposed similar to that of the Internet for the management of the EPC; it consists of EPC Information Systems and a global Object Name Server (ONS), which can be seen as the equivalent to the DNS.

A mapping between EPC and IPv6 is needed in order to integrate EPC over IPv6. This integration is justified, since EPC is not a unique standard for products identification.

Table 1
Comparison of different protocols for the integration of IPv6 in constrained and legacy technologies

Protocol/Feature	Code size optimized (low memory req.)	Header size optimized (overhead level from IPv6 header)	Communication stack independent	Feasible for legacy technologies (application level editable)	Feasible for proprietary technologies (non editable)	Require border router or Gateway	IPv6 address managed by end-device
IPv6 (Base)	×	× ●●●●●	×	×	×	√	√
lwIP	√	× ●●●●●	×	×	×	√	√
uIP	√	× ●●●●●	×	×	×	√	√
6LoWPAN (RFC 4944)	√	√ ●●●○○	×	×	×	×	√
6LoWPAN (RFC 6282)	√	√ ●●○○○	×	×	×	×	√
GLoWBAL IPv6	√	√ ●○○○○	√	√	×	×	√
IPv6 addressing Proxy	√	√ ○○○○○	√	√	√	×	×

A Unique Identifier (UID) has been defined, which consists of a 40-bit identifier hard-coded by the manufacturer to ensure it is really unique. This uniqueness property is being considered by the pharmaceutical industry, because it satisfies its requirements for offering an efficient, trustable, and safely traceable solution.

Finally, the integration of the Digital Objects Identifier (DOIs) should be considered because of its extended use in physical things such as books and movies.

5. Discussion and conclusions

The key contribution of this paper is the proposal of a set of technologies for the extension of legacy technology addressing to the IPv6 address space. This will allow the management of all things around us and access to their information independently of the technology used to convey this information.

For this purpose an integration stack and appropriate hardware platforms have been proposed. In addition, the address space integration has been supplemented by the Half-Gateways, which bridge legacy technologies to the IPv6 world, either at the network layer, or at the application layer as required. We have instantiated this integration with concrete examples for EIB/KNX, X10, CAN, Bluetooth Low Energy, and IEEE 802.15.4. The proposed technology is not limited to the above; additional legacy technologies, protocols and identifiers living in the fringe of the Internet (thus making up the IoT) can be considered such as the DOI, EPC and UID as mentioned. Table 1 summarizes the main features of the existing and the proposed solutions.

Table 1 shows that lwIP and uIP have mainly focused on the reduction of the code footprint, since the stack was defined for wired technologies such as embedded systems with Ethernet Interfaces. For a wireless medium with constrained frame size, 6LoWPAN with header compression mechanism presents a high processing load. For this reason, we propose GLoWBAL IPv6. This presents a reduced overload, based on the reduced overhead from GLoWBAL IPv6 header in relation with the overhead from IPv6 header and even 6LoWPAN header. In addition, GLoWBAL IPv6 would allow the integration in the application layer (payload) of the AAID in order to make it compatible with solutions, which are programmable at the application level but not are able to be modified in the communication stack. Therefore, it is communication stack independent, which is useful for the IPv6 integration over already deployed networks based on closed stacks over technologies such as the Bluetooth Low Energy and IEEE 802.15.4.

Since not all the technologies and solutions are able to be programmed at the application layer, we have defined an IPv6 addressing Proxy, which offers compatibility with proprietary technologies and

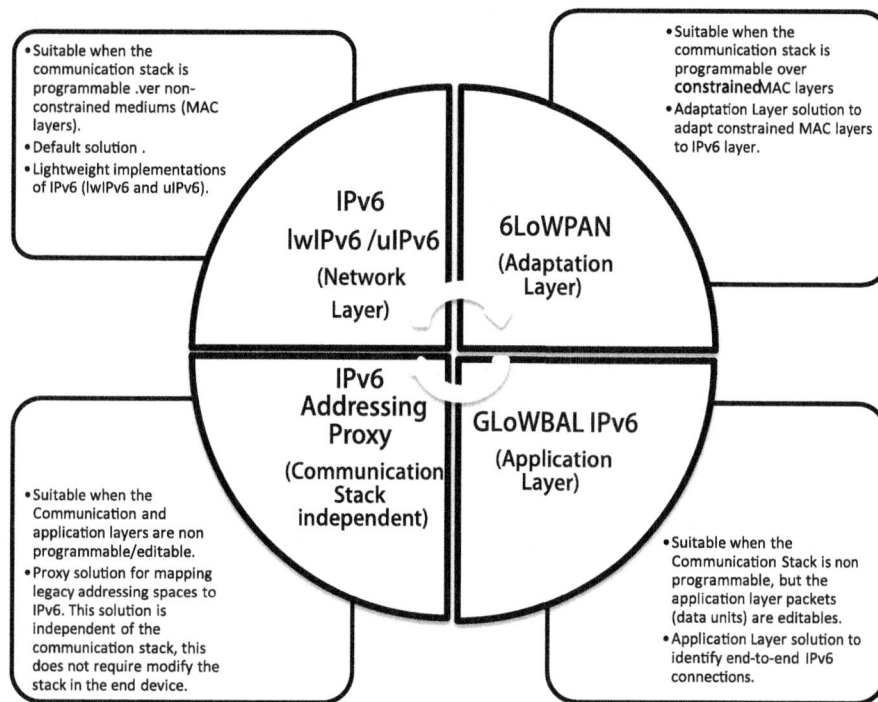

Fig. 6. Classification of the presented solutions in function of the communication layer upgraded.

native address spaces. The main advantage of GLoWBAL IPv6 with respect to the IPv6 addressing Proxy is that with GLoWBAL IPv6 the end device is aware about IPv6, while in the IPv6 addressing Proxy, the end-device is totally agnostic about the version from IP.

The second key contribution is the ability to integrate into the Internet legacy technologies and devices with a new rich range of IPv6-enabled services. This way both real and virtual objects can interoperate and communicate by way of IPv6 end-to-end, if their corresponding services so wish. Thus, homogeneous, transparent and scalable access to devices and services can be achieved. From the Web of Things approach, the CoAP/RestFul methods could be applied directly to the end-device, increasing the scalability of the solution, offering flexibility and allowing for extension of the ubiquitous concept with mobility and global interoperability. This is just what back-end services like Software-as- a-Service would like to see.

Such as presented in the Table 1, the main problem from the proposed adaptations to integrate Internet in constrained devices such as 6LoWPAN, GLoWBAL IPv6, and IPv6 addressing Proxy is that these nodes require of 6LoWPAN Border Routers or gateways to adapt from the lightweight version of the protocol headers and protocols to the common one, in order to interoperate with the rest of the IPv6-enabled entities.

Figure 6 summarizes the suitability of each one of the presented solutions in terms of constrained level for the Medium Access Control (MAC) layer, and the programmability of the different layers from the communication stack.

Our vision is that in the current Internet of Things world everything can be discovered through global resource directories, distributed as desired. These directories would be based on Internet technologies such as the example of mDNS/DNS, and accessed in a homogeneous way through Web Services technologies over IPv6 such as HTTP and CoAP at the application level. This will be complemented with SenML over JSON, RDF, or EXI for the semantic description such as those defined by SPITFIRE [24].

Ongoing work is focused on offering mobility and multi-homing support for the GLoWBAL IPv6 protocol and the IPv6 addressing Proxy. We also need to extend our technology to multiple addressing proxies – themselves accessed by relevant protocols (e.g. using IPv6 Anycast). We believe that applications in mobile environments and the integrated in devices such as smart phones with multiple communications interfaces should not depend on any particular IPv6 sub-network. For this reason, we consider an extension of the application protocols for smart things; this should include session management to define security associations, manage mobility and multi-homing support for the open sessions from each smart object. Presumably the architecture and gateways should address this device context regarding the open sessions such as the synchronization of the AAID among the different GLoWBAL IPv6 gateways.

The final conclusion is that smart phones, personal data terminals, and other mobile computing devices are still far from what a Future Internet of Things will require to connect services, people, and things. But, full IPv6 integration is the first step towards this destination. As next steps one envisages support for mobility, multi-homing, discovery techniques, and management solutions in order to make things more autonomous and to enable a communications era based on the Future Internet of Things, Services and People.

Acknowledgments

The authors would like to thank the European Project "Universal Integration of the Internet of Things through an IPv6-based Service Oriented Architecture enabling heterogeneous components interoperability (IoT6)" from the FP7 with the grant agreement no: 288445, and the Spanish ministry, for education, social politics and sport, for sponsoring this research activity with the grant FPU program (AP2009-3981). Finally, this research has been partially carried out by the Intelligent Systems and Networks group, of the University of Murcia, Espinardo, Spain, awarded for its excellence as a research group in the frame of the Spanish "Plan de Ciencia y Tecnologia de la Region de Murcia" from the "Fundacion Seneca" (04552/GERM/06).

References

[1] L. Atzori, A. Iera and G. Morabito, The Internet of Things: A survey, *Computer Networks* **54**(15) (2010), 2787–2805.

[2] M. Zorzi, A. Gluhak, S. Lange and A. Bassi, From today's INTRAnet of things to a future INTERnet of things: A wireless-and mobility-related view, *IEEE Wireless Communications* **17**(6) (2010), 44–51.

[3] G. Kortuem, F. Kawsar, D. Fitton and V. Sundramoorthy, Smart objects as building blocks for the Internet of things, *Internet Computing, IEEE* **14**(1) (2010), 44–51.

[4] Z. Shelby and C. Bormann, 6LoWPAN: The Wireless Embedded Internet, Wiley, ISBN: 978-0-470-74799-5, 2009.

[5] E. Welbourne, L. Battle, G. Cole, K. Gould, K. Rector, S. Raymer, M. Balazinska and G. Borriello, Building the Internet of Things Using RFID: The RFID Ecosystem Experience, *Internet Computing, IEEE* **13**(3) (2009), 48–55.

[6] A.J. Jara, M.A. Zamora and A.F. Skarmeta, GLoWBAL IP: an adaptive and transparent IPv6 integration in the Internet of Things, *Mobile Information Systems* **8**(3) (2012), 177–197.

[7] Merz, T. Hansemann and C. Hubner, Building Automation: Comm. Systems with EIB/KNX, LON und BACnet, Springer, Series on Signals and Communication Technology. ISBN.978-3-540-88828-4, 2009.

[8] M.A. Zamora, J. Santa and A.F.G. Skarmeta, An integral and networked Home Automation solution for indoor Ambient Intelligence, *IEEE Pervasive Computing* **9** (2010), 66–77.

[9] M. Farsi, K. Ratcliff and M. Barbosa, An overview of controller area network, *Computing and Control Engineering Journal* **10**(3) (1999), 113–120.

[10] D. Uckelmann, M. Harrison and F. Michahelles, *Architecting the Internet of Things*, Springer, ISBN 978-3-642-19156-5, 2011.

[11] Adam Dunkels, Full TCP/IP for 8-Bit Architectures. In Proceedings of the first international conference on mobile applications, systems and services (MOBISYS 2003), San Francisco, May 2003.

[12] R. Gurram, B. Mo and R. Gueldemeister, A Web Based Mashup Platform for Enterprise 2.0, Web Information Systems Engineering, Lecture Notes in Computer Science, pp. 144-151, Vol. 5176, 2008.

[13] D. Guinard, V. Trifa, S. Karnouskos, P. Spiess and D. Savio, Interacting with the SOA-Based Internet of Things: Discovery, Query, Selection, and On-Demand Provisioning of Web Services, *Services Computing, IEEE Trans* **3**(3) (2010), 223–235.

[14] Z. Shelby, Embedded web services, *Wireless Communications, IEEE* **17**(6) (December 2010), 52–57.

[15] A.J. Jara, P. Martinez-Julia and A.F. Skarmeta, Light-weight multicast DNS and DNS-SD (lmDNS-SD): IPv6-based resource and service discovery for the Web of Things, International Workshop on Extending Seamlessly to the Internet of Things, 2012.

[16] A.P. Castellani, M. Gheda, N. Bui, M. Rossi and M. Zorzi, Web Services for the Internet of Things through CoAP and EXI, Communications Workshops (ICC), 2011 IEEE International Conference on, 5–9 June 2011.

[17] B. Carballido, D. Pesch, R. De Paz Alberola, S. Fedor and M. Boubekeur, Constrained Application Protocol for Low Power Embedded Networks: A Survey, International Workshop on Extending Seamlessly to the Internet of Things (es-IoT), IEEE proceedings, 978-0-7695-4684-1, 2012.

[18] R. Roman, P. Najera and J. Lopez, Securing the Internet of Things, *Computer* **44**(9) (Sept. 2011), pp. 51–58.

[19] R. Moskowitz, *HIP Diet EXchange (DEX)*, IETF draft-moskowitz-hip-rg-dex (work in progress), Internet-Draft, Internet Engineering Task Force (IETF), 2012.

[20] S.-D. Lee, M.-K. Shin and H.-J. Kim, *EPC vs. IPv6 mapping mechanism*, Advanced Communication Technology, The 9th International Conference on, Vol. 2, 2007.

[21] W. Colitti, K. Steenhaut, N. De Caro, B. Buta and V. Dobrota, *REST Enabled Wireless Sensor Networks for Seamless Integration with Web Applications* Mobile Adhoc and Sensor Systems (MASS), IEEE 8th International Conference on, 17-22 Oct. 2011, pp. 867–872.

[22] A.J. Jara, M.A. Zamora and A.F.G. Skarmeta, An internet of things–based personal device for diabetes therapy management in ambient assisted living (AAL), *Personal and Ubiquitous Computing* **15**(4) (2011), 431–440.

[23] M. Tatipamula, P. Grossetete and Esaki, IPv6 integration and coexistence strategies for next-generation networks, *Communications Magazine, IEEE* **42**(1) (Jan 2004), 88–96.

[24] D. Pfisterer, K. Romer, D. Bimschas, O. Kleine, R. Mietz, C. Truong, H. Hasemann, A. Kroler, M. Pagel, M. Hauswirth, M. Karnstedt, M. Leggieri, A. Passant and R. Richardson, *SPITFIRE: toward a semantic web of things*, Communications Magazine, IEEE **49** (11) (2011), 40–48.

7

Range assignment problem on the Steiner tree based topology in ad hoc wireless networks

Rashid Bin Muhammad
Department of Computer Science, Kent State University, Kent, OH, USA
E-mail: rmuhamma@cs.kent.edu

Abstract. This paper describes an efficient method for introducing relay nodes in the given communication graph. Our algorithm assigns transmitting ranges to the nodes such that the cost of range assignment function is minimal over all connecting range assignments in the graph. The main contribution of this paper is the $O(N \log N)$ algorithm to add relay nodes to the wireless communication network and 2-approximation to assign transmitting ranges to nodes (original and relay). It does not assume that communication graph to be a unit disk graph. The output of the algorithm is the minimal Steiner tree on the graph consists of terminal (original) nodes and relay (additional) nodes. The output of approximation is the range assignments to the nodes.

Keywords: Connectivity, range assignment, steiner tree, and approximation

1. Introduction

An ad hoc wireless network is a collection of independent devices (transceiver nodes) that have to communicate themselves in the absence of any central authority. The absence of central authority implies that the coordination necessary for communication has to be carried out by the nodes themselves. Ad hoc network nodes allow to communicate directly to each other using wireless transceivers without need for a fixed infrastructure. Any two devices (nodes) in the region achieve communication either directly if they are within each others transmission range (i.e., within one hop) or indirectly via other nodes, if they are out of each others range (multi-hop). Note that in this work we assume that transceiver nodes are portable devices. Typically, portable devices have limited power resource available. One of the main characteristic properties of the ad hoc wireless networks is the ability to vary the power used in communication among nodes, which implies the concept of transmission range, hence the range assignment.

There are many situations in which fixed wired networks, such as the Internet, are not realistic or practicable because it may not physically possible (due to fire, earthquake or flood) to deploy wired network or due to economic infeasibility. In such situations a collection of hosts (emergency workers) with wireless devices may form a temporary network (ad hoc wireless network). Furthermore, it is not hard to imagine that soon after the initial deployment of emergency network, we need to expand this network by adding additional hosts (additional emergency workers). Now the question is where we deploy these additional hosts so that the final network is connected and use minimal communication power. The motivation for our problem arises from this very scenario [19]. Consider a group of emergency workers, e.g. fire fighters, deployed in some geographical region. These workers need to communicate by transmitting and receiving radio signals on an ad hoc network. Workers are equipped with terminal devices. From now on, we call them terminal nodes. A support team with additional transceivers, called relay nodes, are use to form the connected network. The task of the support team is to place relay nodes in the region so that the terminal nodes and the relay nodes together can form a connected communication graph while at the same time minimizing the energy consumption by setting transmitting ranges. Note that the preliminary version of this work appeared in Reference [23].

A range assignment for the finite set N of n network nodes is a function, $RA :\to R^+$ (where R^+ is the set of non-negative numbers), that assigns to every node $u \in N$ such that $0 < RA(u) \leqslant r_{\max}$, where $r_{\max}$ is the maximum transmitting range. The cost of a range Assignment is defined as $\text{cost}(RA) = \Sigma(RA(u))^\alpha$ for some real constant $\alpha > 0$. The underlying intuition is that the element of a finite set N are given transceiver nodes, and one can choose for each transceiver nodes $u \in N$ a corresponding data transmission range i.e., radius, $RA(u)$. Sending data at radius, $RA(u)$, consumes energy proportional to $(RA(u))^\alpha$. The parameter is α called the distance-power gradient. In an ideal environment, i.e., in the empty space $\alpha = 2$ but vary from 1 to more than 6 depending on the environmental conditions of the place where the transceiver nodes are located, see [25].

Let $dist(u, v)$ denote the Euclidean distance between two arbitrary nodes $u, v \in R^2$. We define range assignment RA on the following kind of graph. Let $G_{RA} = (N, E_{RA})$ be directed communication graph of range assignment RA where an edge (communication link) belongs to the set E_{RA} if, and only if, node u can send the data to node u, i.e. the radius of node u is at least as large as the distance between nodes u and v. Formally, we say as follows:

Proposition 1. Given nodes $u, v \in (N, E_{RA})$, $\text{edge}(u, v) \leqslant E_{RA}$ if, and only if, $RA(u) \geqslant dist(u, v)$.

In this work, we demand that the induced communication graph G_{RA} must be strongly connected to ensure all-to-all communication. Note that here by strong connectivity we mean the relevant optimization problem, i.e. the problem to find a range assignment RA that has minimal $cost(RA)$ among all range assignments satisfying the strong connectivity property.

Prosopition 2. Given directed communication graph G_{RA} is strongly connected if every two node $u, v \in G_{RA}$ can send data to each other i.e. u and v are reachable from each other.

The paper is organized as follows. In the next section we define the problem. In the Section 3 we discuss the relevant work. Preliminaries and definition are given in Section 4. Section 5 presents the computational model. Section 6 describes the algorithm based on the Steiner tree. Section 7 presents 2-Approximation algorithm for range assignment problem. Section 8 discuss the open problem. Finally, Section 9 concludes the paper with some final remarks.

2. Problem definition

We shall formulate the problem in the geometric graph, using [15,20], as follows. Given a set N of n points, the problem is to introduce minimal number of additional points such that the graph $G(N \cup N_{RA}, RA)$ is strongly connected and the cost of range assignment function, $\text{Cost}(RA) = \sum(RA(u))^\alpha$ is minimum over all connecting range assignment functions, where α is the distance-power gradient. Note that if range assignment function $RA = 0$ (i.e. transmitting range $r = 0$), this problem reduces to finding the Euclidean Steiner tree problem [21] for the set N and in context of ad hoc wireless networking, the solution of this problem is proposed in [20,22].

Given the location of the terminal nodes, we are interested in the problem of establishing strong connectivity by minimizing the number points (physical loations) where the support team should placed additional relay nodes and minimizing the range assignment, RA, of the resultant graph (tree). In a certain sense, this problem can be seen as a generalization of the problem of determining the critical transmitting range for connectivity, where the constraint that all the nodes have the same transmitting range is dropped. Note that the solution of the problem in which given nodes have the same transmitting range is proposed in [15,19]. Formally, let N be a set of terminal nodes, with $|N| = n$. These nodes are located in a certain bounded convex region S of the Euclidean space R^2. The problem instance is completely defined by the set of locations of n terminal nodes, $N = \{x_i \leqslant S : 1 \leqslant i \leqslant n\}$, and the range assignment function RA. The resultant network topology will be an directed graph $G(N, E(N, RA)) = G(N, RA)$, where $E(N, RA)$ is an edge set defined by $E(N, RA) = \{(x_i, x_j) : x_i, x_j \leqslant N, i \neq j, \|x_i - x_j\| \leqslant RA\}$. A solution to the problem is a set of locations of relay nodes, $N_{RA} = \{y_i \leqslant S : 1 \leqslant i \leqslant N_{RA}\}$ and range assignment function RA and represented in a single maximal connected component in the graph $G(N \cup N_{RA}, RA)$.

3. Relevant work

To our knowledge, the connectivity problem from this practical perspective has been discussed only in [15,19,20]. The MST algorithm proposed in [15] uses Boolean model for the network. The proposed scheme first calculates the MST and then placed the relay nodes on the edges of the MST that are longer than transmitting range r. The overall complexity of the algorithm is determined by the first step which is $O(N \log N)$. In [20], the problem was studied from network lifetime viewpoint. The paper discussed the problem how to increase the network lifetime by choosing the locations of relay sensors such that the Euclidean distances among sensors are at most some δ i.e. the range of the sensors. The algorithm computes the Steiner points and; then sorts the edges and placed relay sensors on the edges larger than the transmitting. The overall complexity of the algorithm is determined by the sorting step, which is $O(N \log N)$.

Kirousis et al. [14] showed that the range assignment problem for one-dimensional networks is in P but NP-hard in the case of three-dimensional networks. For two-dimensional networks, the problem is remains NP-hard [5]. They also gave a minimum spanning tree (MST) based 2-approximation algorithm minimum power symmetric connectivity. In fact, their dynamic programming algorithm solved the complete range assignment problem in $O(n^4)$ for the case of n collinear nodes and some power requirement. Recall, the goal of the complete range assignment problem is to establish a strong connected subgraph of a given graph. Calinescu et al. [3] improved the result [14] and proposed the fully polynomial $1 + ln2$ approximation scheme base on the Steiner tree.

Blough et al. [1] consider a constrained version of the range assignment problem, where the node transmitting ranges must be assigned in such a way that the resulting communication graph is strongly connected and the energy cost is minimum. The result presented in this paper have shown that a weakly symmetric range can reduce the energy cost. Clementi et al. [4] presented the first polynomial-time, approximation algorithm for the Min Assignment problem. The algorithm guarantees a 2-approximation ratio and runs in $O(hn^3)$ time. The Min Assignment problem consists in finding a range assignment of minimum power consumption for a given set of radio stations located on a line and an integer $h \geqslant 1$. Rodoplu et al. [29] proposed a distributed algorithm to construct topology, which is guaranteed to contain the least energy path connecting any pair of nodes in the unit disk graph. Their algorithm relies on a simple radio propagation model for transmit power roll-off. Using this they achieved the minimum power topology, which contains the minimum-power path from each node to a designated node. However, their protocol is not time and space efficient. Li et al. [16] improved the result [29] by proposing an efficient localized algorithm to construct a topology that is sub-graph of the graph constructed by Rodoplu et al. [29] and proved that the sub-graph is sparse, i.e. it has a linear number of links.

Other researchers working in the field of wireless ad hoc and sensor networks have also consider the issue of power efficiency and transmitting range but have taken different approaches. For example, Hou et al. [12] approach was to adjust transmitting ranges of the given node to reduce overall consumption of the network. Particularly, they analyze the effect of adjusting transmission power to reduce interference and hence achieve higher throughput with respect to the fixed transmission ranges of the nodes [32]. On the other hand, Heinzelman et al. [11] approach was to assign the task of maintaining connectivity to energy nodes. They describe an adaptive clustering-based routing protocol that maximizes network lifetime by randomly rotating the role of per-cluster base station among nodes with higher energy reserves.

4. Preliminaries

This paper assumes that underline graph is geometric graph. A geometric graph $G = (V, E)$ is a graph in the plane so that its vertices are points and its edges are straight-line segments connecting pairs of these points. Recall, a geometric graph need not be planar. We represent sensors with points, relay sensors with additional points or Steiner points, and sensors transmission radius or distance between two sensors with edge length or Euclidean distance.

As we have mentioned above, a range assignment for the set N on network nodes is a function RA that assign to $u \in N$ a transmitting range $RA(u)$. Formally, let N be a set of nodes in the Euclidean space R^d, $d = 2$. The problem is to determine a range assignment function RA such that corresponding communication graph is strongly connected and Cost(RA)$= \sum_{u \in N}(RA(u))^\alpha$ is minimum over all connecting range assignment functions, where α is the distance-power gradient.

Let N be a set of n planar points. The Voronoi diagram partitions the plane into regions such that each region contains exactly one point. The region of a point $n_i \in N$ consists of all points closer to n_i than to any other points on the plane. The region of a point $n_i \in N$, $1 \leqslant i \leqslant n$, is called the Voronoi cell or Voronoi polygon and denoted as $VP(n_i)$. The vertices of these polygonal regions are called Voronoi vertices and the polygonal boundaries i.e., edges of the regions, are called Voronoi edges. The collection of Voronoi polygons $VP(n_i)$ for each $n_i \in V$ is called Voronoi diagram and often denoted as $VD(N)$. For further discussion on the properties of the Voronoi diagram, see [18].

The Delaunay triangulation, $DT(N)$, is the straight-line dual of Voronoi diagram $VD(N)$. Each Delaunay face (triangle) corresponds to a Voronoi vertex. The interior of each Delaunay triangle of $DT(N)$ contains no point $n_i \in N$. The Voronoi diagram and the Delaunay diagram can be constructed in $O(n \log(n))$ time [31]. Reader may consult [2].

Lemma 1. Given a set N of $n \geqslant 3$ points. The $DT(N)$ has $E = 3(n-1) - |\partial CH(N)|$ number of edges and $T = 2(n-1) - |\partial CH(N)|$ number of triangles, where $|\partial CH(N)|$ is the number of points on the boundary of the Delaunay triangulation of the point set N.

Lemma 2. $DT(N)$ contains at least one MST for N. The MST for N can be determined in time $O(n)$ from DT(N) [27].

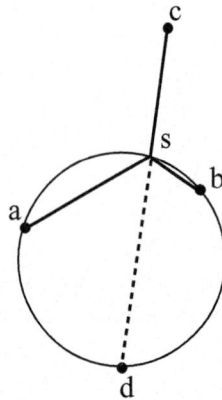

Fig. 1. The addition of Torricelli Point.

The Euclidean Steiner minimal tree (ESMT) problem asks for a shortest tree spanning a given set N of n terminals in the plane. Contrary to the minimum spanning tree (MST) problem, connections in Steiner minimum trees (SMTs) are not required to be between the terminals only. Additional intersections, called Steiner points, can be introduced to obtain shorter spanning tree. The ESMT problem is as follows. For a given set N of points in the plane, construct a minimal length tree, which connects terminals in N. In order to minimize the total length of the tree, additional points $S = s_1, \ldots, s_m$ can be used. Here, the distance metric is the L_2 metric function, known as the Euclidean metric.

4.1. Relevant Steiner tree properties

We shall mention here only some basic properties of SMTs needed in the paper. The reader referred to [13] for a comprehensive survey. (1) Steiner points are incident with exactly three edges making $120°$ with each other (angle condition). (2) SMTs for n terminals have at most n-2 Steiner points. (3) SMTs are union of full Steiner trees (FSTs) terminals spanned by an FST have degree one. FSTs have two Steiner points less than they have terminals. If two FSTs share a terminal N, then the two edges incident make at least $120°$ with each other. Therefore, no terminal can be in more than three FSTs. In other words, each terminal $N_i, 1 \leqslant i \leqslant n$, has at most degree 3 i.e., $deg(n_i) \leqslant 3$. (4) It has been conjectured that the MST is no more than $\frac{2}{\sqrt{3}}$ times as long as the minimum length ESMT [9]. Polak has shown that this conjecture is true for $n = 4$. Graham and Hwang [10] showed that the lower bound on the SMT/MST $\geqslant \frac{1}{\sqrt{3}}$. The problem has been shown to be NP-Complete in [8].

4.2. Relevant Steiner tree results

Now, we present some classical results regarding 3-terminal and 4-terminal Steiner tree problem and discuss their incorporation into the framework of our algorithms.

In case of three terminals, if one of the angles of $\triangle abc$ is at least $120°$, then the Steiner tree consists of simply the two edges subtending the obtuse angle. Therefore, assume that all internal angles of $\triangle abc$ are less than $120°$. The addition of an additional point (or Torricelli point) in $\triangle abc$ is depicted in Fig. 1 [13].

Lemma 3. An additional point in the plane, the sum of whose distances from three given points is minimal, can be located if and only if all internal angles of a triangle are less than $120°$.

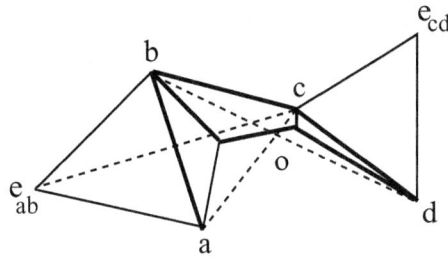

Fig. 2. The Quadrilateral abcd.

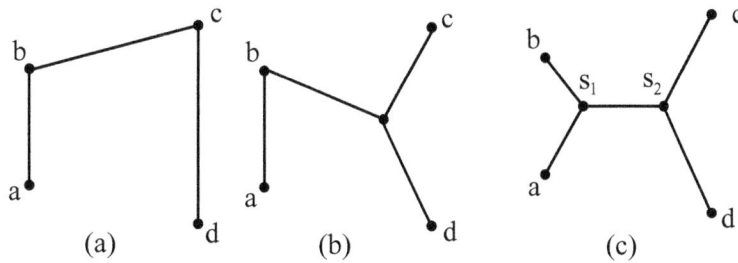

Fig. 3. Illustrates the addition of 2 additional points in quadrilateral. For proof of the fact that (a), (b), and (c) in the Figure are sufficient to define the optimal Steiner tree for 4 terminals see Gilbert and Polak [14].

In case of four terminals, consider the quadrilateral abcd in the Fig. 2. Points e_{ab} and e_{cd} form equilateral triangles with (ab) and (cd) respectively. By Du et al. [7] the necessary and sufficient condition for the existence of FST in that the quadrilateral abcd must be convex. Weng [34] showed that convexity of the quadrilateral abcd (see Fig. 2) is implied by following two conditions.

1. $\angle edaab$, $\angle eabcd$, $\angle edaab$, and $\angle eabca$ are all less than $120°$.
2. $\angle doc < 120°$.

An important insight of Polak's [26] paper on existence of SMT of four terminals can be stated as follows.

Lemma 4. If four terminals a, b, c, and d form a convex quadrilateral, then SMT exists for those terminals.

5. Computational model

The communication graph defines the set of wireless links that the nodes can use to communicate with each other. Based on the above discussion, the existence of the edge (link) between two nodes depends on the distance between the nodes, transmission range assignment, and environmental conditions i.e., distance power gradient, α. Given a transmitting range assignment, we define communication graph G such that directed edge $(u, v) \in G$ if, and only if, $dist(u, v) \leqslant r(u)$ i.e., Euclidean distance from u to v is at most $r(u)$.

This paper assumes the long-distance model [28], which abstracts many characteristics of the environment. In this model, the average long-distance path loss is proportional to the separation distance between transmitting node v and receiving node t raised to a certain exponent, α, which is called the distance power-gradient. The value of α depends on the environmental conditions and it has been experimentally evaluated in many scenarios [30]. Typically, nodes are located in a two-dimensional Euclidean space. These nodes are connected by wireless links that are induced by their energy level. According to the long-distance model, in the given region, each node u transmits a signal with power $P(u)$, which is received by other node v such that $dist(u, v)^\alpha \leqslant P(u)/\gamma$, where $dist(u, v)$ is the Euclidean distance between nodes u and v, $\alpha \leqslant 1$ is the distance power gradient, and $\gamma \leqslant 1$ is the transmission quality parameter. Without loss of generality, in this paper, we assume $\gamma = 1$. The radio coverage region in this model is a disk of radius at most $r(u) \geqslant (P(u)/\gamma)^{1/\alpha}$ centered at the transmitter u. Since $\gamma = 1$, thus we have $r(u) \geqslant (P(u))^{1/\gamma}$. The value $r(u)$ is the transmission range of transmitter u, i.e., the maximum distance at which u can transmit in one hop with power $P(u)$.

6. Overall strategy of the algorithm

For the ease of the presentation, we divide the overall algorithm's strategy into four distinct phases namely, construction phase, test phase, merge phase, and the reduction phase. The preliminary version (Kruskal's version) of this work appeared in the Reference [24]. Our algorithm operates much like Prim's algorithm for finding the minimum spanning tree. This algorithm has the property that the adding of edges in the set always forms a single tree [27]. As illustrated in the example below, the algorithm starts by placing terminal in the priority queue. The tree starts from an arbitrary terminal and grows from there until the tree spans all the vertices in N. At every stage, it selects the shortest available edge that can extend the tree to an additional terminal and mark the terminals up to 4-terminals for which edges are making right turns or left turns. The algorithm ends the construction phase by partitioning the set into subsets of 4-, 3- and 2-terminals using markings. Note that since we are using right turn (left turn) technique to create a subset, at least one terminal is common in adjacent subsets. The performance of this phase depends on how we implement the priority queue. If queue is implemented as a binary heap, building heap requires $O(n)$ time. Since each operation that requires to extract the minimum runs $O(n)$ time, and there are $O(n)$ such operations therefore, the total time is $O(n \log n)$. Selecting the shortest edge from the neighboring edges requires $O(n)$ time. Thus, the total time for construction phase is $O(n \log n)$. However, the asymptotic running time of this phase can be improved by implementing priority differently e.g. Fibonacci heap.

In the test phase, the algorithm first performs the convexity test using lemmas 1 and 2 for 3-terminal subset and 4-terminal subset. After that, it performs the length test to find out if there is any edge in the subsets whose length is greater than $\delta > 0$. If such an edge exists in 3-terminal and 4-terminal subsets, additional points (relay sensors) are required to increase the lifetime of network. Subsets that pass the convexity test are those whose overall lengths can be minimize by adding Steiner points. Hence, increase the lifetime of network. The algorithm includes the additional point in 3-terminal subsets using the algorithm in [33] (or see Fig. 1) and additional two points in 4-terminal subsets using simple geometry in [17] (or see Lemma 4 or Fig. 3). It is easy to see that operations in this phase require constant time. Therefore, this phase require linear time.

In the merging phase, the algorithm simply combines the 2-terminal, 3-terminal and 4-terminal subset by union operation. Since at least one terminal is common in adjacent subsets, union operation will not create the cycles.

Lastly, the reduction phase computes the lengths of each edge in the tree produced by merging phase in $O(n)$ time. Sort the edges in non-decreasing order of lengths using some optimal sorting algorithm (e.g. heap sort) that require $O(n \log n)$ time. For all edges, if the length of edge is greater than δ then add $\left(\left\lceil \frac{|e_i e_{i+1}|}{\delta} \right\rceil - 1 \right)$ 2-degree points which can be done in $O(n)$ time. Thus, the total time for reduction phase is $O(n \log n)$.

The output of the algorithm is the tree that spans all the terminals of the given set N whose each edge has a length at most given positive δ.

6.1. Algorithm ADD_NODE

In this section, we formalize the idea presented in the preceding section. The goal of our algorithm is to abandon long-distance edges (communication links) and establish small edges instead. For this purpose, the algorithm chooses a location of additional nodes or relay node based on the Steiner topology. The following algorithm is based on the Prim's algorithm appeared in [20] in the context of energy conservation in wireless networks and the version based on the Kruskal's algorithm appeared in [24].

1. Given a set N of n terminals. Place $n_i \in N$ in a priority queue. Call this priority queue Q.
2. For each $n_i \in Q$, select the shortest available edge in Adj[n_i].
3. Check whether new edge making right turn (left turn) with previous edge.
4. Mark all terminals (up to 4 terminals) corresponding to consecutive edges that are making right turns (left turns). At the end of this step, we will have connected sub-graphs consisting of 4-, 3-, and 2-terminals.
5. Check the convexity of the 3- and 4-terminal subsets using Lemmas 3 and 4 respectively and include additional points if sub-graphs pass these tests.
6. Merge 2-, 3-, and 4-terminal subsets (sub-graphs).
7. Suppose $e_1, e_2, \ldots, e_{n-1}$ are all edges in the resultant tree. Compute $l(e_i)$ for each edge e_i.
8. Sort the edges $e_i, 1 \leqslant i \leqslant n - 1$, in a non-decreasing order of $l(.)$.
9. For all $n - 1$ edges, If $|e_i e_{i+1}| > \delta$, add $\left(\left\lceil \frac{|e_i e_{i+1}|}{\delta} \right\rceil - 1 \right)$, where $1 \leqslant i \leqslant n - 1$, 2-degree additional points.

6.2. Correctness

In this section, we shall show that the resultant graph produced by the algorithm is a minimum length tree. Following lemma is based on [19,20].

Lemma 5. The tree T produced by the algorithm is minimal if and only if each $(u, v) \notin T$ and for all edges (x, y) in the cyclic graph arising from T by adding (u, v), $C_T(u, v)$, we have $l(u, v) \geqslant l(x, y)$, where $l(u, v)$ denotes the Euclidean distance (edge length) between u and v.

Proof: Suppose that the tree T produced by the algorithm is minimal. If $l(u, v) < l(x, y)$ (i.e., condition given in the Lemma 3 does not satisfy) then there must exist an edge $(u, v) \notin T$ and an edge $(x, y) \in C_T(u, v)$ with $l(u, v) < l(x, y)$. Since edge (x, y) is a bridge edge, removed of (x, y) from T divides T into two components. Adding (u, v) i.e., $(u, v) \cup T - \{x, y\}$ gives a new tree T'. Since $l(u, v) < l(x, y)$, T' has smaller length than T. This contradicts the minimality of T.

Conversely if $l(u, v) \leqslant l(x, y)$. Suppose algorithm produced a tree T which spans given set of points and show that some minimal spanning tree T' produced by same set of points is same as that of T in terms of length i.e., $l(T) = l(T')$. This will prove that T is minimal as well. We use induction on the edge $m \in T' - T$. When $m = 0$, $T = T'$ is trivial. So let $(u', v') \in T' - T$, If we remove (u', v') from

T', the T' is divided in two connected components C_1 and C_2. If we add the path $C_T(u', v') - \{u', v'\}$ to $T' - \{u', v'\}$ i.e., $T' - \{u', v'\} \cup C_T(u', v') - \{u', v'\}$, connected components C_1 and C_2 becomes connected again. Therefore, the graph arising from T by adding (u', v'), $C_T(u, v')$, has to contain an edge of T', because otherwise $T' - \{u', v'\}$ would still be connected. The minimality of T' implies that $l(u, v) \geqslant l(u', v')$: replacing edge (u', v') by (u, v) in T', we get another tree T'' and if $l(u, v) < l(u', v')$, this tree would have smaller length than T' contradicting the minimality of $T'T$. On the other hand, lemma states that $l(u', v') \geqslant l(u, v)$, so that $l(u', v') = l(u, v)$ and $l(T'') = w(T')$. Therefore, T'' is a minimal spanning tree as well. T'' has one more edge in common with T than T', and using induction, we have $l(T) = l(T'') = l(T')$. This completes the proof.   □

7. Approximations for range assignment

In this section, we present an approximation of the optimal solution of range assignment problem by using the algorithm ADD_NODE. In essence, we are defining the range assignment, RA, on the tree produced by the ADD_NODE algorithm.

Approximation

1. ADD_NODE
2. For each node $u \in T$, define range assignment,
   $RA_t = max_{(u,v) \in T} dist(u, v)$.

Now we show that the above "approximation algorithm" produces a 2-approximation of the optimal solution. The following lemma is based on [14].

Lemma 6. Let RA_{Opt} be an optimal range assignment of the problem and RA_T be the range assignment defined above [see line 2 of the Approximation]. Then $\text{Cost}(RA_T) < 2 \times \text{Cost}(RA_{opt})$

Proof: For the proof, we use the technique described in [30]. It consists of two steps. First, we show that $\text{Cost}(RA_{opt})$ is greater than the cost of tree produced by the algorithm i.e., Cost(T). Then, we show that cost of range assignment on the resultant tree, $\text{Cost}(RA_T)$ is less than twice of the cost of resultant tree i.e., $\text{Cost}(RA_T) < 2\,\text{Cost(T)}$. Firstly, $\text{Cost}(RA_{opt}) > \text{Cost(T)}$: For optimal assignment a given set N of n points, construct a shortest path destination tree rooted at u. Given the shortest path destination tree, the correspond tree T' (say) can changing the directed edges. Since each of the $n - 1$ nodes other than the root must be assigned a range, we have $\text{Cost}(RA_{opt}) > \text{Cost}(T')$. Secondly, since each edge of the resultant tree T can be chosen at most by two nodes, we have $\text{Cost}(RA_{opt}) < 2 \times \text{Cost(T)}$.   □

Now we show that the Approximation algorithm satisfies the strong connectivity property (Proposition 2) i.e. the resultant graph (tree) produced by the Approximation is strongly connected.

For this purpose, we define the minimal distance between u and v over all nodes as follows: $dist_u(N) = min\{dist(u, v) : u \in N \text{ and } v \in N \backslash \{u\}\}$. Note that this idea is depend upon the notion of well-spread defined in [6]. We claim that for a given set of nodes N, a range assignment RA in the approximation, the condition $RA(u) \geqslant dist_u(N)$ must hold for all nodes in the set N. We formalize our claim as follows:

Lemma 7. For a given set of nodes N, if a range assignment RA allocated by the Approximation is strongly connected, then $RA(u) \geqslant dist_u(N)$ for all $u \in N$.

Proof: Assume to the contrary, that there exists a range assignment RA in the Approximation such that $RA(u) < dist_u(N)$. That is, condition given in line 2 of the Approximation algorithm does not hold for some particular but arbitrarily chosen nodes u and v. Then, by the Proposition 1, if there was a node u with $RA(u) < dist_u(N)$, it could not send data to any other node (terminal or relay node). This implies that there cannot be an edge (communication link) leaving the node u. Thus, G_{RA} cannot be connected. Since G_{RA} is not connected therefore, it cannot be strongly connected. This contradicts the supposition that $RA(u) < dist_u(N)$. Hence, the supposition is false and the lemma is true. This completes the proof.   □

Indeed, the same node connectivity adopted in the Lemma 7 can be use to prove the following stronger result about the connectivity in the resultant graph produced by the algorithm ADD_NODE. In particular, the following lemma shows that G_{RA} preserves the connectivity of G_{ADD_NODE}.

Lemma 8. The resultant graph, G_{RA}, produced by the approximation algorithm is connected if, and only if, the graph G_{ADD_NODE} produced by the algorithm ADD_NODE is connected.

Proof: For necessary condition, suppose u and v are any particular but arbitrarily chosen nodes in G_{RA}. Then by line 2 of the approximation algorithm, the algorithm will consider only those edges that are connected in the graph G_{ADD_NODE}. Therefore, nodes u and v are connected and this is what to be shown.

For sufficient condition, assume to the contrary, that there exists at least one edge in G_{ADD_NODE} that is connected i.e. $edge(u,v) \in G_{ADD\_NODE}$ and not connected in G_{RA} i.e., edge $(u,v) \notin G_{RA}$. Consider the $edge(u,v)$ with minimum range assignment i.e., Euclidean distance $dist(u,v)$, among all pairs of nodes u and v that are connected in G_{RA} but not connected in G_{ADD_NODE}. Recall, by definition $edge(u,v) \in E_{RA}$ iff $R(u) \geqslant dist(u,v)$ so, minimum range assignment means the edge connecting node u and node v in the G_{ADD_NODE}. The node u and v must be connected directly by $edge(u,v)$ in G_{ADD_NODE} (see line 2 of ADD_NODE algorithm). Otherwise, a different $edge(x,y)$ must exist on the path connecting nodes u and v would have a Euclidean distance (range assignment) $dist(x,y)$ less than $dist(u,v)$ and nodes u and v would not be the edge with minimum range assignment i.e. minimum Euclidean distance, $dist(u,v)$. Since the $edge(u,v)$ is in G_{ADD_NODE} by line 2 of the algorithm ADD_NODE, the cost of the shortest path connecting nodes u and v is their Euclidean distance i.e. $RA = dist(u,v)$. Since $dist(u,x)$ is less than $dist(u,v)$. Also, since $dist(u,v)$ is the least such value for any non-connected nodes in G_{RA} and u and x must also be connected in G_{RA} (by line 2 in approximation algorithm). For the same reason, node v and node x must be connected in G_{RA}. This contradicts the supposition that nodes $u$and v are not connected in the graph G_{RA} produced by algorithm approximation. Hence, the supposition is false and the sufficient condition holds. And this completes the proof.   □

8. Open question

The goal of our proposed scheme is to discard long-distance communication links and instead create small links (range assignments). For this purpose, the algorithm chooses a location of additional nodes or relay node using the notion of Steiner tree construction. Clearly, we cannot throw away nodes that are too far from neighbor nodes otherwise, the resultant graph will be partitioned. In general, there is a trade-off between network connectivity and sparseness. Recall, sparseness means that the number of edges in the graph should be in the order of the number of nodes. Although the graph constructed by the approximation algorithm has a nice feature of being at most twice the cost of an optimal cost while maintaining the connectivity, the question of whether the constructed graph is actually a sparse remains open. Despite of the observation that there exists a node placement and transmitting range setting such that some long-distance links can be removed from the resultant communication graph without increasing the cost of the algorithm. We strongly suspect that answer to the question is negative (at least in the context of our problem setting). Our argument is as follows. There always exists a node configuration and transmitting range setting such that the number of communication links in the graph constructed by ADD_NODE algorithm equals to the number of links in the resultant graph.

9. Conclusion

The paper discussed the algorithmic issues concerning the problem of introducing additional nodes (relay nodes) in the existing wireless network from practical viewpoint. That is, setup a communication links in the emergencies by introducing additional or relay nodes. The paper also presented a 2-approximation algorithm to assign the transmitting ranges in such a way that the resultant network is connected.

References

[1]   D.M. Blough, M. Leoncini, G. Resta and P. Santi, *On the Symmetric Range Assignment Problem in Wireless Ad Hoc Networks*, in Proceedings of 2nd IFIP International Conference on Theoretical Computer Science: Foundations of Information Technology in the Era of Networking and Mobile Computing, pp. 71–82, 2002.

[2]   B. Boots, A. Okabe, K. Sugihara and S. Chiu, *Spatial Tessellations: Concepts and Applications of Voronoi Diagrams*, Wiley, New York, 2000.

[3]   G. Calinescu, I. Mandoiu and A. Zelkovsky, *Symmetric connectivity with minimum power consumption in radio networks*, in Proceedings of 2nd IFIP International Conference on Theoretical Computer Science: Foundations of Information Technology in the Era of Networking and Mobile Computing, pp. 119–130, 2002.

[4]   A.E.F. Clementi, A. Ferreira, P. Penna, S. Perennes and R. Silvestri, The mimimum range assignment problem on linear radio networks, *Algorithmica* **35**(2)(2003), 95–110.

[5]   A.E.F. Clementi, P. Penna and R. Silvestri, *Hardness Results for the Power Range Assignment Problem in Packet Radio networks*, in Proceedings of 2nd Workshop on Approximation algorithms for Combinatorial Optimization Problems (RANDOM/APPROX'99), pp. 197–208, 1999.

[6]   A.E.F. Clementi, P. Penna and R. Silvestri, On the power assignment problem in radio networks, *Mobile Networks and Applications* **9**(2) (2004), 125–140.

[7]   D. Du, F. Hwang, G. Song and G. Ting, Steiner minimal trees on sets of four points, *Discrete Computational Geometry* **2**(1) (1987), 401–414.

[8]   M.R. Garey, R.L. Graham and D.S. Johnson, The complexity of computing Steiner minimal trees, *SIAM Jornal on Applied Mathematics* **32**(4) (1977), 835–859.

[9]   E.N. Gilbert and H.O. Polak, Steiner minimal trees, *SIAM Journal on Applied Mathematics* **1**(16) (1968), 1–29.

[10] R.L. Graham and F.K. Hwang, Remarks on Steiner minimal trees, *Bulletin of the Institute of Mathematics Academia Sinica* **4**(1) (1976), 177–182.

[11] W.R. Heinzelman, A. Chandrakasan and H. Balakrishnan, *Energy-Efficient Communication Protocol for Wireless Micro-Sensor networks*, in Proceedings of IEEE Hawaii International Conference on System Sciences, pp. 4–7, 2000.

[12] T.C. Hou and V.O.K. Li, Transmitting range control in multihop packet radio networks, *IEEE Transactions on Communications* **34**(1) (1986), 38–44.

[13] F. Hwang, D. Richards and P. Winter, The Steiner tree problem, Annals of Discrete Mathematics 53, Elsevier, Amsterdam, 1992.

[14] L.M. Kirousis, E. Krankis, D. Krizanc and A. Pelc, Power consumption in packet radio, *Theoretical Computer Science* **243**(1–2) (2000), 289–305.

[15] H. Koskinen, J. Karvo and O. Apilo, *On Improving Connectivity of Static Ad-Hoc Networks By Adding Nodes*, In Challenges in Ad Hoc Networking, pp. 169-178. Springer, Boston, 2006.

[16] X.Y. Li and P.J. Wan, *Constructing Minimum Energy Mobile Wireless Networks*, in Proceedings of the 2nd ACM International Symposium on Mobile Ad Hoc Networking and Computing (MobiHoc 2001), pp. 283–286, 2001.

[17] Z.A. Melzak, *Companion to Concrete Mathematics*, Wiley, New York, 1973.

[18] R.B. Muhammad, *A parallel Algorithm for Planar Voronoi Diagram on Hypercube Model of Computation*, in Proceedings of the 16th IASTED International Conference on Parallel and Distributed Computing and Systems (PDCS'04), Cambridge, Massachusetts, USA, pp. 542–547, 2004.

[19] R.B. Muhammad, *Connectivity Setup using Steiner Tree in Ad Hoc Wireless Networks*, in Proceedings of Eight Annual IEEE Wireless and Microwave Technology Conference (WAMICON'06), Clear Water, Florida, USA, pp. five pages, full text on CD-ROM, 2006.

[20] R.B. Muhammad, *Deterministic Energy Conserving Algorithms for Wireless Sensor Networks*, in Proceedings of the IEEE International Conference on Networking Sensing and Control (ICNSC'06), Ft. Lauderdale, Florida, USA, pp. 324–329, 2006.

[21] R.B. Muhammad, *A Parallel Local Search Algorithm for Euclidean Steiner Tree Problem*, in IEEE Proceedings of the 7th International Conference on Software Engineering, Artificial Intelligence Networking and Parallel/Distributed Computing (SNPD'06), Las Vegas, USA, pp. 157–164, 2006.

[22] R.B. Muhammad, *Parallelization of Local Search for Euclidean Steiner Tree Problem*, in Proceedings of The 44th ACM Southeast Conference (ACMSE'06), Melbourne, Florida, USA, pp. 233–238, 2006.

[23] R.B. Muhammad, *Range Assignment Approximation Based on Steiner Tree in Ad Hoc Wireless Networks*, in Proceedings of 22nd IEEE International Conference on Advanced Information Networking and Applications (AINA'08), Okinawa, Japan, pp. 100–105, 2008.

[24] R.B. Muhammad, *Transmitting Range Assignment using Steiner Tree in Ad Hoc Wireless Networks*, in IEEE Proceedings of the 5th International Conference on Information Technology: New Generations (ITNG'08), Las Vegas, USA, pp. 408–413, 2008.

[25] K. Pahlavan and A. Levesque, *Wireless Information Networks*, Wiley-Interscience, New York, 1995.

[26] H.O. Polak, Some remarks on the Steiner problem, *Journal of Combinatorial Theory* **24**(3) (1978), 278–295.

[27] F.P. Preparata and M.I. Shamos, *Computational Geometry: An Introduction*, Springer-Verlag, New York, 1988.

[28] T. Rappaport, *Wireless Communications: Principle and Practice*, Prentice Hall, New Jersey, 2002.

[29] V. Rodoplu and T. Meng, Minimum energy mobile wireless networks, *IEEE Journal on Selected Areas in Communications* **17**(8) (1999), 1333–1344.

[30] P. Santi, *Topology Control in Wireless Ad Hoc and Sensor Networks*, Wiley, Hoboken, NJ, 2005.

[31] M.I. Shamos, *Geometric Complexity*, in Proceedings of the 7th Annual ACM Symposium on Theory of Computing, pp. 224–223, 1975.

[32] H. Takagi and L. Kleinrock, Optimal transmission ranges for randomly distributed packet radio terminals, *IEEE Transactions on Communications* **23**(3) (1984), 246–257.

[33] E.A. Thomson, The method of minimum evolution, *Annals of Human Genetics London* **36** (1973), 333–340.

[34] J. Weng, Restudy of Steiner minimal trees on four points, Preprint, 1991.

8

A friendly location-aware system to facilitate the work of technical directors when broadcasting sport events

Sergio Ilarri[a,*], Eduardo Mena[a], Arantza Illarramendi[b], Roberto Yus[a], Maider Laka[c] and Gorka Marcos[c]

[a]*Department of Computer Science and Systems Engineering, University of Zaragoza, Zaragoza, Spain*
[b]*Basque Country University, San Sebastián, Spain*
[c]*Vicomtech Research Center, San Sebastián, Spain*

Abstract. The production costs of broadcasting sport events that require tracking moving objects are continuously increasing. Although those events are very demanded by the audience, broadcasting organizations have economical difficulties to afford them. For that reason, they are demanding the development of new professional (software and hardware) equipments that lead to a considerable reduction of the production costs.

In this paper, we present a software system that takes into account these needs. This system allows a technical director to indicate his/her interest about certain moving objects or geographic areas in run-time. The system is in charge of selecting the cameras that can provide the types of views requested on those interesting objects and areas. So, it decreases the human effort needed to produce (create, edit and distribute) audiovisual contents, giving at the same time the opportunity to increase their quality. For this, the system provides a friendly interface to specify requirements and obtain which monitoring video cameras attached to moving or static objects fulfill them, along with a query processor to handle those requests in a continuous and efficient way. We illustrate the feasibility of our system in a specific scenario using real data of a traditional rowing race in the Basque Country.

Keywords: Mobile multi-camera management, location-aware systems, location-dependent queries, video broadcasting, sport events

1. Introduction

In the last years, different factors are provoking a deep revolution in the broadcast industry. First of all, there has been a strong decrease of the advertisement rates, mainly due to the economic crisis and the audience fragmentation caused by the digitalization and optimization of the spectrum (channel multiplication in satellite and terrestrial television) and the appearance of new communication platforms (e.g., video-blogs, Internet video platforms, mobile broadcasting, and so on). Secondly, continuous needs related to the upgrade of the technology (e.g., digitalization, development of media asset management systems, HD, 3D, etc.) are implying huge economic efforts. Finally, it is also evident that the expectations

---

*Corresponding author: Sergio Ilarri, University of Zaragoza, Department of Computer Science and Systems Engineering, Edificio Ada Byron, María de Luna 1, E-50018 Zaragoza, Spain. E-mail: silarri@unizar.es.

of the audience are increasing. The quality and richness of the content (e.g., amazing views, last generation graphics, and so on) demanded by the audience and the advertisers is much bigger than in other distribution platforms such as Internet video platforms.

In such a context, organizations focus their efforts on the enrichment and the diversification of their offers but trying to reduce their production costs. On the one hand, the enrichment is tackled by the generation of very attractive and high-quality material, including in many cases technological support (e.g., the Obama hologram in the US presidential election in 2008). On the other hand, the reduction is considered by the acquisition of new professional products that, apart from begin able to deal with the last technology (e.g., IP interoperability, HD resolution, remote control mechanisms), either have a lower price due to the inclusion of hardware coming from the consumer electronic field or add new features that decrease the human effort required for content production. An example of this is the combination of professional cameras with low-cost cameras controlled remotely. This leads to an enrichment of the content consuming experience without having a big impact on the total number of cameramen required. However, that solution has a serious impact in one of the most complex and critical tasks in a live content production environment: the more cameras are employed, the more images are available, and therefore more complicated is for technical directors (people responsible for the content production of an event) to select the best video stream to broadcast.

The audiovisual production of rowing races in the Basque Country is a relevant paradigm of this situation and it will serve us as sample scenario in this paper. The live broadcasting of such rowing races requires a very complex infrastructure: one helicopter, sailing boats with cameras and GPS transmitters, more cameras in the harbor, and a production mobile unit (usually a trailer). In such a complex context, the technical director is responsible for the selection and coordination on the fly of the graphical material, video signals, and views that are finally broadcasted to the TV audience. These tasks become especially difficult when different unexpected events happen at different geographic areas or when different moving objects become interesting at the same time.

In order to help technical directors to obtain the best broadcasting results, we present in this paper a system that helps them to select the best candidate video signals coming from static or mobile cameras (i.e., cameras installed on rowing boats, other sailing boats, fixed locations, etc.). Our proposal, based on the LOQOMOTION system [19] extended with videocamera management, relies on mobile agent technology to bring the processing to the best place in the distributed wireless scenario, at any time. Thus, the camera selections provided by the system are updated continuously in an efficient manner, and it is possible to deal with different geographic areas of interest at the same time. The system can even alert about upcoming situations defined previously by the technical director (e.g., a certain object is within an area), or when some event happens (e.g., a camera gets close to a certain location). A preliminary version of our proposal appeared in [23], which has been improved and extended in this paper; among other extensions, we consider cameras that can both pan and tilt, and we analyze the precision of our system by testing a complete prototype using real GPS location data captured during a rowing race celebrated in September 2010.

The main contributions of our proposal are:

- We extend a general architecture for location-dependent query processing with videocamera management to help technical directors that broadcast sport events to deal with the multimedia information coming from different (static or moving) videocameras.
- The system enables technical directors to indicate his/her interest about certain (static or moving) objects or (fixed or moving) geographic areas predefined or defined in run-time. The system is in charge of selecting the cameras that can provide the types of views requested by the technical director on those interesting objects and areas.

Fig. 1. "Kaiku" boat in a rowing race (image provided by courtesy of the Kaiku club).

- A flexible approach is followed, which can be applied to different distributed scenarios and requirements, and therefore new functionalities can be added to the system without compromising the main architecture.

The rest of the paper is structured as follows. In Section 2 we detail the features of the sample sport event used as motivating context and some interesting multimedia location-dependent queries that we would like to process automatically. In Section 3 we summarize the basic features of cameras and the work of a technical director. In Section 4 we introduce the concept of location-dependent queries, a general architecture for location-based query processing, and our proposal to extend it with the modeling and management of videocameras. In Section 5 we explain how queries in the sample scenario are processed by the proposed architecture. In Section 6 we present our prototype and some experiments that we have performed to validate our proposal. In Section 7 we review some related works. Finally, conclusions and future work are included in Section 8.

2. Motivating context: Rowing races in San Sebastian

Rowing racing (see Fig. 1) is a very popular sport in every seaside town of fishing tradition along the north of Spain. In the Basque Country this sport is very important, and almost every town has its own team, which always counts with the unconditional local support.

Although there are multiple competitions, the most important one is celebrated in the San Sebastian bay, once a year, since 1879. The boats leave from the harbor where they come back after making a turn in the sea, outside the protection of the bay, covering a total distance of 3 nautical miles. There are four lanes, although those lanes are only physically distinguished in the starting and turning points (see Fig. 2).

We have chosen this race due to the fact that it is a paradigmatic example of a potential live broadcasting event that can benefit from our work. In this race, celebrated during two consecutive weekends every September, the city of San Sebastian is crowded with many visitors, and the audience of the Basque TV broadcaster is very high. The technology and equipment involved in the event have been evolving during

Fig. 2. Start of a rowing race (image provided by courtesy of the EiTB Media Group).

the last years and nowadays there are multiple cameras (on sailing boats, on a helicopter, in the harbor, on an island nearby, etc.), a GPS transmitter on every boat, a software application for the panel of judges to help them to determine the distance between the boats and the location of a boat with respect to its lane, a software tool to show on TV the real positions of the boats in a 3D reconstruction of the bay, microphones to capture the atmosphere sounds in different places, and so on.

In such a scenario, it is possible to attach a remotely controlled camera to each rowing boat. We would like to highlight that in many cases these cameras will provide innovative and interesting views from the content production perspective. For instance, these cameras may capture the exciting moment of the turning in the sea, the view of the audience perceived by the rowers, overtaking maneuvers, unusual ocean-side views of an island that is located in the middle of the bay, etc. Those views or points of interest can be classified into predefined ones (e.g., the view of spectators in the harbor) or defined while broadcasting (e.g., the view of an overtaking maneuver between two boats captured by the camera of a third boat). Figure 3 shows graphically some of the predefined points of interest on top of a 3D reconstruction of the race scenario.

However, it would be crucial to help the technical director to select among the available cameras those that can provide nice views, considering that the cameras could be remotely controlled and be continuously moving. It is important to emphasize that receiving the video signals is not a problem in this scenario, but the selection among the multiple video signals is a challenge. The system described in this paper helps the technical director in the identification and management of the best candidate videocameras that can view a specific (static or moving) area or object of interest; such target objects and areas can be predefined or specified during the broadcasting.

In the following, we enumerate some motivating queries that we would like to be answered and updated continuously:

1. *Query 1: View a certain boat.* The technical director could focus on a particular boat due to many reasons (it is the local team, it is leading the race, etc.), some of them caused by unexpected situations (an accident, a broken oar, etc.) which obviously should be captured quickly from any camera.

2. *Query 2: Capture a close side view of any boat.* The technical director could want to broadcast a close side view of some boat to show the big effort performed by its team of rowers during the race.

3. *Query 3: Capture a wide view of the island from the ocean side.* This is interesting for technical directors because this view of the island is usually very spectacular (it is a sheer cliff full of

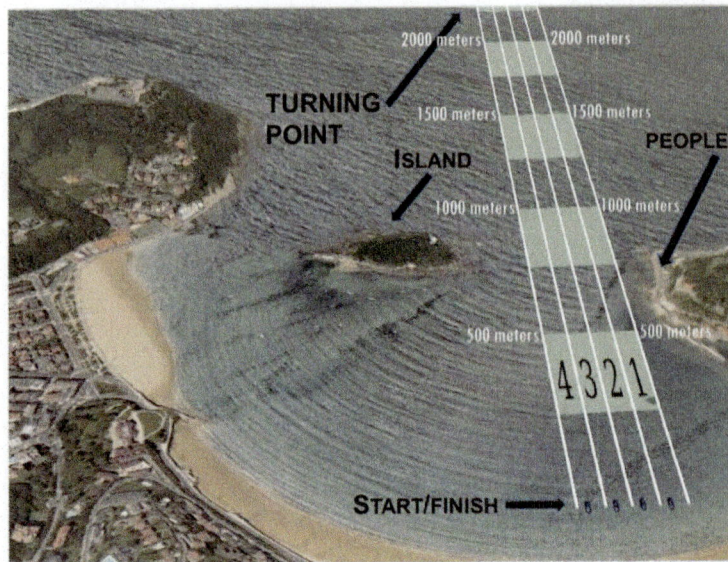

Fig. 3. Predefined points of in a rowing competition interest.

seagulls) and not as typical and easy-to-access as the well-known view of the island from the bay. For this query, technical directors are interested in cameras located far from the island (to capture it completely) but within a certain range (to get also a detailed picture). So, for this query, the locations of the cameras play an important role.

Other areas that are very interesting for technical directors, and therefore could be the target of similar queries, are: the *ciaboga* area (the turning point, which is a key part of the competition), the area of the harbor or the promenade at the seaside (usually crowded with people watching the race), etc. It would be beneficial to technical directors to be able to predefine areas that are usually relevant at some time during the race. However, the system should also allow technical directors to select, while broadcasting, any area of interest by dragging the mouse on a map.

3. Technological context: Cameras and technical direction

In this section, we describe the features of cameras and some basic ideas about the work of a technical director.

3.1. The features of cameras

Today, most of the low-cost/mid-range cameras employed in scenarios similar to the one described (e.g., F1 car races, sailing, etc.) offer a fixed view (e.g., a front view from the driver's perspective). From the production point of view, this implies important difficulties for the generation of attractive and rich content, mainly due to the limitation or lack of control of the rotation (pan and tilt) and zoom of the camera.

However, the electronics consumer sector is providing new cameras with very competitive prices that offer an acceptable image quality while providing rich remote control functionalities. Figure 4 shows some examples of these cameras, that allow a remote control of their parameters.

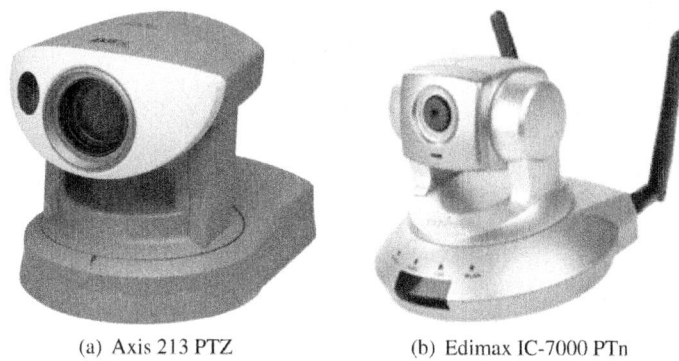

(a) Axis 213 PTZ   (b) Edimax IC-7000 PTn

Fig. 4. Remote motorized cameras.

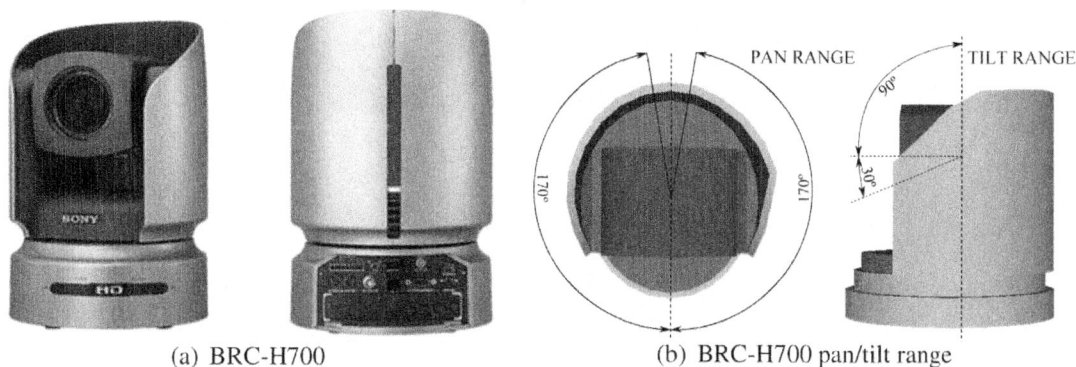

(a) BRC-H700   (b) BRC-H700 pan/tilt range

Fig. 5. Rotation ranges of a camera.

PTZ (pan-tilt-zoom) cameras [8], and particularly those that can be controlled remotely, are used in many fields, such as corporate training, distance learning, videoconferencing, and even broadcasting. Their main specifications are the *field-of-view* (*FOV*), image quality, and remote control distance.

The field-of-view of a camera is defined based on the rotation range of both the vertical and the horizontal axis. The rotation of the axis perpendicular to the ground is called *pan* (cameras can usually pan 360° or slightly less; a left rotation is denoted by a negative value), and the vertical rotation is called *tilt* (cameras can usually tilt 180°; a downward rotation is denoted by a negative value). Figure 5 shows an example of these parameters for a robotic camera (*Sony BRC-H700*), where the pan could take values between −170° and +170° from the reference point and the tilt between −30° and 90°.

Other specifications of a camera are the *zoom* and the *focus* (they depend on the lens of the camera). The zoom is the adjustment of the focal length of the lens to make the target appear close-up or far away, and the focus is related to the clarity of the objects captured in the image. Usually the cameras have *AutoFocus*, which means that the camera is able to adjust the lens so that the target is always in focus.

This kind of robotic cameras are remotely controlled and their remote control units are usually connected to them (or other devices) by a serial port, for instance to an optical multiplex unit using optical fiber. Depending on the specific remote control unit, it is possible to program different configurations for several cameras. These configurations can store values for the pan, tilt, zoom, and focus.

The integration of the cameras in the workflow of a live broadcasting must be supported by camera operators, who need to invest an important effort to manage the cameras and keep their concentration

Fig. 6. Production-control room at the EiTB TV station.

continually. In many cases, according to Vicomtech's experience (http://www.vicomtech.es/), this is a very important problem, especially when the number of cameras increases significantly. In this sense, the work presented in this paper aims to provide a tool that simplifies the management of such cameras, in order to enable the production of innovative and attractive content using wireless (static or mobile) cameras controlled remotely.

3.2. The work of the technical director

In television, a fundamental aspect of the job of a technical director is to manually select between the available video sources, perform live edits, and overlay text on the images when it is necessary. Figure 6 shows an example of the workspace of a technical director, called *Production Control Room* (*PCR*). There are several facilities in a PCR, such as:

- A *video monitor wall*, where multiple sources are displayed, such as cameras, video tape recorders (VTR), graphics, and other video sources.
- A *vision mixer*, which is a control panel used to select the video sources that will be broadcasted or to manipulate them (e.g., adding digital effects).
- *Intercom and interruptible feedback* (*IFB*), used to communicate with crew and camera operators.

Technical directors have overall responsibility for the operations in a PCR. They coordinate the work of the whole crew, select the video source that will be on air, and look into technical problems. Technical directors have to make quick decisions in live broadcast scenarios such as the one described in this paper.

4. Monitoring proposal

As the underlying architecture for a location-aware system for monitoring sport events, we advocate the adoption of a general-purpose location-dependent query processing system [19], that will be extended

```
SELECT   B.id
  FROM   RowingBoats AS B
 WHERE   inside(0.2 miles, boat38, B)
```

Fig. 7. Sample location-dependent query.

with the needed new functionalities. In order to get a better understanding of that system, in this section we present some basic concepts related to it. So, we first define and explain the concept of location-dependent query. Then, we summarize the general architecture proposed for location-dependent query processing. Finally, we indicate the new functionalities added to this system to manage references to videocameras in location-dependent queries.

4.1. Location-dependent queries

Location-dependent queries are queries whose answer depends on the locations of the objects involved. For example, a user with a smartphone may want to locate available taxi cabs that are near him/her while he/she is walking home in a rainy day. These queries are usually considered as *continuous queries* [42], whose answer must be continuously refreshed. For example, the answer to the previous query can change immediately due to the movements of the user and the taxi cabs. Moreover, even if the set of taxis satisfying the query condition does not change, their locations and distances to the user could change continuously, and therefore the answer to the query must be updated with the new location data.

To express location-dependent queries, we will use an SQL-like syntax with the following structure:

```
    SELECT   projections
      FROM   sets-of-objects
     WHERE   boolean-conditions
[ ORDER BY   sorting-criteria ]
```

where *projections* is the list of attributes that we want to retrieve from the selected objects, *sets-of-objects* is a list of object classes that identify the kind of objects interesting for the query, *boolean-conditions* is a boolean expression containing objects from *set-of-objects* that must satisfy the specified *location-dependent constraints*, and *sorting-criteria* is the ordering criteria that will be used for presentation of the results. The *ORDER BY* clause is optional, as in standard SQL. However, the sorting criteria can be particularly important when dealing with queries that retrieve cameras, as several cameras may satisfy the query constraints and some criteria is needed to show the most promising results first.

As an example of a location-dependent query, the query in Fig. 7 asks for rowing boats that are within 0.2 miles around *boat38*. This query includes an *inside* constraint expressed with the general syntax *inside(r, obj, target)*, which retrieves the objects of a certain *target* class (such objects are called *target objects*) within a specific distance *r* (which is called the *relevant radius*) of a certain moving or static object *obj* (that is called the *reference object*). Thus, in the sample query in Fig. 7 the radius of the inside constraint is 0.2 miles, and there is one reference object (*boat38*) and one target class (*RowingBoats*).

It is important to emphasize that, although we use this SQL-like language to express the queries, higher-level facilities will be available for end users (e.g., predefined queries will be posed just by clicking on certain GUI buttons), as described in Section 6.

Moving and static objects in a scenario are not single points but have a certain geographic extension, depending on their size. Thus, objects and areas are managed in the same way (an object is characterized by a volume, which is its *extent* [38]). The techniques described in [18] to manage queries that take into account the 2D extent of the objects involved have been adapted in this paper to deal with 3D extents.

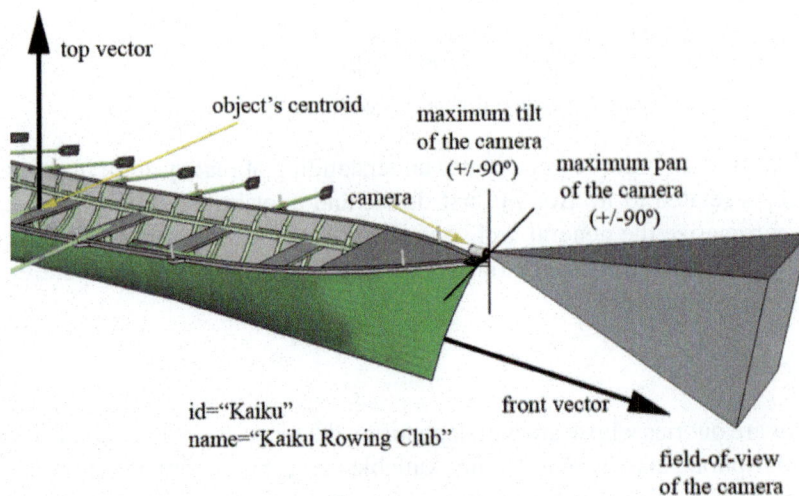

Fig. 8. Main elements of a rowing boat.

Class *Objects* is the set of entities in the scenario, which could be equipped with one or more cameras. *RowingBoats* is a subclass of *Objects*. An individual of *Objects* can be represented by the tuple:

<id, name, extent, centroid, frontVector, topVector, cams>

where *id* is a unique identifier of the object, *name* is the name of the object, *extent* is the volume occupied by the object, *centroid* is the centroid of the extent of the object, *frontVector* and *topVector* are vectors pointing towards the frontal and the top part of the object, and *cams* is the list of cameras attached to the object (if any). The *frontVector* and *topVector* enable distinguishing different kinds of views of an object that could be provided by cameras (e.g., a front view of the object or a view from the top). Other interesting vectors (e.g., a vector pointing towards the bottom or the rear of the object) do not need to be defined because they can be obtained from the *frontVector* and *topVector*. For an example of the main elements characterizing a rowing boat, see Fig. 8.

4.2. Processing location-dependent queries

To process location-dependent queries in a mobile environment, we have proposed in previous works the system LOQOMOTION [19] (*LOcation-dependent Queries On Moving ObjecTs In mObile Networks*), a distributed location-dependent query processing system whose architecture is based on mobile agents. Mobile agents [3,43] are programs that execute in contexts called *places*, hosted on computers or other devices, and can autonomously travel from *place* to *place* resuming their execution there. Thus, they are not bound to the computer where they were created; instead, they can move freely across different computers and devices. Mobile agents provide interesting features for distributed and wireless environments (e.g., see [40]), thanks to their autonomy, adaptability, and capability to move to remote computers.

LOQOMOTION deploys a network of agents to perform the query processing over a distributed set of objects which can detect other objects moving within their range. Notice that a certain object could only detect a subset of objects in a scenario because of its limited range. The basic idea is that mobile agents move among objects in order to detect the target objects that are relevant for a query. As an example,

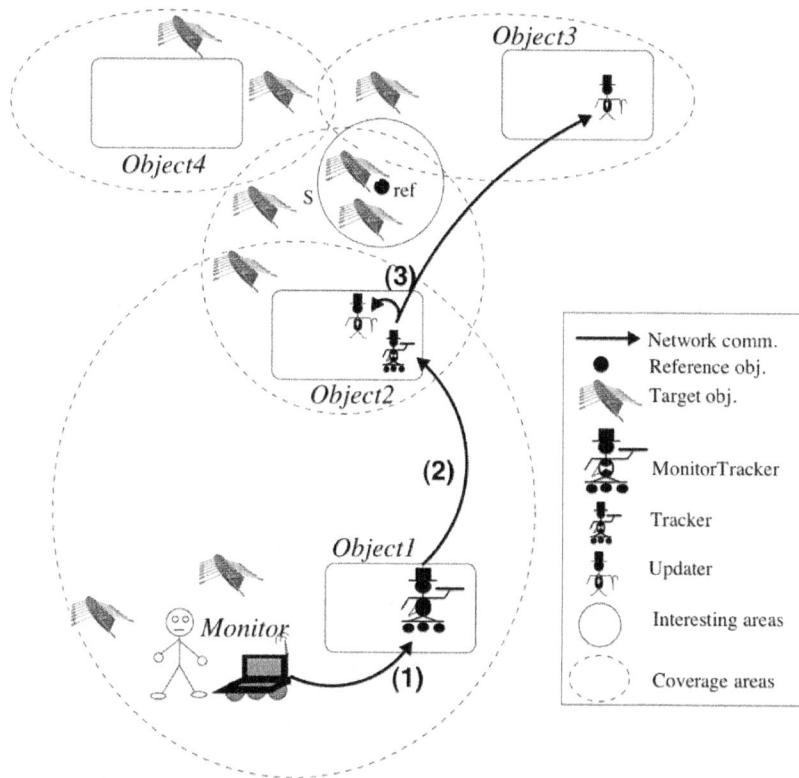

Fig. 9. Architecture of LOQOMOTION.

let us assume that in the scenario shown in Fig. 9 the object *Monitor* wants to retrieve the rowing boats within the area S centered on the black object *ref*.

In the figure, we represent with rectangles the objects that have the capability to detect other objects within its range (*coverage area*). The query will by processed by an agent called *MonitorTracker*, that will execute on a certain object (step 1 in Fig. 9). To do that, it first needs to know the location of *ref*. Notice that the object *ref* is beyond the coverage area of that object, and so the system will need to query some other object that is able to detect the *ref* object. In particular, the location of *ref* is known by *Object2* (as *ref* is within its area). So, a mobile agent called *Tracker* travels to *Object2* to retrieve the current location of *ref* (step 2 in Fig. 9). Once the system knows the location of *ref*, it also knows exactly the circular area of interest S. Then, the rowing boats within that area are obtained, by using agents called *Updaters* (step 3 in Fig. 9). Thus, one Updater executing on each object whose range intersects with S (in the example, *Object2* and *Object3*) will keep track of the rowing boats entering S. Of course, as the interesting objects move, the network of agents will re-organize itself as needed (e.g., notice that when *ref* moves the area S moves as well). For more details, see [19,20].

4.3. Extension to monitor multimedia data: Management of videocameras

In our context, one of the most interesting attributes of the objects in the scenario are obviously the videocameras, which play a key role for us. Thus, we do not only need to retrieve the objects that satisfy certain location-dependent constraints, but also filter out those objects whose cameras do not satisfy

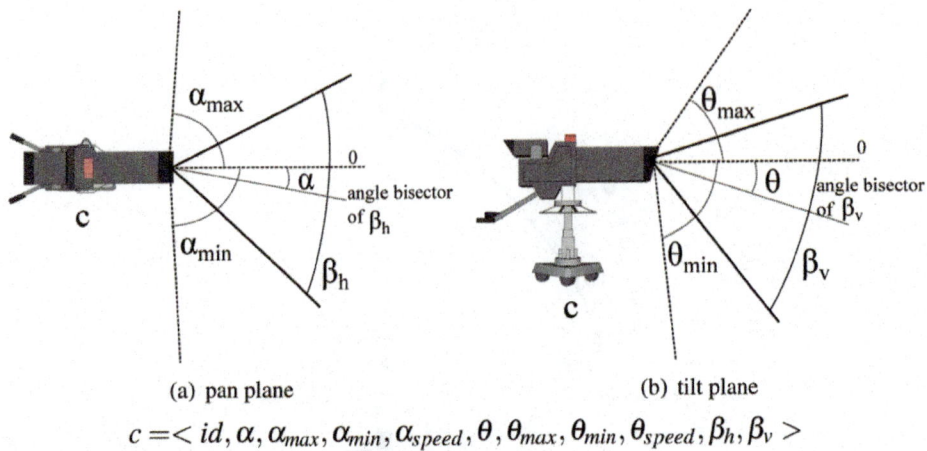

(a) pan plane   (b) tilt plane

$$c = <id, \alpha, \alpha_{max}, \alpha_{min}, \alpha_{speed}, \theta, \theta_{max}, \theta_{min}, \theta_{speed}, \beta_h, \beta_v>$$

Fig. 10. Modeling a videocamera.

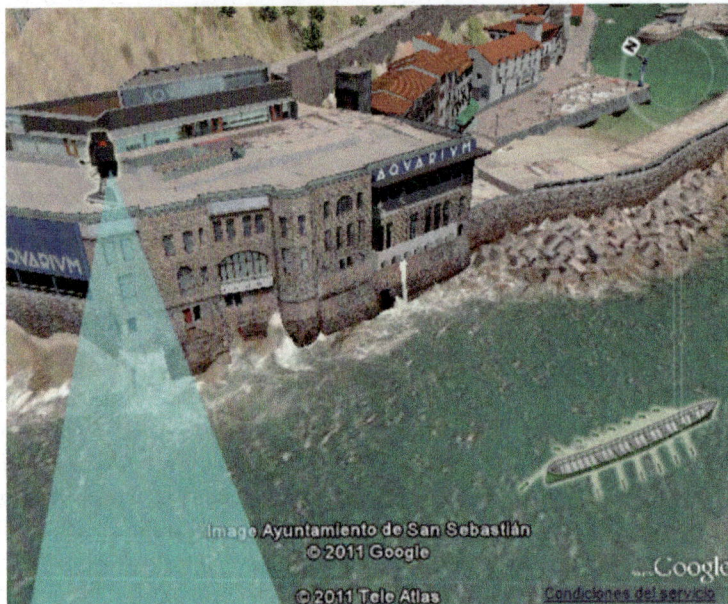

Fig. 11. 3D representation of a camera and a rowing boat.

certain conditions/features needed for a suitable viewing of the target (i.e., a static or moving object or area). For this purpose, we model a camera c as shown in Fig. 10.

In the figure, id is a unique identifier, β_h and β_v are the horizontal and vertical angle of view, respectively. Concerning the horizontal turn (pan), α, α_{max}, α_{min}, and α_{speed} are the current pan, the maximum pan possible, the minimum pan possible, and the pan speed (degrees/second) of such a camera, respectively. Concerning the vertical turn (tilt), θ, θ_{max}, θ_{min}, and θ_{speed} are the current tilt, the maximum tilt possible, the minimum tilt possible, and the tilt speed (degrees/second), respectively. Thus, by considering both the pan and the tilt of a camera, the system is able to manage a 3D space containing 3D objects (Fig. 11 shows a 3D representation of a scene).

We define some functions that abstract us from the specific calculations needed, based on the features

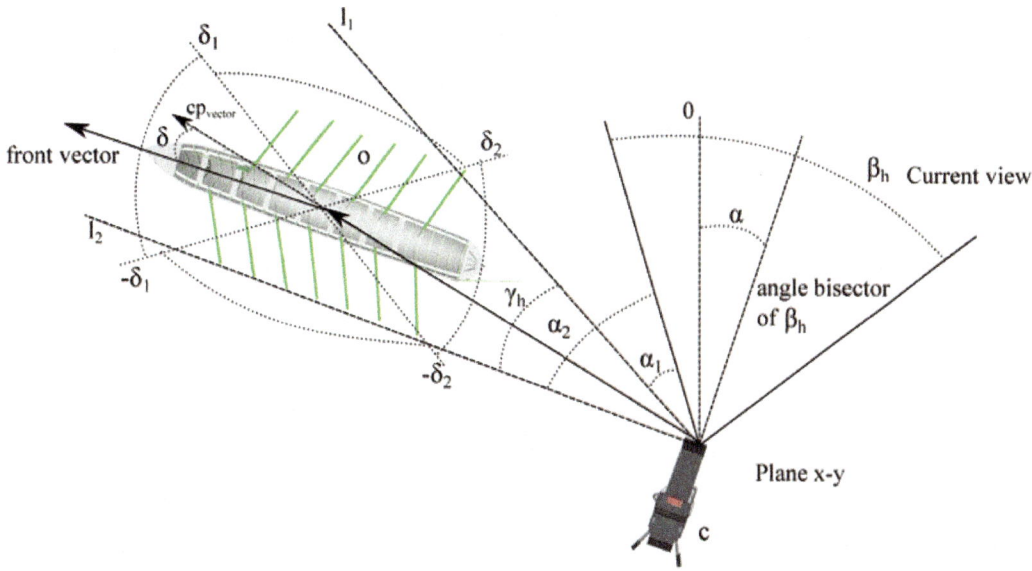

$$
panToView(o,c,v,cov) = \begin{cases} \alpha_1 + \varepsilon & \begin{aligned} & cov \in \{incomplete, any\} \\ & \wedge\, (view(c,o) = v \vee v = any) \\ & \wedge\, c.\alpha_{min} \leq \alpha + \alpha_1 \leq c.\alpha_{max} \end{aligned} \\[2em] \alpha_2 & \begin{aligned} & cov = full \wedge \gamma_h \leq \beta_h \\ & \wedge\, (view(c,o) = v \vee v = any) \\ & \wedge\, c.\alpha_{min} \leq \alpha + \alpha_2 \leq c.\alpha_{max} \end{aligned} \\[2em] -999 & otherwise \end{cases}
$$

Fig. 12. Definition of the $panToView$ function.

of the cameras, to verify certain conditions. Particularly, we define the functions $panToView$ (see Fig. 12) and $tiltToView$ (see Fig. 13).

These functions return the signed angle that the videocamera c should pan or tilt, respectively, to view an object o from a certain view v (*front, rear, side, top, bottom, middle*, or *any*) with the specified coverage cov (*full, incomplete, any*). Positive values for $panToView$ mean "right pan" and negative ones mean "left pan". Similarly, for $tiltToView$ positive values mean "upward tilt" and negative ones mean "downward tilt". The sign of the angles α, α_1, α_2, θ, θ_1, and θ_2, also indicate the direction of the corresponding rotation. Notice that sometimes we would need to do both movements (pan and tilt) to be able to view an object, due to the fact that the camera and the object could be at different heights and horizontal locations. To compute the $panToView$, lines $l1$ and $l2$ (see Fig. 12) are traced from the location of the camera to both sides of the geographic volume associated to the object o (in this case, a rowing boat, including its oars); similarly, lines $l3$ and $l4$ are traced to compute the $tiltToView$ (see Fig. 13). It is important to consider that, as explained in Section 3.1, cameras usually have mechanical limitations about the possible turn angle ranges ($[\alpha_{min} - \alpha_{max}]$ for pan and $[\theta_{min} - \theta_{max}]$ for tilt). So, when the camera cannot pan or tilt to view the object o, satisfying the specified conditions, the above functions return a special value (in our prototype, -999). In the previous formulas for $panToView$ and $tiltToView$, ε is a small value greater than 0, which must be added to the angle in order to start having the target

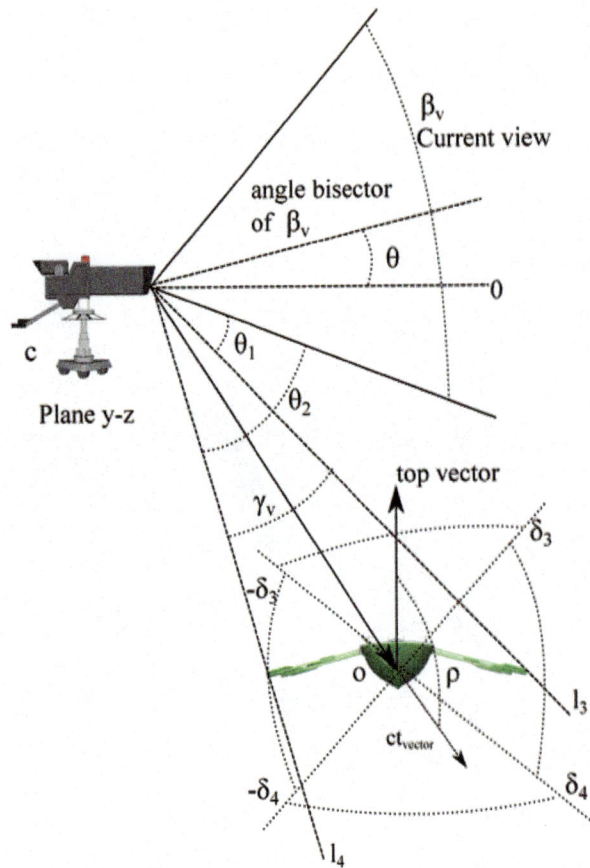

$$tiltToView(o,c,v,cov) = \begin{cases} \theta_1 + \varepsilon & cov \in \{incomplete, any\} \\ & \wedge\, (view(c,o) = v \vee v = any) \\ & \wedge\, c.\theta_{min} \leq \theta + \theta_1 \leq c.\theta_{max} \\ \\ \theta_2 & cov = full \wedge \gamma_v \leq \beta_v \\ & \wedge\, (view(c,o) = v \vee v = any) \\ & \wedge\, c.\theta_{min} \leq \theta + \theta_2 \leq c.\theta_{max} \\ \\ -999 & otherwise \end{cases}$$

Fig. 13. Definition of the $tiltToView$ function.

object within the view of the camera.

The function $view(c,o)$ returns values in $\{front, rear, side, top, bottom, middle\}$ by considering in which of the quadrants defined by the angles $\pm\delta_1$, $\pm\delta_2$, $\pm\delta_3$, and $\pm\delta_4$ (the definition of front, rear, side, top, bottom, and middle views could be defined by any other partition) falls the angle between these two elements: 1) the vector defined by the location of the camera c and the centroid (of the volume) of the target object o and origin at the centroid of o (cp_{vector} in Fig. 12 and ct_{vector} in Fig. 13); and 2) a predefined vector which indicates the *front* or the *top* of the object o, depending on whether we are considering the horizontal dimension (pan) or the vertical dimension (tilt), respectively (see angle δ in Fig. 12 and angle

ρ in Fig. 13). Notice that angles δ_1 and δ_2 are measured from the front vector (see Fig. 12) and angles δ_3 and δ_4 are measured from the top vector (see Fig. 13). Besides, angles that represent a turn to the right from the corresponding vector are considered positive and those to the left are considered negative; so, $\delta_1, \delta_2, \delta_3,$ and δ_4 are values that range from $0°$ to $180°$. The *view* function is defined as follows:

$$view(c, o) = \begin{cases} front & |\delta| \geqslant \delta_2 \\ rear & |\delta| \leqslant \delta_1 \\ side & (\delta_1 < |\delta| < \delta_2) \\ top & |\rho| \geqslant \delta_4 \\ bottom & |\rho| \leqslant \delta_3 \\ middle & (\delta_3 < |\delta| < \delta_4) \\ & 0° \leqslant \delta, \delta_1, \delta_2, \rho, \delta_3, \delta_4 \leqslant 180° \end{cases}$$

Notice that *view*(*c, o*) has to be applied for both the horizontal and the vertical plane. So, according to the location of the object and the camera in Figs 12 and 13, in that scenario *view*(*c, o*) returns "rear" and "top". Moreover, it should be noted that if a camera must pan to focus on the target it will need some time to perform the turning. We define the function *timeToPan*, which returns the time (in seconds) needed to pan the camera α degrees:

$$timeToPan(o, c, v, cov) = \frac{panToView(o, c, v, cov)}{c.\alpha_{speed}}$$

$$timeToPan(c, pan) = \frac{pan}{c.\alpha_{speed}}$$

Similarly, we define the *timeToTilt* function using *tiltToView* and $c.\theta_{speed}$. The cameras can pan and tilt at the same time, and therefore the time needed to pan and tilt a camera would be *MAX(timeToPan, timeToTilt)*.

These functions will be used in the next section to take into account the features of videocameras in location-dependent queries. By defining appropriate location-dependent queries, we benefit from the power of SQL to specify in a flexible way the features of the cameras required to capture different types of views.

5. Using LOQOMOTION in the rowing boats scenario

In this section, we first indicate how the different elements of the query processing architecture apply in the context of rowing races in San Sebastian:

- *Distributed Query Processing.* In the context of rowing races in San Sebastian there is a single object (a TV trailer) that is able to access all the objects (boats, people, etc.) in the scenario. Although LOQOMOTION is particularly adapted to perform well in a distributed infrastructure where there are different objects that can monitor different geographic areas, it can obviously also work when there is a single object covering the whole area of interest.
- *Inside Constraints.* As cameras that are very far from their target are usually of little interest, an *inside* constraint with an appropriate relevant radius can be used to retrieve the candidate cameras for capturing the kind of view (close, wide open, etc.) we want (in this work we do not deal with the possibility of zooming).

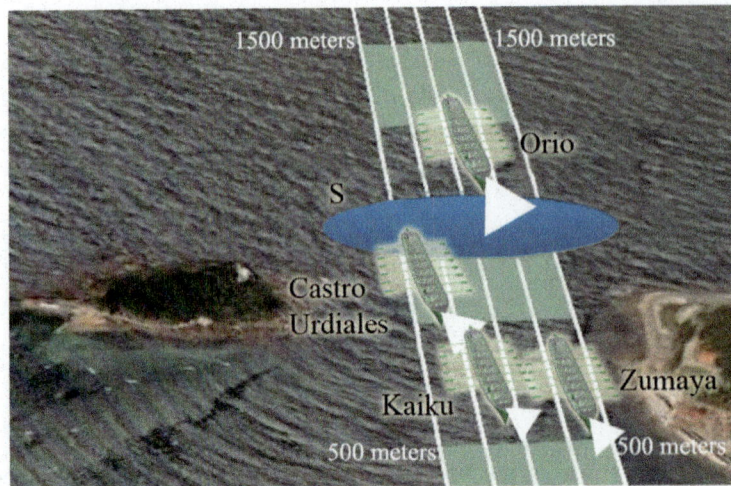

Fig. 14. Sample scenario for the example queries.

- *Reference Objects.* In the context of rowing races, the reference object is the interesting object (e.g., a particular boat, the island, etc.) that must be viewed. As described in Section 4.1, the extent of the objects is considered when processing the queries.
- *Target Objects.* In the context of rowing races, the target objects are the objects (e.g., boats) that have cameras that may satisfy the conditions required to view the area of interest. For example, cameras located on boats or fixed cameras in the harbor are part of the answer to location-dependent queries.

As seen above, the proposed architecture fits the context of rowing races in San Sebastian. In the rest of this section, we will analyze how the motivating queries described in Section 2 can be expressed (using an SQL-like syntax) in a way that allows their processing with the proposed architecture. For illustration, we will consider the scenario shown in Fig. 14 (the boats are not done to scale, as they have been enlarged for clarity), where we assume that all the boats have a camera situated in their bow, which can pan $+/-90°$. We will use *o.cam* to reference the camera of an object o.

5.1. Query 1: View a certain boat

Let us suppose that we want to retrieve the cameras that can focus on a particular boat, for example the one named "Kaiku". The query would be expressed as follows:

```
SELECT   O.cam.id, pan, tilt, time
  FROM   Objects AS O
 WHERE   pan=panToView(Kaiku, O.cam, any, full)
         AND tilt=tiltToView(Kaiku, O.cam, any, full)
         AND pan<>-999 AND tilt<>-999
         AND time=MAX(timeToPan(O.cam, pan), timeToTilt(O.cam, tilt))
ORDER BY   time ASC
```

Besides the identifiers of the cameras, it returns the number of degrees that they should pan (*pan* value) and tilt (*tilt* value) to focus on the boat "Kaiku" (from any angle but providing a full view) as well as the time needed by that camera to do it (*time*). The cameras that currently have a full view of the "Kaiku"

Fig. 15. Virtual view of "Kaiku" provided by "Castro Urdiales" in the scenario of Fig. 14.

boat (independently of the combination of front/rear/side and top/middle/bottom that defines the view, as for example a "front and top" view of the object) appear first in the answer ($pan = 0$, $tilt = 0$, $time = 0$), and the more time they need to turn (to fully view the reference object "Kaiku") the later they appear in the answer because of the *ORDER BY* clause. For example, in Fig. 14 the camera of the boat "Castro Urdiales" would be retrieved first (as it has a full view of the rear of "Kaiku", see Fig. 15) and the one of "Orio" would be ranked in the second position (as it can turn slightly to have a full view of the rear of "Kaiku"). Cameras that cannot view that object (e.g., the one of the boat "Zumaya" in Fig. 14) are not included in the answer because of the $pan <> -999$ and $tilt <> -999$ constraints (see Section 4.3). However, as this query is evaluated continuously, the answer is always up to date. Thus, the information retrieved by the continuous query will help technical directors to select the best views.

5.2. Query 2: Capture a close side view of any boat

Let us imagine that we want to retrieve only cameras which are currently streaming a side view of a boat captured from a nearby location. This query would be expressed as follows:

```
SELECT    O.cam.id, dist
  FROM    Objects AS O, RowingBoats AS B
 WHERE    pan=panToView(B.id, O.cam, side, any)
          AND tilt=tiltToView(B.id, O.cam, any, any)
          AND pan=0 AND tilt=0
          AND inside(250 meters, B.id, O) AND dist=distance(B.id, O.id)
ORDER BY  dist
```

where we assume that a camera must be located no further than 250 meters in order to provide a close (full or incomplete) view of that boat. Moreover, we order the cameras according to the distance to the target, as closer cameras are more likely to be able to provide detailed views. Thus, for example, in the

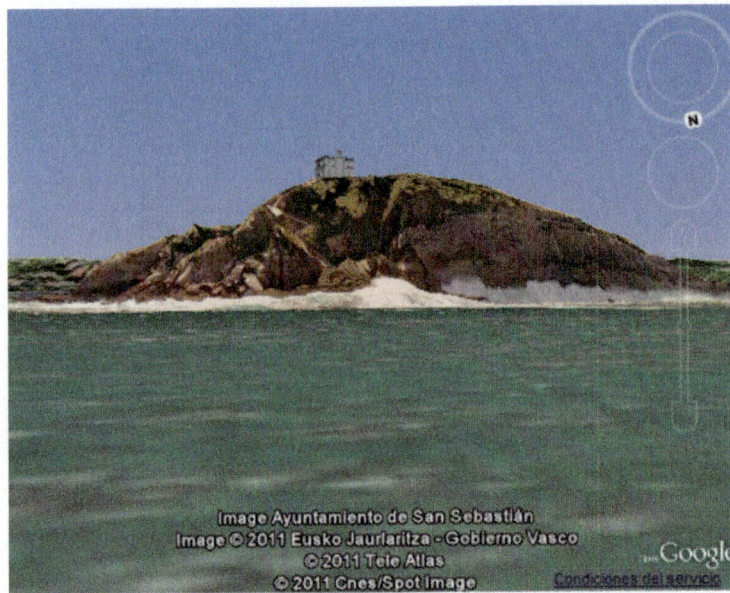

Fig. 16. Virtual view of the island from the ocean side.

scenario of Fig. 14 the camera of "Castro Urdiales" would be a good candidate to view the boat "Kaiku" (see Fig. 15).

5.3. Query 3: Capture a wide view of the island from the ocean side

For this query, let us assume that the technical director previously defined an area S (see Fig. 14), on the ocean side, which represents where a camera should be located in order to have the wanted wide view of the island from the ocean side (see Fig. 16). Thus, the query could be expressed as follows:

```
SELECT      O.cam.id, pan, tilt, time
  FROM      Objects AS O
 WHERE      inside(0 meters, S, O)
            AND pan=panToView(island, O.cam, any, full)
            AND tilt=tiltToView(island, O.cam, any, full)
            AND pan<>-999 AND tilt<>-999
            AND time=MAX(timeToPan(O.cam, pan), timeToTilt(O.cam, tilt))
ORDER BY    time ASC
```

where with the constraint *inside(0 meters, S, O)* we require objects no further than 0 meters from S (that is, objects within S). In the scenario shown in Fig. 14 no camera is currently within the area S. However, the "Orio" boat will enter soon S and so it will become part of the answer to the continuous query. It should be noted that the technical director may want to change the size and location of the area S at any time, in order to focus on a different area (e.g., an accident that has happened). The system is able to adapt to changes in the monitoring requirements.

In Table 1 we summarize the current answers to the sample queries discussed in this section, according to the scenario shown in Fig. 14.

Table 1
Answers for the sample queries

Query	Answer (cameras)
1.– View a certain boat ("Kaiku")	–"Castro Urdiales" –"Orio"
2.– Capture a close side view of any boat	For the "Kaiku" boat: –"Castro Urdiales"
3.– Capture a wide view of the island from the ocean	No answer for the moment

6. Prototype and experimental evaluation

In order to test the system proposed, a prototype has been developed. This prototype consists of three modules: 1) a simulator, 2) a query processor, and 3) a Graphical User Interface (GUI) for technical directors. Testing this kind of system in a real-life environment is difficult because there are many real objects, devices, and wide-area scenarios involved; therefore, we have had no chance to test it on the real rowing race in San Sebastian, which is celebrated only once a year. So, the simulator of wireless and distributed environments described in [21] has been adapted to deal with the sample scenario of this paper. This simulator used a file containing the real GPS location data of each rowing boat recorded every second during the rowing race celebrated in September 2010. Therefore, real trajectories were used to move objects in the simulator. Moreover, the simulator allows us to dynamically change other parameters, such as the current pan and tilt of each camera, to simulate the real movements of the cameras.

The technical director needs a simple GUI that simulates his/her work environment, where the query processor is integrated. As we are not using any real camera in the test, cameras are simulated using the Google Earth API (http://code.google.com/apis/earth/), to show an approximate view of what the cameras would view in the real scenario. This was inspired by the work presented in [28], where the data coming from GPS receivers in the boats during the broadcasting of live rowing events are integrated into a 3D terrain model. Figure 17 shows a screenshot of this GUI in our prototype. The technical director could see up to four camera streams at the same time and use the buttons on the bottom left panel to interact with the query processor, as explained in Section 5. This interface is inspired by mobile production units where the technical director has several input video sources to select, a screen where the source selected is processed (Preview screen), and another screen (Program screen) to see what it is being broadcasted; in our GUI, by clicking on the button "air", we "broadcast" the content of the Preview screen into the Program screen.

The system has been tested using the query described in Section 5.1. In that query, the technical director wants to retrieve all the cameras that could provide a full coverage and any view of the rowing boat "Kaiku". The parameters used in the tests are the following ones:

– There are four rowing boats with a camera each; they are moving according to the real GPS location information captured during the race in 2010.
– There are three fixed cameras, one on the island, another one on the promenade, and the last one on a sailing boat near the rowing boats.
– The cameras are set with horizontal focus $70°$, vertical focus $45°$, pan range $\pm130°$, tilt range $\pm90°$, pan and tilt speed 5.5 degrees/second.

In the rest of this section, we evaluate the precision and recall of the system, as well as the precision of the estimation of the time needed by a camera to focus on a target.

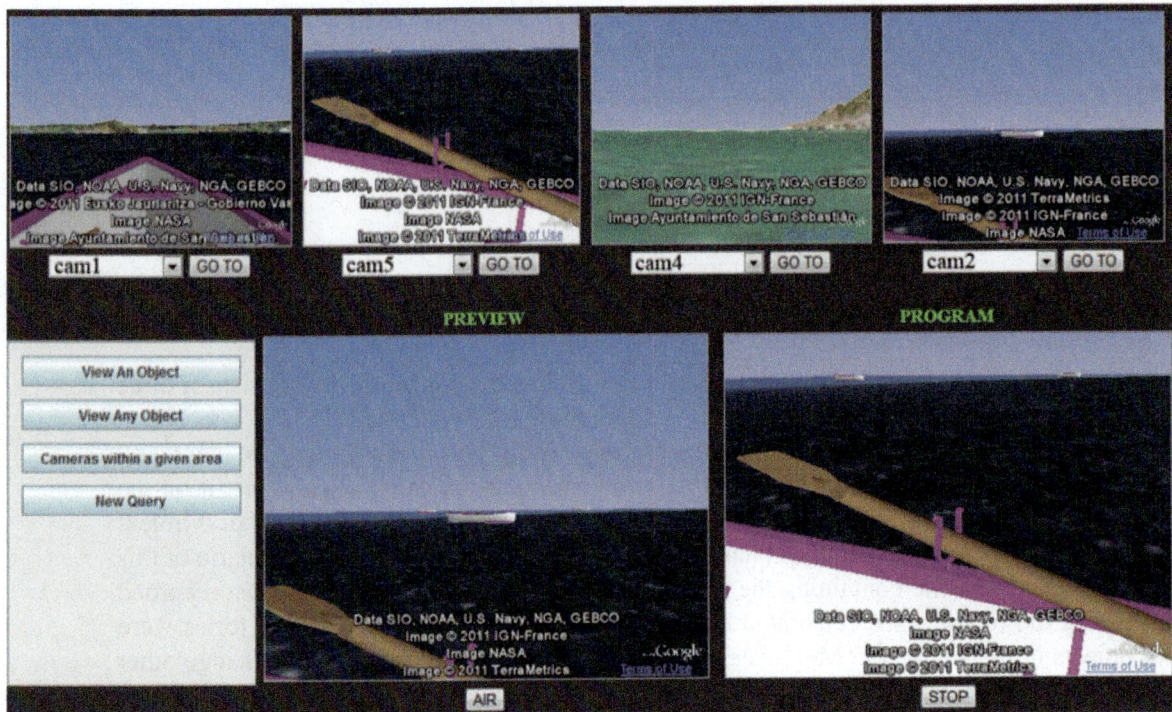

Fig. 17. GUI for the technical director.

6.1. Testing the precision and recall of the cameras in the answer

The first test is conceived to obtain the difference between the number of cameras in an answer given by the system in real time and the optimal answer. Figure 18 shows the results of this test, where we are representing the number of cameras given in the answer (which is refreshed every second). Remember that there exist seven cameras in the scenario. An error is indicated in the graph with an arrow at the specific time instant when such an error occurs. When an answer includes, for example, one extra camera (a camera that will not obtain the required view), it is denoted by "+1 Cam" and when the answer misses one valid camera (one that could obtain the required view), it is denoted by "−1 Cam". Along with the number of extra/missing cameras, we specify between brackets the amount of time (in seconds) that this error lasts.

During the race, the graph shows that the coverage of "Kaiku" is quite high compared with the number of existing cameras: the maximum number is six (because the camera of "Kaiku" obviously cannot fully focus on itself) and at least three cameras obtain the required view on the target rowing boat. Notice also that the errors committed by the system never last more than 2 seconds. The reason is the efficient continuous refreshment of the answer.

It can also be observed that the number of cameras able to capture the target varies quickly once the turning point is passed (around time=10:00). This variation is due to the query asking for a full view of "Kaiku". As the distance and speed of the rowing boats increase, some cameras are unable to maintain the target completely focused all the time (this would require exceeding the pan limit), and so they disappear for a short time from the answer set even when they can provide a view that covers most of the target object.

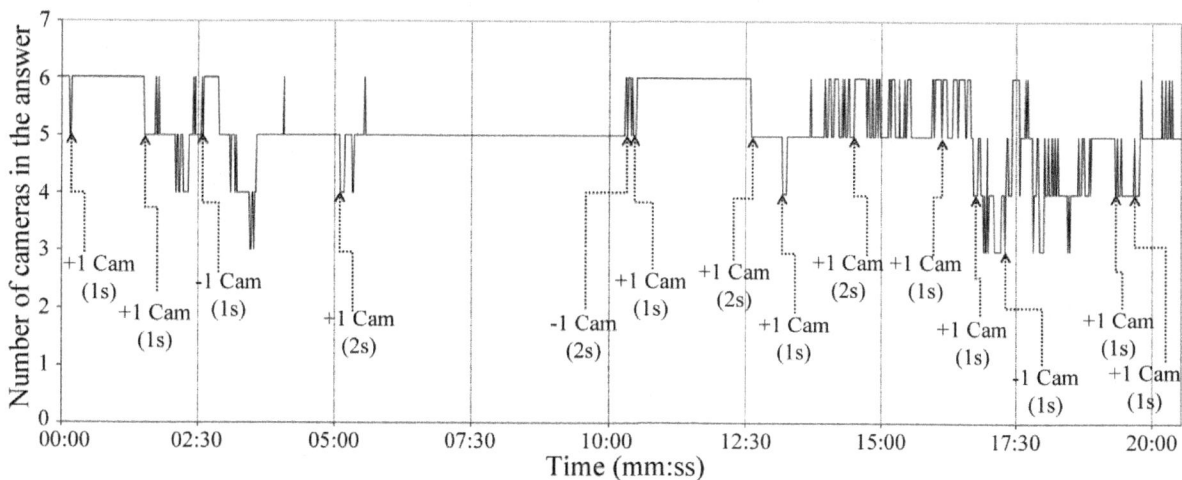

Fig. 18. Error in precision/recall of the cameras.

Even when the query processor must perform its calculations every second, the results are quite satisfactory for a period of twenty minutes, and the errors are corrected quickly enough to not be taken into account by the user, i.e., the technical director.

6.2. Testing the precision of the estimated time to view

The goal of the second test is to evaluate the precision of the answer given by the query processor in terms of the time information provided. When the query processor lists a camera in the answer, the time needed by the camera to turn (pan and/or tilt) to focus on the rowing boat is computed and shown to the technical director. This estimation could be wrong because the rowing boats, and therefore their cameras, could move unexpectedly, although these errors will be corrected quickly as the query is continuously processed.

In Fig. 19, we show the total error committed by all the cameras (the difference between the time given by the query processor to turn the cameras and the real time needed) at every instant. In the figure, we use triangles to mark time instants when a camera is included in the answer but its error is too big (i.e., that camera should not be included in the answer).

Initially, all the cameras (pointing at the same direction as the front vector of their rowing boats) have to pan to focus on "Kaiku" and the query processor estimates the time needed for the movement with a small error. After that, the error is zero because the rowing boats are moving at approximately the same speed. Around time 02:30, the rowing boats are leaving the bay and they get closer to each other. "Kaiku" is too close to its neighbor boat whose camera onboard is not able to fully focus on "Kaiku". The next interesting time instant is around 10:00, when the rowing boats are in the turning point. Here the error is caused by the neighbor boat of "Kaiku" because the system estimates that the time to view "Kaiku" is much greater than the real time needed, due to the quick turning maneuver. Once the turning point is passed, the speed of the rowing boats and their relative distance increase. That means that the fastest rowing boats are not going to be able to focus on "Kaiku" during a few seconds (around time 14:00). Finally, the maximum distance between the rowing boats is achieved, and so there are some cameras that cannot view "Kaiku" anymore and the system incurs several errors, indicated by the triangles around time 17:30–20:00.

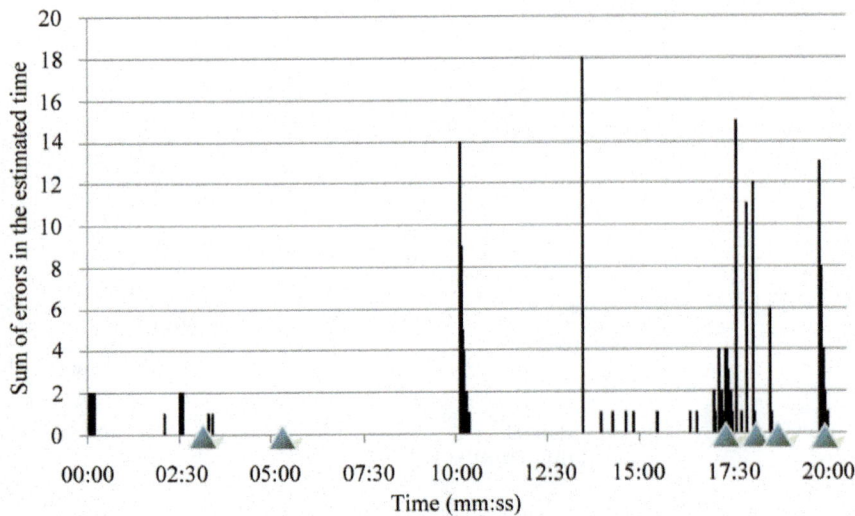

Fig. 19. Error in the estimated time.

This test shows that the system commits small errors concerning the estimated time to view the target. Besides, those errors are fixed very quickly (in the test, in five seconds at the most), since the queries are continuously processed. The answer presented to the user contains too high errors only six times along the twenty minutes of race (remember that an answer is updated every second), and those errors only last one second.

7. Related work

In this section, we analyze works that focus on the use of multiple cameras to produce contents (for video lectures, sport events, and others) as well as works that are closer to the field of location-based services. Works performing object tracking by using multiple cameras [24,50] are not explicitly considered, as they pursue a different goal: estimating/monitoring the trajectory of an object from images, and in our proposal the objects are detected by using sensors [30] (specifically, GPS receivers).

7.1. Camera management for video lectures

Several works in the literature have considered the problem of automatic camera management for recording and broadcasting lectures, talks and seminars [4,5,17,29–31,39,41,51]. In this section, we provide an overview of some proposals:

- One of the first works in the area is the *AutoAuditorium* system [4,5] (see http://www.autoauditorium. com/), which uses two cameras and microphones to obtain information about what is happening on the stage and perform an automatic audio mixing, tracking of the people on stage, and automatic camera selection.
- Another interesting work is [30,39], which also proposes a two-camera system that is able to track the presenter and the audience and generates automatically the produced video. Several production rules have been implemented in this system, inspired by the way professional video producers work, in order to take the appropriate recording decisions. According to the tests performed, the system passes the Turing test successfully.

- The system *FlySPEC*, proposed in [29], combines a PTZ camera and a panoramic camera and integrates requests from users, along with a mechanism to solve conflicting requests. In this way, the system benefits from the involvement of the audience to reduce the probability of unsatisfactory recordings.
- As another example [51], presents the *Microsoft Research LeeCasting System* (*MSRLCS*), that supports a scripting language to facilitate the customization of production rules for different room configurations and production styles.
- The *Virtual Director* described in [31] considers an *automation specification language* (*ASL*) that directors can use to express broadcast specifications.

Finally, it is interesting to mention the *Virtual Videography* approach described in [17]. Whereas the previous works perform a live broadcast, this last work advocates an offline processing, as this provides more time and more information to perform the video production (e.g., it is possible to look into the future).

Although the context and purpose of these works is different than the ones considered in this paper, they highlight the interest of the development of automatic video production techniques to save production costs and enable fast access to multimedia information.

7.2. Camera management for sport events

Several other works have focused on helping producers of sport videos [34], as we describe in this section.

Some of these works [6,7,12] are part of the *APIDIS* (*Autonomous Production of Images based on Distributed and Intelligent Sensing*) project (http://www.apidis.org/). For example, in [7] a system that helps the video production and video summarization in the context of basketball games is presented. This system is based on a divide-and-conquer strategy and considers the trade-off among three desirable properties: *closeness* (resolution or level of detail of the images), *smoothness* (continuity in the movement and in the story-telling contents, despite the switching of camera views), and *completeness* (displaying as many objects and interesting actions as possible). Besides, user preferences are taken into account in order to provide personalized videos.

Outside the APIDIS project, the *My eDirector 2012* project, presented for example in [36], advocates extracting semantic metadata from the video streams to enable the personalization of the multimedia information broadcasted. Several problems are considered in this project, such as the detection and tracking of athletes and the automatic detection and tracking of high-level patterns. For more information, the web page of the project can be consulted (http://www.myedirector2012.eu/).

Besides basketball and athletic competitions, producing suitable soccer videos has also attracted the interest of research. As an example, the work presented in [14] tackles the problem of summarizing videos (i.e., obtaining video abstracts) of soccer games by applying different image processing algorithms to analyze the input videos. Soccer videos were also considered as an evaluation scenario within the context of the APIDIS project (see [6]). The work in [2] also focuses on the analysis and summarization of soccer videos by using image recognition techniques and presents an experimental evaluation based on criteria such as the video quality, the intelligibility, the quality of the zooming and panning performed, the view size, and the view duration. As another example, a multi-camera tracking approach based on a belief propagation algorithm is proposed in [13], where thanks to the collaboration among the cameras it is possible to compensate the effects of poor observations due to phenomena such as occlusions. There are other works and patents that deal with the problem of summarizing soccer videos, such as [35].

Other sports have also been considered, such as tennis, baseball, and cricket. For example, in the *TennisSense* platform [9], nine IP cameras around the tennis court are used to automatically infer what is happening in a tennis game. The extracted metadata is then used for indexing and browsing the stored tennis videos. In [33] the problem of video summarization is considered in the context of baseball games. The proposed method is able to generate abstracts based on metadata describing the semantic content of videos encoded in the MPEG-7 format. Moreover, it takes into account the preferences of the users to build a personalized summary of the input video. Finally, in [26] excitement clips (e.g., the close-up of a player or the referee, the gathering of players, etc.) from sports videos are extracted by exploiting audio features (such as an increase in the audio level of the voice of the commentators or the cheers of the audience). The proposal is validated by using soccer and cricket videos.

All the works described in this section tackle sport events, like the proposal in this paper, but they have a different purpose. Thus, our system helps video producers to locate cameras that can provide certain images in an environment with moving objects and moving cameras, whereas the purpose of these works is usually to perform an automatic video production in an offline setting (so, for example, achieving a good performance for real-time processing is not an issue) by using image processing techniques. In other words, the focus of these works is on the operation and exploitation of the cameras, rather than on the complexity of managing multiple video signals in real time.

Although some works explicitly mention the possibility of live broadcasting, such as [47] (that focuses on soccer games and indicates that even though their system is currently performing off-line it can be improved to enable on-line processing), the context of these works is different from the one presented in this paper. Thus, for example, moving cameras are not used in those related works. However, according to [47], the automatic generation of broadcast video is facilitated when a fixed set of static cameras is considered, since the types of views available (e.g., far view, close-up view, etc.) is limited by the set of cameras used. According to [11], which presents a multi-camera selection technique based on features of objects and frames and considers a basketball game scenario (among others), an efficient mechanism for detection, tracking and feature extraction is a key element to enable view selection in a short time.

7.3. Other works on multi-camera management

In this section we briefly indicate some other interesting works for multi-camera management. An interesting and representative work concerning the management of multiple cameras and screens can be found in [10], which proposes different approaches for concurrent access and management of multiple static cameras. It is also worth mentioning the proposal in [46], where the authors work on the optimal monitoring of multiples video signals. In [1] the authors use an array of cameras to acquire a 3D scene.

However, it is important to emphasize that these representative works on multi-camera management and monitoring consider only cameras that are static (i.e., at fixed locations). On the contrary, the cameras considered in our proposal can move.

7.4. Works on location-dependent queries

Developing query processing techniques that manage location-dependent queries efficiently is an intensive area of research [22,45], due to the great interest in Location-Based Services (LBS). The existing proposals differ in several aspects, such as the types of queries that they can process, the assumptions that they rely on, and/or the underlying infrastructure required. For example, *MobiEyes* [15] requires the cooperation of the moving objects that are the targets of the queries to process the queries distributively, *LOQOMOTION* [19] performs a distributed processing on certain objects, the proposal in [25] considers a multi-cell distributed environment to process range queries over static target objects, *DOMINO* [48] exploits some knowledge about the trajectories of the objects, *SEA-CNN* [49] focuses on the processing of continuous kNN queries, the work presented in [32] considers soft timing requirements that may exist for the processing of location-dependent queries, and [44] tackles location-dependent queries in a wireless broadcast environment through a global indexing scheme.

However, we are not aware of any work that has applied an architecture for processing location-dependent queries in the context of sport events to retrieve relevant multimedia data. Although there are works that emphasize the importance of location-dependent query processing for multimedia computing (e.g., [27]), they do not consider that the queries themselves can have as a final goal to retrieve multimedia data that are relevant for the user. Works that propose location-dependent multimedia services for mobile users (e.g., [37], where a middleware based on mobile agents is presented) are not directly related either to the research presented in this paper.

Moreover, as far as we know, the processing of location-dependent queries has not been studied before in relation to any kind of multimedia data.

8. Conclusions and future work

In this paper, we have shown the usefulness of location-dependent query processing in a real-world sport event: the rowing races in San Sebastian. With that purpose, we have extended a general architecture for location-dependent query processing with the capabilities needed to detect suitable (static or moving) cameras to view a particular moving object or area. Besides, we have shown the precision of our system by testing our prototype using real GPS location data captured during a rowing race celebrated in September 2010. The main features of our proposal are:

- The features of videocameras are considered to help technical directors to access the multimedia information coming from different (static or moving) videocameras.
- Predefined or dynamically built geographic areas or static/moving objects can be set as targets for which suitable cameras can be found by the system.
- The proposal is flexible, and so new functionalities can be added to the system without important changes to the main architecture.

The underlying location-dependent query processing system used benefits from the use of mobile agents to carry out the processing tasks wherever they are needed. Thus, agents are in charge of 1) tracking the location of interesting (static or moving) objects and the field-of-view of all the cameras, and 2) refreshing the query answers continuously and efficiently.

As future work, we plan to extend the system to enable automatic tracking of targets by videocameras, in order to further help the technical directors with the broadcasting process. It would be interesting to extend this study to other sport scenarios too. Particularly, we think that the context of classic cycling competitions, such as the Tour of France, could be very challenging (the scenario could extend along many kilometers in mountain landscapes where a centralized solution could be unsuitable). In other scenarios, some useful multimedia information could also be provided by mobile devices carried by people (e.g., spectators or tourists), and so exploiting the possibilities of Mobile Ad Hoc Networks (MANETs) for content dissemination [16] and retrieval of multimedia information would be interesting. In these highly distributed and dynamic scenarios, the whole potential of LOQOMOTION could be better exploited.

Acknowledgments

This research work has been supported by the CICYT project TIN2010-21387-C02. We also thank David Antón and Aritz Legarretaetxebarria for their help with the implementation of our prototype and technical support, respectively.

References

[1] D.G. Aliaga, Y. Xu and V. Popescu, Lag camera: A moving multi-camera array for scene-acquisition, *Journal of Virtual Reality and Broadcasting* **3**(10) (December 2006).

[2] Y. Ariki, S. Kubota and M. Kumano, Automatic production system of soccer sports video by digital camera work based on situation recognition. In *Eighth IEEE International Symposium on Multimedia (ISM'06)*, IEEE Computer Society, 2006, pages 851–860.

[3] F.L. Bellifemine, G. Caire and D. Greenwood, *Developing Multi-Agent Systems with JADE*, Wiley, 2007.

[4] M.H. Bianchi, AutoAuditorium: A fully automatic, multi-camera system to televise auditorium presentations. In *1998 Joint DARPA/NIST Smart Spaces Technology Workshop*, 1998.

[5] M.H. Bianchi, Automatic video production of lectures using an intelligent and aware environment. In *Third International Conference on Mobile and Ubiquitous Multimedia (MUM'04)*, ACM Press, 2004, pages 117–123.

[6] F. Chen and C.D. Vleeschouwer, A resource allocation framework for summarizing team sport videos. In *16th IEEE International Conference on Image Processing (ICIP'09)*, IEEE Computer Society, 2009, pages 4293–4296.

[7] F. Chen and C.D. Vleeschouwer, Personalized production of basketball videos from multi-sensored data under limited display resolution, *Computer Vision and Image Understanding* **114**(6) (June 2010), 667–680.

[8] I.-H. Chen and S.-J. Wang, An efficient approach for the calibration of multiple PTZ cameras, *IEEE Transactions on Automation Science and Engineering* **4**(2) (April 2007), 286–293.

[9] C.O. Conaire, P. Kelly, D. Connaghan and N.E. O'Connor, TennisSense: A platform for extracting semantic information from multi-camera tennis data. In *16th International Conference on Digital Signal Processing (DSP'09)*, IEEE Computer Society, 2009, pages 1062–1067.

[10] G.W. Daniel and M. Chen, Interaction control protocols for distributed multi-user multi-camera environments, *Systemics, Cybernetics and Informatics* **1**(5) (2003), 29–38.

[11] F. Daniyal, M. Taj and A. Cavallaro, Content and task-based view selection from multiple video streams, *Multimedia Systems* **46**(2–3) (January 2010), 235–258.

[12] D. Delannay, N. Danhier and C.D. Vleeschouwer, Detection and recognition of sports(wo)men from multiple views. In *Third ACM/IEEE International Conference on Distributed Smart Cameras (ICDSC'09)*. IEEE Computer Society, 2009.

[13] W. Du, J. bernard Hayet, J. Piater and J. Verly, Collaborative multi-camera tracking of athletes in team sports. In *Workshop on Computer Vision Based Analysis in Sport Environments (CVBASE'06)*, 2006, pages 2–13.

[14] A. Ekin, A.M. Tekalp and R. Mehrotra, Automatic soccer video analysis and summarization, *IEEE Transactions on Image Processing* **12**(7) (July 2003), 796–807.

[15] B. Gedik and L. Liu,. MobiEyes: A distributed location monitoring service using moving location queries, *IEEE Transactions on Mobile Computing* **5**(10) (October 2006), 1384–1402.

[16] J. Haillot and F. Guidec, A protocol for content-based communication in disconnected mobile ad hoc networks, *Mobile Information Systems* **6**(2) (April 2010), 123–154.

[17] R. Heck, M. Wallick and M. Gleicher, Virtual videography, *ACM Transactions on Multimedia Computing, Communications, and Applications* **3**(1) (February 2007), 4:1–4:28.

[18] S. Ilarri, C. Bobed and E. Mena, An approach to process continuous location-dependent queries on moving objects with support for location granules, *Journal of Systems and Software* **84**(8) (August 2011), 1327–1350.

[19] S. Ilarri, E. Mena and A. Illarramendi, Location-dependent queries in mobile contexts: Distributed processing using mobile agents, *IEEE Transactions on Mobile Computing* **5**(8) (August 2006), 1029–1043.

[20] S. Ilarri, E. Mena and A. Illarramendi, Using cooperative mobile agents to monitor distributed and dynamic environments, *Information Sciences* **178**(9) (May 2008), 2105–2127.

[21] S. Ilarri, E. Mena and A. Illarramendi, A system based on mobile agents to test mobile computing applications, *Journal of Network and Computer Applications* **32**(4) (July 2009), 846–865.

[22] S. Ilarri, E. Mena and A. Illarramendi, Location-dependent query processing: Where we are and where we are heading, *ACM Computing Surveys* **42**(3) (March 2010), 1–73.

[23] S. Ilarri, E. Mena, A. Illarramendi and G. Marcos, A location-aware system for monitoring sport events. In *Eight International Conference on Advances in Mobile Computing & Multimedia (MoMM 2010)*, ACM, Austrian Computer Society (OCG), 2010, pages 305–312.

[24] O. Javed and M. Shah, *Automated Multi-Camera Surveillance: Algorithms and Practice*, volume 10 of *The Kluwer International Series in Video Computing*, chapter 5 "Tracking in Multiple Cameras with Disjoint Views", Springer, 2008, pages 59–84.

[25] J. Jayaputera and D. Taniar, Data retrieval for location-dependent queries in a multi-cell wireless environment, *Mobile Information Systems* **1**(2) (April 2005), 91–108.

[26] M.H. Kolekar, Bayesian belief network based broadcast sports video indexing, *Multimedia Tools and Applications* **54**(1) (August 2011), 27–54.

[27] A. Krikelis, Location-dependent multimedia computing, *IEEE Concurrency* **7**(2) (April/June 1999), 13–15.

[28] M. Laka, I. García, I. Macía and A. Ugarte, TV sport broadcasts: Real time virtual representation in 3D terrain models. In *3DTV Conference: The True Vision – Capture, Transmission and Display of 3D Video (3DTV-Con'08)*, IEEE Computer Society, 2008, pages 405–408.

[29] Q. Liu, D. Kimber, J. Foote, L. Wilcox and J. Boreczky, FlySPEC: a multi-user video camera system with hybrid human and automatic control. In *Tenth ACM International Conference on Multimedia (MULTIMEDIA'02)*, ACM Press, 2002, pages 484–492.

[30] Q. Liu, Y. Rui, A. Gupta and J.J. Cadiz, Automating camera management for lecture room environments. In *SIGCHI Conference on Human Factors in Computing Systems (CHI'01)*, ACM Press, 2001, pages 442–449.

[31] E. Machnicki and L.A. Rowe, Virtual director: automating a webcast. In *Multimedia Computing and Networking* **4673**, SPIE, 2002, pages 208–225.

[32] Z. Mammeri, F. Morvan, A. Hameurlain and N. Marsit, Location-dependent query processing under soft real-time constraints, *Mobile Information Systems* **5**(3) (August 2009), 205–232.

[33] N. Nitta, Y. Takahashi and N. Babaguchi, Automatic personalized video abstraction for sports videos using metadata, *Multimedia Tools and Applications* **41**(1) (January 2009), 1–25.

[34] J. Owens, *Television Sports Production*. Focal Press, 2007. Fourth Edition.

[35] H. Pan and B. Li. Summarization of soccer video content. United States Patent 7657836, 2004.

[36] N. Papaoulakis, N. Doulamis, C. Patrikakis, J. Soldatos, A. Pnevmatikakis and E. Protonotarios, Real-time video analysis and personalized media streaming environments for large scale athletic events. In *First ACM Workshop on Analysis and Retrieval of Events/Actions and Workflows in Video Streams (AREA'08)*, ACM Press, 2008, pages 105–112.

[37] M.H. Raza and M.A. Shibli, Mobile agent middleware for multimedia services. In *Ninth International Conference on Advanced Communication Technology (ICACT'07)*, IEEE Computer Society, 2007, pages 1109–1114.

[38] P. Rigaux, M. Scholl and A. Voisard, *Spatial databases with application to GIS*, Morgan Kaufmann Publishers Inc., 2002.

[39] Y. Rui, L. He, A. Gupta and Q. Liu, Building an intelligent camera management system. In *Ninth ACM International Conference on Multimedia (MULTIMEDIA'01)*, ACM Press, 2001, pages 2–11.

[40] C. Spyrou, G. Samaras, E. Pitoura and P. Evripidou, Mobile agents for wireless computing: the convergence of wireless computational models with mobile-agent technologies, *Mobile Networks and Applications* **9**(5) (October 2004), 517–528.

[41] A.A. Steinmetz and M.G. Kienzle, e-Seminar lecture recording and distribution system. In *Multimedia Computing and Networking* **4312**, SPIE, 2001, pages 25–36.

[42] D. Terry, D. Goldberg, D. Nichols and B. Oki, Continuous queries over append-only databases, *ACM SIGMOD Record* **21**(2) (June 1992), 321–330.

[43] R. Trillo, S. Ilarri and E. Mena, Comparison and performance evaluation of mobile agent platforms. In *Third International Conference on Autonomic and Autonomous Systems (ICAS'07)*, IEEE Computer Society, 2007, pages 41:1–41:6.

[44] A.B. Waluyo, B. Srinivasan and D. Taniar, Global indexing scheme for location-dependent queries in multi channels mobile broadcast environment. In *19th International Conference on Advanced Information Networking and Applications (AINA'05)*, IEEE Computer Society, 2005, pages 1011–1016.

[45] A.B. Waluyo, B. Srinivasan and D. Taniar, Research on location-dependent queries in mobile databases, *International Journal of Computer Systems: Science and Engineering* **20**(2) (March 2005), 77–93.

[46] J. Wang, Experiential sampling for video surveillance. In *First ACM International Workshop on Video Surveillance (IWVS'03)*, ACM Press, 2003, pages 77–86.

[47] J. Wang, C. Xu, E. Chng, H. Lu and Q. Tian, Automatic composition of broadcast sports video, *Multimedia Systems* **14**(4) (September 2008), 179–193.

[48] O. Wolfson, A.P. Sistla, S. Chamberlain and Y. Yesha, Updating and querying databases that track mobile units, *Distributed and Parallel Databases* **7**(3) (July 1999), 257–287.

[49] X. Xiong, M.F. Mokbel and W.G. Aref, SEA-CNN: Scalable processing of continuous k-nearest neighbor queries in spatio-temporal databases. In *21st International Conference on Data Engineering (ICDE'05)*, IEEE Computer Society, 2005, pages 643–654.

[50] A. Yilmaz, O. Javed and M. Shah, Object tracking: A survey, *ACM Computing Surveys* **38**(4) (2006), 13:1–13:45.

[51] C. Zhang, Y. Rui, J. Crawford and L.-W. He, An automated end-to-end lecture capture and broadcasting system, *ACM Transactions on Multimedia Computing, Communications and, Applications* **4**(1) (January 2008), 6:1–6:23.

9

Balanced bipartite graph based register allocation for network processors in mobile and wireless networks

Feilong Tang[a,b], Ilsun You[c,*], Minyi Guo[a], Song Guo[b] and Long Zheng[b]

[a]*Department of Computer Science and Engineering, Shanghai Jiao Tong University, Shanghai 200240, China*

[b]*School of Computer Science and Engineering, The University of Aizu, Fukushima 965-8580, Japan*

[c]*School of Information Science, Korean Bible University, 16 Danghyun 2-gil, Nowon-gu, Seoul, South Kroea*

Abstract. Mobile and wireless networks are the integrant infrastructure of mobile and pervasive computing that aims at providing transparent and preferred information and services for people anytime anywhere. In such environments, end-to-end network bandwidth is crucial to improve user's transparent experience when providing on-demand services such as mobile video playing. As a result, powerful computing power is required for networked nodes, especially for routers. General-purpose processors cannot meet such requirements due to their limited processing ability, and poor programmability and scalability. Intel's network processor IXP is specially designed for fast packet processing to achieve a broad bandwidth. IXP provides a large number of registers to reduce the number of memory accesses. Registers in an IXP are physically partitioned as two banks so that two source operands in an instruction have to come from the two banks respectively, which makes the IXP register allocation tricky and different from conventional ones. In this paper, we investigate an approach for efficiently generating balanced bipartite graph and register allocation algorithms for the dual-bank register allocation in IXPs. The paper presents a graph uniform 2-way partition algorithm (FPT), which provides an optimal solution to the graph partition, and a heuristic algorithm for generating balanced bipartite graph. Finally, we design a framework for IXP register allocation. Experimental results demonstrate the framework and the algorithms are efficient in register allocation for IXP network processors.

Keywords: Register allocation, network processor, bipartite graph, mobile and wireless network, register bank

1. Introduction

Mobile and pervasive computing (MPC) is an exciting new paradigm that provides "disappeared" services for people even while moving around, where mobile and wireless networks with powerful computing and communication abilities are the integrant infrastructure to make this vision a reality. More and more applications, for example, mobile video playing, require a high end-to-end network bandwidth to ensure the specified QoS requirements in such environments [2,30].

Powerful computation ability is requisite to high bandwidth in order to meet on-demand basic network services, e.g. packet classification and forwarding, as well as intrusion detection and firewalls. Routers are the core components of high-speed networks [10,14,23]. In general, packets are received and sent

---

*Corresponding author. E-mail: ilsunu@gmail.com.

though input and output buffer which reside in on-chip memory that is characterized by large capacity and long access latency. A router should process packets at line rates of high speed, and at the same time be sufficiently programmable and flexible to support current and potential services. Traditional ASIC processors cannot meet such requirements because of their limited processing ability, and poor programmability and scalability.

Network processors (NP) emerge to address such high-speed challenges that are beyond the ability of general-purpose processors, while providing flexibility and programmability. It is a new breed of packet forwarding engine that is designed to meet the simultaneous demands of high speed and great flexibility of today's network equipments [35]. The goals are to provide the performance of traditional ASICs and the programmability of general-purpose processors [16].

Keeping up with external line rate without losing packets makes it imperative that all latencies be kept to a minimum. For this purpose, network processors, for example, Intel IXP [4,25], in general contain a large number of registers to decrease the number of memory accesses [26]. In such a circumstance, efficient register allocation plays an extremely important role for the network applications to achieve high throughput.

Register allocation is the process of multiplexing a large number of target program variables onto a small number of CPU registers. The goal is to keep as many operands as possible in registers to maximize the execution speed of application programs. The register file of Intel' IXP is built using two-ported register files: one read port and one write port. The area efficiency of two-ported registers relative to multiport registers is important in allowing the large number of registers to fit in the allocated silicon area. On the other hand, the use of two-port registers places some restrictions on which combinations of registers source operands can use legally for an instruction [21]. These restrictions make the IXP register allocator tricky and different from conventional ones. Two main differences are band conflict and bank imbalance.

This paper focuses on how to effectively allocate registers for IXP network processors to reduce the bank conflict and balance the bank loads as much as possible, based on the balanced bipartite graph. Our motivation is to increase routing speed through allocating tasks on registers of the IXP-based network process efficiently to improve the performance of mobile and wireless networks.

The rest of the paper is organized as follows. In the next section, we review related work carefully. Section 3 briefly describes the register file and register allocation problems on two banks based IXP. In Section 4, we present a fixed-parameter tractable algorithm (FPT) for uniform graph bipartite problem and a heuristic algorithm for balanced bipartite graph. The algorithms will be used in the register allocation framework presented in Section 5. Experiments and evaluation are reported in Section 6. Section 7 concludes our paper with the discussion on the future work.

2. Related work

Many significant results on register allocation have been reported. In this section, we carefully review related work on register allocation algorithm and architecture, as well as related proposals for IXP network processors.

Register allocation algorithms. Most schemes on the register allocation used the graph coloring algorithm, which was proposed by Chaitin [9] in 1982. In the graph coloring algorithm, a graph vertex represents a variable that will be assigned a register; a line between two vertexes means a conflict between the two variables so that the two variables can not be assigned to the same register. Such a graph is called the conflict graph. The graph coloring algorithm converts the register allocation into the graph coloring problem and becomes the base of register allocation. Briggs et al. [24] improved Chaitin's work by an optimistic graph coloring algorithm. The basic idea is to put the conflicted variable to a temp stack when the conflict graph cannot be simplified. Optimistically, the conflict may be solved a little later. Chow and Hennessy used a frequency-sensitive heuristic graph coloring algorithm, which assigns the priority based on the usage frequency. This method is very effective to programs with loops. Lueh [11] proposed a fusion-based register allocation approach, assigning registers in units of basic blocks, where potential conflicts are limited in individual basic blocks. Appel et al. [3] divided the register allocation into two sub-problems: optimization of the insert of overflowed codes and optimization of the register merging. They solved the first sub-problem using integer linear programming (ILP) to achieve the optimal scheme of register allocation.

Zhuang' proposal [33] is the most relevant to our work. They proposed three heuristic approaches, which differ in the order of register allocation (RA) and bank assignment. The first one is Pre-RA Bank Allocation Approach, simplified Pre-RA. This approach assigns banks before registers. It breaks odd cycles through live range splitting rather than deleting edges to make the graph bipartite. An algorithm is invoked to near-balance the vertexes into two banks. The well-know RA algorithm [9] is used to allocate registers for each bank in Pre-RA. However, it suffers from difficulty to judge physical register pressure in the two banks before RA and often results in imbalance and large spilling cost. The second one is Post-RA Bank allocation Approach, simplified Post-RA. This approach firstly allocates registers with conventional RA algorithms. Post-RA is similar to the approach that IXP products often employ. The rest parts are almost the same as Pre-RA. Finally, the third approach is the mixed register allocation. Their experimental results show that Post-RA has better runtime performance but inserts the more instructions to resolve bank conflicts and balance bank load. Nevertheless, it splits vertexes regardless of how it will affect the rest of the graph and incurs unnecessary move instructions between banks. Unfortunately, these approaches only considered the register allocation for general processors, without involving the features of registers in IXP. Moreover, Monreal et al. [32] presented a novel physical register management scheme that allows for a late allocation (at the end of execution) of registers to reduce the time accessing to register files, which significantly saves the number of registers and shortens the access time to register files. The approach used the virtual-physical registers, together with an on-demand register allocation policy and a stealing mechanism that prevents older instruction from being delayed by younger ones. The main idea is to allocate physical registers at the end of the execution stage, rather than at decode time.

Register allocation architectures. Jang et al. [31] studies banked register file under VLIW architecture, where register banks are partitioned such that one bank is associated with each functional unit or by associating a cluster of functional units with each bank. Abella and González [13] proposed an adaptive micro-architecture that achieves significant dynamic and static power savings in the register file, at the expense of a very small performance loss. The proposed mechanism dynamically limits the number of in-flight instructions in order to save dynamic and static power in register files or rename buffers, and reduces register pressure, based on monitoring how much time instructions spend in both the issue queue and the reorder buffer and limit their occupancy, and taking resizing decisions based on these observations. Wagner et al. [20] proposes a special register allocation technique, where the register allocator maintains two sets of virtual registers: one for scalar values and one for register arrays. All virtual registers are indexed by a unique number. Any element of a register array can be accessed in two different ways: first by direct addressing or indirectly by the use of a bit-packet pointer. In case of insufficient physical registers using indirect access, spill code is generated for all virtual registers within a register array; otherwise, only the particular virtual register is spilled. Park et al. [19] studies dual bank allocation problem. Nevertheless, they don't have the bank constraint like ours. Collins et al. [26] proposes a dynamic register allocation algorithm on IXP, but it requires modification to the hardware. Very Wide Register [29] is an asymmetric register file architecture, which has single ported cells and asymmetric interfaces to the memory and to the datapath, for low power embedded processors. The basic components in this organization are the interface to the memory, single ported cells and the interface to the datapath. The interface of this foreground memory organization is asymmetric: wide towards the memory and narrower towards the datapath. This foreground memory is similar to a register file which is incorporated in the datapath pipeline cycles. The wide interface enables to exploit the locality of access of applications through wide loads from the memory to the foreground memories (registers). At the same time the datapath is able to access words of a smaller width for the actual computations. ParShield [28] is a register architecture that provides cost-effective protection for register files against soft errors, which selectively protects a subset of the registers by generating, storing, and checking the ECCs of only the most vulnerable registers while they contain useful data. ParShield also adds a parity bit for all the registers and re-uses the ECC circuitry for parity generation and checking as well. In particular, ParShield has no performance impact and little area requirements. Kolson et al. [6] proposed a register allocation architecture to loop in multiple register files, where the available registers have been partitioned into multiple banks. A distinguishing characteristic of the approach is that the register assignment may span multiple iterations of the original loop. Hiser et al. [18] proposed a register assignment method for code generation for instruction-level parallelism (ILP) architectures with partitioned register banks, through the register component graph that abstracts away machine specifics within a single representation, especially for instruction-level parallelism architectures. The approach can produce good code, which increases the parallelism of operations, by separating partitioning from scheduling and register assignment.

Register allocation for IXP network processors. There also have been reports on register allocation for Intel IXP network processors. Shangri-La compiler [22] used a C-like high-level programming language developed by Intel, and provides the functions of optimal register allocation. George et al. [17] researched the single thread-based register allocation for IXP. They took many features of IXP into account, however, did not consider the important feature of IXP—multiple threads. Zhuang et al. [34] studies register allocation across threads on IXP and doesn't deal with the bank conflicts problem. Zhou [8] proposed a framework for allocation of registers of IXP, improving the work in [33] to some extent.

3. The IXP register allocation problems

The General Purpose Registers (GPRs) in an IXP network processor are physically split into two banks: Bank A and Bank B. The dual banks cause *bank conflict* and *bank imbalance* problems that have to been considered in register allocation for IXPs.

3.1. Bank conflict

For each instruction with two GPR source operands, one must come from Bank A and another from Bank B. The destination is not constrained. Certain allocation of GPR source operands may cause a conflict. For example, consider the three instructions sequence in Fig. 1(a). The first instruction above requires that A and B must be on different banks. The second instruction requires that A and C must also be on different banks. The above two allocations constrain that B and C must be allocated to the same

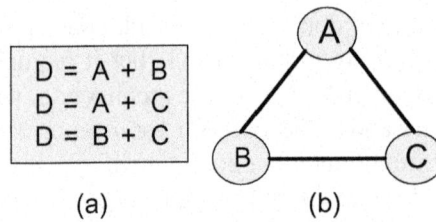

Fig. 1. Bank Conflict in IXP.

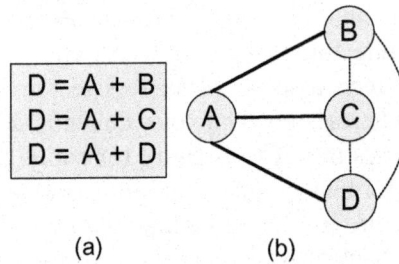

Fig. 2. Bank Imbalance in IXP.

bank since they are both in a bank different from that of A. However, the third instruction conflicts with this since it requires B and C to be allocated on different banks. Two variables are said to *conflict* with each other if they are specified in the same instruction as source operands.

These allocation problems can be understood and solved by some simple graphical analysis. An example of such a graph, called bank allocation graph (BAG) [12], is shown in the Fig. 1(b), corresponding to instruction sequence in Fig. 1(a). The edges in BAG indicate that the variables conflict with each other. These edges are called *conflict edges*. Whenever a cycle is formed with an odd number of edges, dual bank allocation is impossible [12], like in Fig. 1(a). It means that we have to resolve some bank conflicts. If the BAG is *odd-cycle-free*, we say it's *conflict-free*.

A simple solution provided in the IXP assembler is to inserts an instruction to copy one of the registers to a temporary register and replaces the register reference with a reference to the temporary register. The assembler attempts to minimize the number of inserted instructions, but it is error-prone and ineffective. A compiler solution is more efficient.

3.2. Bank imbalance

Another issue of register allocation in IXP is how to balance register workload in the two banks. The graph in Fig. 2(b) is the conflict graph of code segment in Fig. 2(a). The bold lines represent a conflict edge, while the normal solid lines represent that the two variables it connects merely interference with each other but do not conflict with each other. The BAG consists of the bold lines and the vertices incident with them. Obviously, BAG is a subgraph of the IG.

Assume there be two registers in each bank and B, C and D interfere with each other. The instruction sequence in Fig. 2(a) requires A in one bank and B, C and D in another. Nevertheless, the clique formed by B, C and D is not 2-colorable. So they cannot be allocated in the same bank with only 2 registers. This may make the assembler to issue a "*no enough registers*" error message although there may be a register available in the opposite bank. To deal with this, a compiler solution is preferred as well.

4. Fixed-parameter uniform 2-way graph partition

4.1. Bipartite graph and balanced bipartite graph

A bipartite graph is a basic conception in various register allocation algorithms. A bipartite graph is described as follows.

Bipartite Graph:

Given an undirected graph $G = (V, E)$. G is a bipartite graph if the set of vertexes E can be cut into two disjoint subsets A and B, and two vertexes, connected by each edge, locate at the subsets A and B respectively.

Bipartite graph prevents register allocation from bank conflicts because conflicted variables are assigned in different banks. To balance tasks in different banks, however, register allocation for IXP needs to find the balanced bipartite graph, which minimizes the tasks' difference between two subsets of a graph. A balanced bipartite graph is described below.

Balanced bipartite graph:

Given a partite graph $G = (V, E)$ with two subset G1$\subseteq$G and G2$\subseteq$G such that G1$\cup$G2=G. Let $|V1|$ and $|V2|$ be the numbers of vertexes in G1 and G2, respectively. We call G is a balanced bipartite graph if and only if $||V1| - |V2|| \leqslant 1$.

4.2. Fixed-parameter tractable graph bipartition

Many problems in compiler technique are NP-complete, e.g. register allocation via graph coloring. Approximate solutions are given by heuristic algorithms. Known algorithms for these problems to achieve an *optimal* solution require exponential time related to the size of input. Existence of efficient and exact algorithms is considered unlikely. However, this view has been challenged by *parameterized complexity* [27]. Some problems can be solved by algorithms that are exponential only in the size of a fixed parameter while polynomial to the input size. Such algorithms are called fixed-parameter tractable algorithms. These algorithms are of high practical interest for they are able to provide *optimal* solutions within small runtime for small parameter values.

The classic problem, *graph bipartition*, is NP-complete. Guo et al. [15] in 2005 provided an algorithm solving the graph bipartition problem by deleting vertices edge deletion version that runs in $O(2^k \cdot m^2)$ time, where m, n and k is the number of edges, vertices and number of edges to delete respectively. It means that *graph bipartite* problem is FPT. Their result show that the algorithm finishes within minutes for some graph instances with $n = 300, m = 1500$ and $k = 40$. The graph bipartition problem can be formally described as below:

Graph Bipartition:

Given an undirected graph $G = (V, E)$ and a nonnegative integer k, find a subset $C \subseteq E$ of edges (or vertexes) with $|C| \leqslant k$ such that after removal of C, the resulting subgraph $G' \subseteq G$ is bipartite. C is called *cut set*.

The Guo's algorithm gives a C with $|C| \leqslant k$ or proves that such a sub set C does not exist. The idea is to use a routine to compress the know solution iteratively. Given a size-$(k + 1)$ solution, the routine either computes a size-k solution or proves that there is no size-k solution.

We present a graph uniform 2-way partition algorithm by improving Guo's algorithm so that it can deal with a weighted graph and give a minimum weighted cut set, which is the total weights of edges in it is minimal, with size not more than k. The resulting graph is called *maximal bipartite graph*. We call the algorithm as FPT because it is a fixed-parameter tractable graph bipartition, where the key idea is how to maximize the uniform 2-way graph partition.

Maximal uniform 2-way graph partition:

Given a graph $G = (V, E)$ with weights on its edges and the number of vertexes, partition the vertexes of G into two subsets of equal size (or the difference is within 1) that the sum of weights of edges connecting different subset is maximal. We call the process for finding a balanced bipartite graph as maximum uniform 2-way graph partition.

It is equivalent to make the graph bipartite into two equal size with minimal cost since the remaining weights of edges will be maximal. The proof is trivial and we ignore it here. We just show how to get *uniform 2-way partite graph* from a *maximal bipartite graph*.

As for a bipartite graph, it has the property that its vertexes can be divided into 2 disjoint subsets such that all its edges have one endpoint in one set and another endpoint in the other set. Assume the maximum bipartite graph we get in last subsection is divided into such 2 disjoint subset, say $LEFT = \{L_1, L_2, \ldots, L_i\}$ and $RIGHT = \{R_1, R_2, \ldots, R_j\}$ as shown in Fig. 3. The bold lines represent edges in E', which called *connecting set*, and the normal solid lines represent edges in C. Note that C is not part of the maximum bipartite graph. Then we have the following definitions and claim:

Definitions:

A *flip* is a movement of a vertex from *LEFT* to *RIGHT* or vice versa.

$W\{L_x, \{L^*\}\}$ is the sum of weights of edges between L_x and all the nodes in set *LEFT*.

$W\{L_x, \{R^*\}\}$, $W\{R_x, \{L^*\}\}$ and $W\{R_x, \{R^*\}\}$ is defined similarly.

$W\{E\}$ is the sum of weights of edges in set E.

Claim: $\forall \quad L_x \in LEFT$ and $R_x \in RIGHT$, we have

$W\{L_x, \{R^*\}\} - W\{L_x, \{L^*\}\} \geqslant 0$ and

$W\{R_x, \{L^*\}\} - W\{R_x, \{R^*\}\} \geqslant 0$.

Proof: Suppose for an L_x in *Left*, $W\{L_x, \{R^*\}\} - W\{L_x, \{L^*\}\} < 0$, then if we flip L_x to *Right*, the edges $\{L_x, \{R^*\}\}$ become part of the new cut set C' and $\{L_x, \{L^*\}\}$ become part of the connecting set, $E' = E - C'$. We have $W\{C'\} = W\{C\} + W\{L_x, \{R^*\}\} - W\{L_x, \{L^*\}\} < W\{C\}$. It contradicts with the fact $W\{C\}$ is the minimum cut set. $W\{R_x, \{L^*\}\} - W\{R_x, \{R^*\}\} \geqslant 0$ can be proved similarly.

This means that any flip will not decrease the weight of the new cut set.

Assume $|LEFT| \leqslant |RIGHT|$. To balance the two sets, i.e. to make $||RIGHT| - |LEFT||$ minimal, we have to flip $q = (|RIGHT| - |LEFT|)/2$ vertexes from *RIGHT* to *LEFT*.

We can only choose the q vertexes $\{N_1, N_2, \ldots \ldots, N_q\}$ such that

$\{W\{R_x, \{L^*\}\} - W\{R_x, \{R^*\}\} < W\{R_y, \{L^*\}\} - W\{R_y, \{R^*\}\} \mid x \in [1, q], y \in [q + 1, |RIGHT|]\}$

because every flip will no decrease the weight of new cut set. This approach assures the weight of new cut set to be minimal. This operation takes $O(q) < O(|V|)$ time.

4.3. A heuristic algorithm for balanced bipartite graph

The FPT algorithm presented in the last subsection can get optimal solutions, however, it will spend much more time if values of input parameters become large because the uniform 2-way graph partition is a NP-complete problem. So, we also design a heuristic algorithm for efficiently generating bipartite graph by modifying the K-way balanced graph algorithm proposed by Lee [5]. The key idea is to how to convert the balanced bipartite graph into the maximal cut set, which can be reduced to balanced bipartite graph with ease. We formally define the maximal cut set as follows.

Maximal cut set:

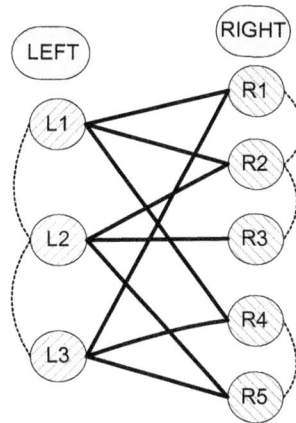

Fig. 3. A bipartite graph.

Given a undirected graph $G = (V, E)$ and *weights of all edges. We partition the set of* vertexes into two disjoint subsets A and B to maximize the sum of weights of edges such that two vertexes of each edge are in different subset, i.e.,

Maximal cut set $C = \{E_{ij} | \max \sum_{i \in A, j \in B} w_{ij}\}$, where W_{ij} is the weight of the edge between vertexes i and j.

The heuristic algorithm uses the conception gain. Let a graph G be partitioned into two subsets S_1 and S_2, and $W(V_i, V_j)$ be the weight of the edge between vertexes V_i and V_j. The gain that vertex V_i moves from S_1 to S_2 is formally defined as:

$$g(v, S_2) = \sum_{v_i \in S_1} W(v, v_i) - \sum_{v_j \in S_2} W(v, v_j), \forall v \in S_1 \tag{1}$$

The heuristic algorithm is illustrated in Fig. 4. At the beginning, states of all vertexes in S_1, and S_2 are empty. The algorithm runs in a number of passes. In each pass, we take the result of last pass as the input, and move a vertex to S_2 such that the movement will increase the cut set. The selected candidate is the vertex with the maximal gain. A pass finishes if all vertexes are tried. Then, we take the result with the maximal cut set as the input of the next pass. The algorithm repeats until no new gain can get. Whenever a vertex V_r is moved, we need to recalculate the gains of all non-moved vertexes. The time complexity of each pass is $O(2|V|^2)$.

$$g(v_a, S_2) = g(v_a, S_2) - 2W(v_a, v_r), \forall v_a \in S_1$$
$$g(v_b, S_1) = g(v_b, S_1) + 2W(v, v_r), \forall v_b \in S_2 \tag{2}$$

4.4. Graph separation

At this point, we get a maximal uniform 2-way graph partition algorithm that runs in $O(2^k \cdot |V|^2) + O(|V|) = O(2^k \cdot |V|^2)$ time, where k is the most number of edges need to be deleted to make the graph bipartite. However, there remains one more issue. Generally, the input graphs are large and may not be able to be made bipartite with a small value of k. Fortunately, network processing programs are relative small. Moreover, the BAGs are typically sparse so there is good chance that we can apply the algorithm

```
input:  G = {V, E}
output: Part[|V|], i.e., vertex set of bipartite graph
variable:  Part[|V|]: integer
        Gain[|V|][2]: real // Gain[i][j] = g(v_i, V_j}
        State[|V|]: bool // false = unused, true = used
        History[|V|][2]: integer
        Temp[|V|]: integer
Algorithm:
```

1. $Part[i] = 0$

2. $for\ i\ =\ 0\ to\ |V| - 1\ do$
   $State[i] = 0;$
   $for\ j\ =\ 0\ to\ 1\ do$
      $Gain[i][j]\ =\ g(v_i, V_j)$
   od
od

3. $for\ i\ =\ 0\ to\ |V| - 1\ do$
   3.1 select unused v_d such that $g(v_d, V_l) = \max_{j,p}(Gain[j][p]); // l \in \{0,1\} v_d \in V_m$
   3.2 $State[d] = 1;$
   3.3 $Part[d] = l;$
   3.4 $History[i][0] = d;$
       $History[i][1] = m;$
   3.5 $Temp[i] = Gain[d][l];$
   3.6 $for\ j\ =\ 0\ to\ |V| - 1\ do$
        $if\ State[j] = 1$
           $continue;$
        $if\ Part[j] = m$
           $Gain[j][l] = Gain[j][l] - 2W(v_d, v_j);$
        $else\ if\ Part[j] = l$
           $Gain[j][m] = Gain[j][m] + 2W(v_d, v_j);$
       od

4. select t to maximize $G = \sum_{j=1}^{t} Temp[j];$

5. $if\ (G > 0)$
      $for\ j\ =\ t + 1\ to\ |V|$
         $Part[History[j][0]] = History[j][1];$
      $goto\ 2;$

Fig. 4. Heuristic balanced bipartite graph algorithm.

on them after pruning the input graph according to the properties of bipartite graph. The basic idea is that we delete an edge e if deleting the edge e separates the graph into 2 disconnected components. For example, deleting the edge *e* in Fig. 5 will divide the graph into 2 disconnected components G1 and G2. These edges are called *edge separators*. Such edges certainly may not induce an odd cycle. The optimal solution of the original graph will be the union of those of G1 and G2.

The graph separation is very useful to get balanced bipartite graph because smaller graphs will have

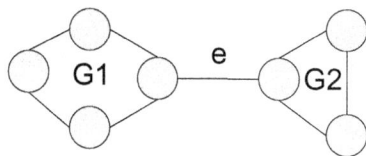

Fig. 5. Example of a graph separator.

a smaller size of cut set so that our FPT algorithm will possibly may used. At the same time, the runtime spent on separated graphs will be significantly reduced. Supposed the graph is equally divided, the runtime will decrease by 50% of the original time, i.e., from $O(2^k \cdot |E|^2)$ to $O(2 \cdot 2^k \cdot (|E|/2)^2) = O(2^{k-1} \cdot |E|^2)$. Moreover, the subgraphs may contain less edges than the k we predefine, which further reduces the runtime.

To find such separators, we choose an edge, delete it from the graph and then use a depth-first-search (DFS) algorithm to test if there is a path from its one end to another. If there is no such a path, the edge is a separator. It takes $O(|E^2|)$ time to test all the edges. We color the graph with 2 colors during the DFS process. If we find a bipartite connected subgraph, we delete it from G since it is conflict-free. This can decrease the size of the graph. If the graph is still too big that there may be little chance we can make it bipartite by a small number of k edges, we will perform more complex pruning proposed in [7]. At the worst, if a cut set of size smaller than k cannot be found, we use the heuristic [1] to provide an approximate solution at polynomial time instead.

5. Balanced bipartite graph based register allocation for IXP

The algorithms presented in Section 4 can generate balanced bipartite graphs for the register allocation. The FPT algorithm is able to work out an optimal solution but spends more time. Therefore, the FPT can work well only for programs with small sizes. For large programs, instead, we may use the heuristic algorithm. In this section, we investigate how to assign registers for IXPs.

Register allocation for IXPs needs to consider not only assigning registers but also assigning banks. In summary, there are three kinds of methods: first registers next banks (FRNB), first banks next registers (FBNR) and mixed registers and banks (MRB). FBNR is easy to implement and can resolve bank conflicts effectively. However, FBNR can not judge the register pressure in advance so that it has a difficulty in balancing the register pressure, which often causes a lot of spilling codes [33]. On the other hand, MRB cannot effectively coordinate the synchronization between the register allocation and the bank allocation. The worst situation is all adjacent vertexes of a vertex are conservatively assigned in a bank, which also cause a lot of unnecessary spilling codes.

We use the first method, i.e., FRNB, because it can reduce spilling codes as much as possible. Possibly, FRNB allocation will result in more bank conflicts than FBNR. But FRNB outperforms to FBNR because code spilling needs to access memory and the cost for resoling bank conflicts is much less than that for code spilling. The framework of register allocation is shown in Fig. 6. Each phase will be presented in the following subsections, where we assume the total register number be N and each bank have N/2 registers.

5.1. Register allocation

We firstly allocate registers since it reduce the total number of code spilling, which is more expensive than the bank conflict. Interference graph is built then virtual registers are mapped into physical register

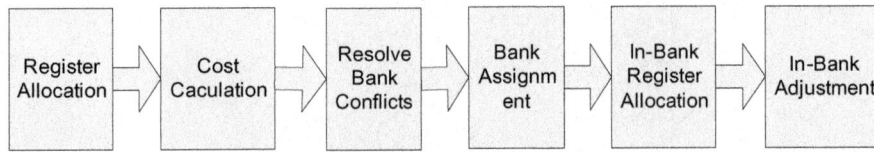

Fig. 6. Register allocation framework for IXP.

$$1.\ B1 = A1$$

$$A2 = A1 + A2 \Rightarrow 2.\ A2 = B1 + A2$$

Fig. 7. Insert a move instruction.

$$1.\ A1 = A1 \oplus B1$$
$$2.\ B1 = A1 \oplus B1$$
$$//now\ B1 = A1_{org}\ ,\ A1 = A1_{org} \oplus B1_{org}$$
$$A2 = A1 + A2 \qquad 3.\ A2 = B1 + A2$$
$$4.\ B1 = A1 \oplus B1$$
$$5.\ A1 = A1 \oplus B1$$
$$//\ now\ A1 = A1_{org}\ ,\ B1 = B1_{org}$$

Fig. 8. In-place bank exchange.

assuming a single monolithic register bank with traditional register allocation algorithm [9]. The resulting interference graph is N-colorable and each vertex presents a physical live range (PLR). Here physical live range is the maximal du-ud chain with respect to the physical register names. They can contain different virtual live ranges that reside in the same physical register.

5.2. Cost model

Before present the cost model, we first introduce two methods for resolving bank conflict.

1) If the register pressure of the program point at which the PLRs conflict is less than N/2, we use a move instruction to resolve the conflict. For example, in the left instruction in Fig. 7, A1 conflicts with A2. A move instruction, $B1 = A1$, is inserted, and B1, which is a register on the opposite bank of A, replaces A1 in the original instruction. Such a B1 can certainly be found since the register pressure is less than N/2, which means there is at less one register available in either bank.

2) When the number of registers is more than N/2-1, one of the bank may not have unused registers. If both the banks have at least one available register, we can resolve conflicts by a move instruction as in case (a). If not, it is possible that the opposite bank, against which the PLRs reside in, has no available register. In this case, we use a tricky technique called in-place bank exchange to avoid spills, instead of inserting a move instruction. For example, in the left instruction in Fig.8, A1 conflicts with A2. Two XOR instructions are inserted before the original one, exchanging the values of A1 and B1. Then B1 replaces A1 in the original instruction. The last two instructions restore the values of A1 and B1.

3) As to BAG, these are equivalent to delete an edge between two vertexes, since they no longer conflict. Theoretically, a new node should be added into the BAG. Nevertheless, it is not necessary since the new node only conflict with one node and such a node will never induce an odd cycle.

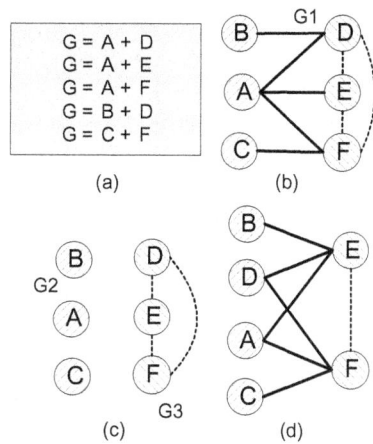

Fig. 9. Flip vertex between banks.

To estimate the cost of resolving PLRs conflicts, we define the following quantities:

1. $conf_i$ is the ith conflict of two PLRs
2. $confwt_i$ is the cost of resolving $conf_i$. In the case (a) mentioned above, it is taken to be the cost of the move instruction, say 1. In the case (b), we don't simply set the cost to be the 4 XOR instructions, say 4, since PLRs conflict with less register pressure have better changes to be resolved as in case (a). Instead, the cost is taken to be $2^{(2p/N-2)}$, where p is the register pressure of the conflict point. Its range is [1,4] when p is within [N/2, N].

The cost to resolve all the conflicts within two PLRs is

$$\sum_i confwt_i \cdot 8^{depth(conf_i)}$$

5.3. Conflict resolving and bank balance

In this step, we build the weighted BAG, whose edges have weights representing the cost to remove it from the graph, i.e. the cost to resolve the conflict between two PLRs. To make a graph bipartite, we can delete vertexes and edges or split nodes. However, we cannot delete nodes here since PLRs cannot be deleted. We employ the uniform 2-way graph partition algorithm presented in Section 4 to make the graph bipartite and balance with the minimal cost. If k passes the threshold, we invoke a heuristic algorithm.

5.4. Bank assignment

At this time, bank conflict graph has been converted into balanced bipartite graph. A connected graph will keep connected after it is reduced into a balanced bipartite graph. As a result, we can use just one way to color the BAG with 2 colors. The vertexes get the same color will be assigned to the same bank.

5.5. Register assignment in a bank

Now the conflicts are resolved and the banks are assigned. The next step is to assign registers in-bank. We perform it with traditional register allocator. Though the graph is balanced, there still may be a few

$$\begin{array}{lllll}
 & 1.\ B1 = A1 & & 1.\ B1 = A1 \\
A2 = A1 + A2 & 2.\ A2 = B1 \oplus A2 & & 2.\ \mathbf{A2} = B1 + \mathbf{A2} \\
\dots\ \dots \quad\Rightarrow & \dots\ \dots \quad\Rightarrow & \dots\ \dots \\
A3 = A1 + A3 & 3.\ B2 = A1 & & 3.\ \mathbf{A3} = B1 + \mathbf{A3} \\
 & 4.\ A3 = B2 \oplus A3
\end{array}$$

Fig. 10. Eliminate a move instruction.

$$\begin{array}{lll}
 & 1.\ A1 = A1 \oplus B1 \\
 & 2.\ B1 = A1 \oplus B1 \\
 & 3.\ \mathbf{A2} = B1 + \mathbf{A2} & 1.\ A1 = A1 \oplus B1 \\
 & 4.\ B1 = A1 \oplus B1 & 2.\ B1 = A1 \oplus B1 \\
A2 = A1 + A2 & 5.\ A1 = A1 \oplus B1 & 3.\ \mathbf{A2} = B1 + \mathbf{A2} \\
\dots\ \dots \quad\Rightarrow & \dots\ \dots \quad\Rightarrow & \dots\ \dots \\
A3 = A1 + A3 & 6.\ A1 = A1 \oplus B2 & 4.\ \mathbf{A3} = B1 + \mathbf{A3} \\
 & 7.\ B2 = A1 \oplus B2 & 5.\ B1 = A1 \oplus B1 \\
 & 8.\ \mathbf{A3} = B2 + \mathbf{A3} & 6.\ A1 = A1 \oplus B1 \\
 & 9.\ B2 = A1 \oplus B2 \\
 & 10.\ A1 = A1 \oplus B2
\end{array}$$

Fig. 11. Eliminate 4 XOR instructions.

cases that the registers cannot be successfully allocated in-bank. For example, for the instructions in Fig. 9(a), assume vertexes A, B and C get bank A while vertexes D, E and F get bank B in Fig. 9(b). Each bank has two registers. The original graph is divided into two subgraphs induced by the vertexes in the same bank respectively. For example, G1 is divided into G2 and G3 as shown in Fig. 9(c). We try to color each subgraph with two colors but fail to do so in bank B. Then some PLRs have to be "spilled" to the opposite bank, which will induce new conflicts. The vertex with least spill cost is to be spilled with the highest priority. The vertex to be spilled is the one with least flip cost as calculated in Section 4.2 and the new conflicts are resolved as described in Subsection 5.2. Vertex B is spilled since it has the least spill cost = 1. The final interference graph is shown in Fig. 9(d), assuming D be "spilled".

5.6. Redundant instruction elimination

A insert instruction at every conflict point may results in redundancy. For example, assume there are unused register in the opposite bank of bank A, and A1 and B1 is not redefined between the two instructions in the left of Fig. 10. In this case, two move instructions may be inserted as shown in the middle of Fig. 11. We can remove the third instruction and rename B2 to B1 in the fourth instruction.

If the register pressure is high, there is more redundancy. Also assume A1 and B1 are not redefined between the original two instructions. Four instructions can be eliminated and the result is shown in the right of Fig. 11. Simple dataflow algorithm can perform the eliminations easily.

6. Experiments and evaluation

We developed a simulation system oriented to mobile and wireless applications using Intel IXP simulation workbench. Each mobile node with an IXP processor connects with others using wireless

Table 1
System configuration parameters

Parameters	Value
K	32
The number of total threads	4
The number of registers in a bank	32
The number of used threads	1

Table 2
Benchmark properties

Applications	Number of instructions
IPSec_forward	6194
10GB_Ethernet	1781
L2_forward	2405
Firewall	566
MPLS	1025

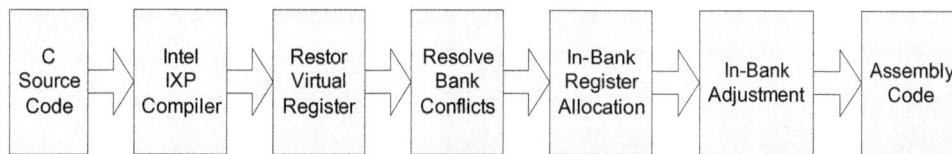

Fig. 12. Compilation flowchart.

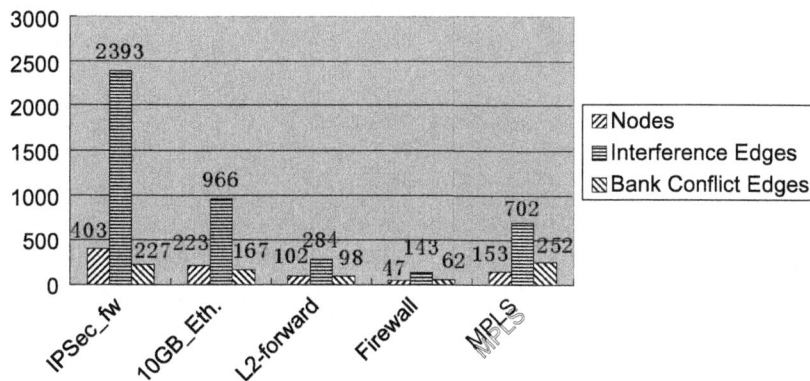

Fig. 13. Number of conflict edges and bank conflict edges.

mesh communication. The node was configured according to Table 1. The benchmarks were compiled using the C compiler for Intel IXP to generate assembly codes, restored in all the virtual registers, and assigned registers with our algorithm, as shown in Fig. 12. Then conflict graph and BAG was constructed from the assembly codes. The construction of BAG is easy since IXP only has 80 instructions. Table 2 shows the properties of the benchmark applications used in our experiments.

We used five real network applications, whose parameters are listed in Table 2 and features are shown in Fig. 13. The average vertex degree is 9.6 in the interference graph and 1.2 in bank conflict graph, respectively.

We tested our algorithms FPT and Heuristic algorithm using the above five benchmarks in terms of

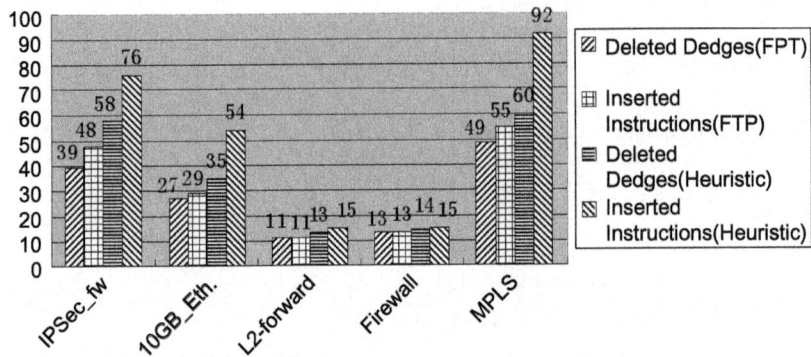

Fig. 14. Results without redundancy elimination.

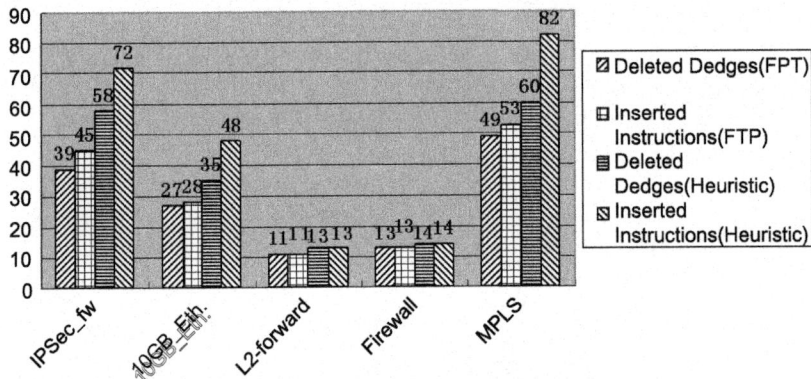

Fig. 15. Results with redundancy elimination.

the following performance metrics.

The number of deleted edges and inserted instructions. In the experiments, we tested the number of edges and instructions that need to be deleted and inserted respectively for IXP register allocation. In Fig. 14, redundant instructions in tested programs were not deleted, while Fig. 15 shows the experimental results after the redundant instructions were removed from the programs. The difference between Figs 14 and 15 is a little because live range in network applications is small, and there are only 2 or 3 conflicts. By comparison, the algorithm FPT outperformed the Heuristic one in terms of whether the number of deleted edges or inserted instructions. In FPT, the number of deleted edges always approximately equals to the number of inserted instructions, which means that FPT can solve conflicts by inserting less instructions. At the same time, Heuristic algorithm needs approximate number of deleted edges and inserted instructions for small graph such as Firewall. For Heuristic algorithm, however, the inserted instructions grow more quickly than the deleted edges with the increase of the size of programs.

Throughput. Typically, a network application consists of an unconditional loop, which continuously handles and forwards incoming packets, as shown in Fig. 16. We used OC48 (2.5 Gb/s) engine to test the throughput of benchmarks using different register allocation algorithms. The results are shown in Fig. 17 and Fig. 18. We found that the two algorithms all provide around 3.5 Gb/s network bandwidth. The processing speed may improve by 2.7% after deleting redundant codes, and the FPT is 2.5% better than the Heuristic algorithm.

```
for ( ; ; )
{
    Rx_Packet(); //receive data packets;
    . . .
    IPv4_Fwd(); //forward data packets;
    . . .
    Tx_Packet(); //send data packet;
}
```

Fig. 16. A typical network processing application.

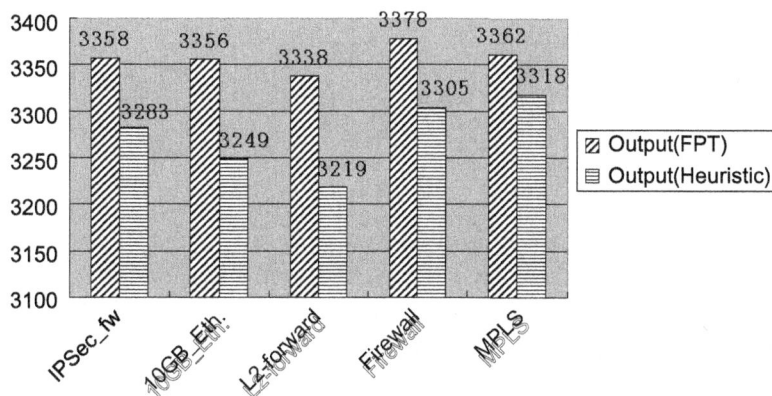

Fig. 17. Throughput without redundancy elimination.

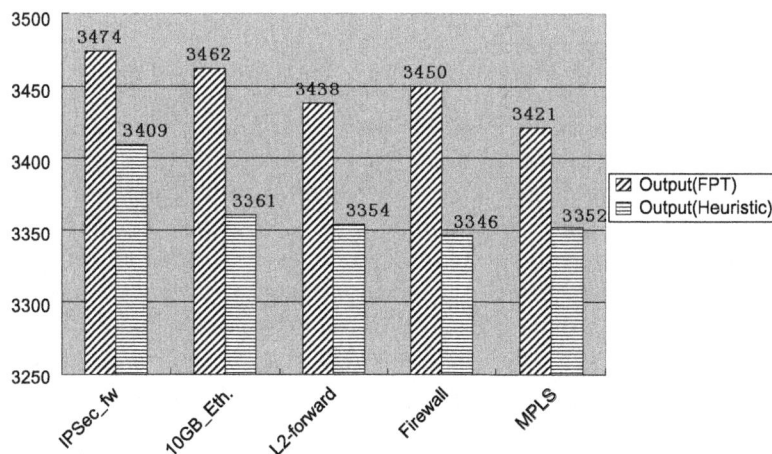

Fig. 18. Throughput with redundancy elimination.

Comparison with Post-RA. We tested the inserted instructions for our FPT algorithm, Heuristic algorithm and Zhuang's Post-RA [9], which firstly assigns registers and then banks. We tested them using the ratio of inserted instructions to the number of conflict edges in order to avoid the affects of different

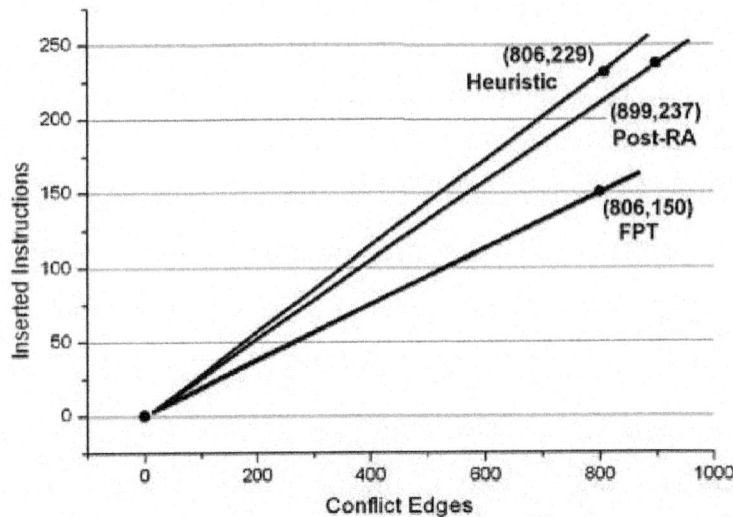

Fig. 19. Comparison among FPT, heuristic and post-RA.

benchmarks. The results are shown in Fig. 19, where Y-axis represents the sum of inserted instructions of all the benchmarks and X-axis represents the sum of total conflict edges. Our FPT algorithm needs fewer insert instructions to resolve more conflicts since FPT takes the interaction of edges as a whole and provide optimal solution instead of approximate solution. Besides, the bank pressure is almost balanced since we take a uniform graph partition algorithm. Note that the data in Fig. 19 is the sum of inserted instructions and conflict edges for all five benchmarks.

From the Fig. 19, we can draw a conclusion that the FPT is the best one in all three algorithms in terms of the ratio of inserted instructions to conflict edges. Post-RA is better than Heuristic, but worse than our FPT. Heuristic is the worst one in this experiment.

7. Conclusion

Mobile and wireless networks need to provide sufficient bandwidth to improve users' transparent experience and meet increasingly QoS demands of the users. Network processors, e.g., IXP, oriented to mobile and wireless networks typically have many registers, where effective register allocation is very important for the efficient computing ability and thus high network throughput. For this purpose, we have presented a graph uniform 2-way partition algorithm (FPT), which can work out an optimal solution to the graph partition, and a heuristic algorithm, which can get approximate results, for the generation of balanced bipartite graph. Based on the algorithms, we designed a framework for two banks based IXP register allocation. Experimental results demonstrate the framework and the algorithms are efficient in register allocation for IXP network processors. Further, our FPT algorithm is practical and effective in solving certain NP-complete problems in embedded systems. By employing such an algorithm to give an optimal solution, we can consider the BAG as a whole and reduce the inserted instructions for conflict resolving.

Acknowledgement

Feilong Tang would like to thank The Japan Society for the Promotion of Science (JSPS) and The University of Aizu (UoA), Japan for providing the excellent research environment during his JSPS Post-doctoral Fellow Program (ID No. P 09059) in UoA, Japan.

This work was supported by the National High Technology Research and Development Program (863 Program) of China (Nos. 2006AA01Z172 and 2008AA01Z106), and the National Natural Science Foundation of China (Nos. 60773089, 60533040 and 60725208).

References

[1]   A. Agarwal, M. Charikar, K. Makarychev and Y. Makarychev, $O(\sqrt{\log n})$ approximation algorithms for min UnCut, min 2CNF deletion, and directed cut problems, *Proceedings of the 37th STOC*. ACM Press, 2005, pp. 573–581.

[2]   A. Durresi and M. Denko, Advances in mobile communications and computing, *Mobile Information Systems* 5(2) (2009), 101–103.

[3]   A.W. Appel and L. George, Optimal spilling for CISC machines with few registers, *Proceedings of the ACM SIGPLAN 2001 Conference on Programming Language Design and Implementation*, 2001, pp. 243–253.

[4]   B.C. Cheng, H. Chen and R.Y. Tseng, Context-Aware Gateway for Ubiquitous SIP-Based Services in Smart Homes, *Proceedings of International Conference on Hybrid Information Technology (ICHIT '06)*, 2006, pp. 374–381.

[5]   C.H. Lee, M. Kim and C.I. Park, An efficient k-way graph partitioning algorithm for task allocation, in parallel computing systems, Proceedings of the First International Conference on Systems Integration, 1990, pp. 748–751.

[6]   D.J. Kolson, A. Nicolau, N. Dutt et al., A method for register allocation to loops in multiple register filearchitectures, *Proceedings of The 10th International Parallel Processing Symposium (IPPS '96)*, pp. 28–33.

[7]   F. Hüffner, N. Betzler and R. Niedermeier, Optimal edge deletions for signed graph balancing, *Proceedings of the 6th Workshop on Experimental Algorithms (WEA'07)*, 2007, pp. 297–310.

[8]   F. Zhou, J.C. Zhang, C.Y. Wu and Z.Q. Zhang, A Register Allocation Framework for Banked Register Files with Access Constraints, *Proceedings of The 10th Asia-Pacific Computer Systems Architecture Conference*, 2005, pp. 269–280.

[9]   G.J. Chaitin, Register allocation and spilling via graph coloring, *Proceedings of the SIGPLAN Symposium on Compiler Construction* 39(4) (1982), 66–74.

[10] G. Mino, L. Barolli, F. Xhafa, A. Durresi and A. Koyama, Implementation and performance evaluation of two fuzzy-based handover systems for wireless cellular networks, *Mobile Information Systems* 5(4) (2009), 339–361.

[11] G.Y. Lueh, T. Gross and A. Adl-Tabatabai, Fusion-based register allocation, *ACM Transactions on Programming Languages and Systems*, 2000, pp. 431–470.

[12] IXP 2800 Network Processor: Programmer's Reference Manual, Dec. 2001. http://www.intel.org.

[13] J. Abella and A. González, On Reducing Register Pressure and Energy in Multiple-Banked Register Files, *Proceedings of IEEE International Conference on Computer Design (ICCD'03)*, 2003, pp. 14–20.

[14] J. Ahn, Novel directory service and message delivery mechanism enabling scalable mobile agent communication, *Mobile Information Systems* 4(4) (2008), 333–349.

[15] J. Guo, J. Gramm, F. Hüffner, R. Niedermeier and S. Wernicke, Improved fixed-parameter algorithms for two feedback set problems, *Proceedings of 9th Workshop on Algorithms and Data Structures (WADS'05)*, 2005, pp. 158–168.

[16] J. Fu, O. Hagsand and G. Karlsson, Queuing Behavior and Packet Delays in Network Processor Systems, *Proceedings of the 15th International Symposium on Modeling, Analysis, and Simulation of Computer and Telecommunication Systems*, 2007, pp. 217–224.

[17] L. George and M. Blume, Taming the IXP network processor, *ACM SIGPLAN Conference on Programming Language Design and Implementation*, 2003, pp. 26–37.

[18] J. Hiser, S. Carr, P. Sweany and S.J. Beaty, Register assignment for software pipelining with partitioned register banks, *Proceedings of 14th International Parallel and Distributed Processing Symposium (IPDPS 2000)*, pp. 211–217.

[19] J. Park, J. Lee and S. Moon, Register Allocation for Banked Register File, *Proceedings of the 2001 ACM SIGPLAN workshop on Optimization of middleware and distributed systems*, pp. 39–47.

[20] J. Wagner and R. Leupers, C compiler design for a network processor, *IEEE Transactions on Computer-Aided Design of Integrated Circuits and Systems* 20(11) (2001), 1302–1308.

[21] M. Adiletta, M. Rosenbluth, D. Bernstein et al., The Next Generation of Intel IXP Network Processors, *Intel Technology Journal* 6(3) (2003), 6–18.

[22] M.K. Chen, X.F. Li, R. Lian et al., Shangrila: Achieving high performance from compiled network applications while enabling ease of programming, *ACM SIGPLAN Conference on Programming Language Design and Implementation*, 2005, pp. 224–236.

[23] M. Safar, H. Sawwan, M. Taha and T. Al-Fadhli, Virtual social networks online and mobile systems, *Mobile Information Systems* 5(3) (2009), 233–253.

[24] P. Briggs, K. Cooper and L. Torczon, Rematerialization, *Proceedings of ACM SIGPLAN Conference on Programming Language Design and Implementation*, 1992, pp. 311–321.

[25] P. Brink, M. Castelino, D. Meng, C. Rawal and H. Tadepalli, Network processing performance metrics for IA- and IXP-Based Systems, *Intel Technology Journal* **7**(4) (2003), 77–91.

[26] R. Collins, F. Alegre, X.T. Zhuang and S. Pande, Compiler assisted dynamic management of registers for network processors, *Proceedings of 20th International Parallel and Distributed Processing Symposium (IPDPS 2006)*, 2006, 10 pages.

[27] R.G. Downey and M.R. Fellows, Parameterized Complexity, *Springer*, 1999.

[28] P. Montesinos, W. Liu and J. Torrellas, Using register lifetime predictions to protect register files against soft errors, *Proceedings of the 37th Annual IEEE/IFIP International Conference on Dependable Systems and Networks (DSN '07)*, 2007, pp. 286–296.

[29] P. Raghavan, A. Lambrechts, M. Jayapala et al., Very Wide Register: An Asymmetric Register File Organization for Low Power Embedded Processors, *Proceedings of Design, Automation & Test in Europe Conference & Exhibition (DATE '07)*, 2007, pp. 1–6.

[30] S. Izumi, K. Yamanaka, Y. Tokairin et al., Ubiquitous supervisory system based on social contexts using ontology, *Mobile Information Systems* **5**(2) (2009), 141–163.

[31] S. Jang, S. Carr, P. Sweany and D. Kuras, A Code Generation Framework for VLIW Architectures with Partitioned Register Files, *Proceedings of International Conference on Massively Parallel Computing Systems*, 1998.

[32] T. Monreal, A. Gonzalez, M. Valero et al. Delaying physical register allocation through virtual-physical registers, *Proceedings of the 32nd Annual International Symposium on Microarchitecture* (1999), 186–192.

[33] X. Zhuang and S. Pande, Resolving Register Bank Conflicts for a Network Processor, *Proceedings of the 12th International Conference on Parallel Architectures and Compilation Techniques (PACT 2003)*, page. 269.

[34] X. Zhuang and S. Pande, Balancing register allocation across threads for a multithreaded network processor, *Proceedings of ACM SIGPLAN 2004 Conference on Programming Language Design and Implementation (PLDI 2004)*, 2004, pp. 289–300.

[35] Z. Liu, H. Che, K. Zheng et al. A trace driven study of the effectiveness of cache mechanism for network processors, *Proceedings of International Conference on Wireless Communications, Networking and Mobile Computing* **2** (2005), 978–981.

10

Comprehensive protection of data-partitioned video for broadband wireless IPTV streaming

Laith Al-Jobouri, Martin Fleury* and Mohammed Ghanbari
Department of CSEE, University of Essex, Essex, UK

Abstract. This paper examines the threat to video streaming from slow and fast fading, traffic congestion, and channel packet drops. The proposed response is a combination of: rateless channel coding, which is adaptively applied; data-partitioned source coding to exploit prioritized packetization; and duplicate slice provision, which is the focus of the evaluation in this paper. The paper also considers the distribution of intra-refresh macroblocks as a means of avoiding sudden data rate increases. When error bursts occur, this paper shows that duplicate slices are certainly necessary but this provision is more effective for medium quality video than it is for high quality video. The percentage of intra-refresh macroblocks can be low and still reduce the impact of temporal error propagation.

Keywords: Data-partitioning, error resilience, FEC, IPTV, rateless coding, redundant packets, video streaming, WiMAX

1. Introduction

Broadband wireless delivery of video streaming for IPTV is under active investigation, though this environment is challenging due to signal noise and interference, leading to error bursts. In the time division duplex (TDD) form of multiplexing favored by broadband wireless access networks such as IEEE 802.16e (mobile WiMAX) [10], Automatic Repeat reQuests (ARQs) come almost for free by virtue of the return sub-frame. Therefore, a form of type-1 hybrid ARQ is possible. To achieve this, the paper introduces an adaptive variant of rateless channel coding which utilizes such retransmissions. To avoid the long start-up delays that can arise from packet erasure correction, error correction in the proposed protection scheme is at the byte level. However, byte-level channel coding is powerless against packet drops, as obviously a packet can be lost through buffer overflow before forward error correction (FEC) can be applied. Outright packet drops on the channel can also occur, for example if the signal strength fails to reach a desired receiver threshold. Therefore, further protection is required, which in this paper is provided by duplicate video packets. The paper tests the effectiveness of different levels of duplication according to the source coding scheme used.

We have also employed video data-partitioning [19] which is a form of error-resilient source coding in which the compressed video data are reorganized according to decoding priority. In an H.264/AVC (Advanced Video Coding) codec [35], the data are partitioned between packets rather than re-ordered within a packet (refer to Section 2 for details). In particular, motion vectors (MVs) are packed into a partition-A slice packet, allowing error concealment to reconstruct missing partition-C slices containing

*Corresponding author: Martin Fleury, Department of CSEE, University of Essex, Wivenhoe Park, Colchester CO4 3SQ, Essex, UK. E-mail: fleum@essex.ac.uk.

texture data (quantized transform coefficient residuals). One advantage of this arrangement is that, as partition-A packets can be relatively smaller than the other packets (according to the compression ratio) then these packets are less likely to be corrupted by channel noise or dropped at buffers.

It is also possible to use redundant picture slices [27] rather than duplicate slices. Redundant picture slices employ a higher quantization parameter (QP) and, hence, coarser quantization than the original slices. However, to use redundant picture slices would cause additional drift between the encoder and the decoder, if (say) a partition-A packet was matched with a partition-C packet's data with a different QP. Besides, there is an implementation issue. Though both redundant slices and data-partitioned slices co-exist in the Extended profile, they are not jointly implemented in the JM implementation of H.264/AVC [31] and, in fact, appear not to be implemented at all in most other software codec implementations such as QuickTime, Nero, and LEAD randomly to name a few. However, it is possible with data-partitioning through repeated runs of the encoder to create a duplicate stream of all partition-A slice packets or a duplicate stream consisting of partition-A and partition-B packets or, indeed, a replica of the original stream. Unequal error protection (UEP) is also associated with data partitioning, for example by means of hierarchical modulation in [4]. Nevertheless, the combined congestion and channel error protection in the proposed scheme negates the need for UEP and Equal Error Protection (EEP) is employed instead.

The feasibility of video streaming for broadband access has been established [11] by means of a live WiMAX testbed. However, that study [11] concentrated on varying the configuration parameters and used UDP-transported streams, seemingly without congestion or error control. Other work in [18] was primarily concerned with combining true- and near-video-on-demand, providing a solution to how content should be allocated between the two services. In [9], adaptive multicast streaming was proposed using the Scalable Video Coding (SVC) extension for H.264. Fixed WiMAX channel conditions were monitored in order to vary the bitrate accordingly. Unfortunately, the subsequent decision of the JVT standardization body for H.264/AVC *not* to support fine-grained scalability (FGS) implies that it will be harder to respond to channel volatility. Other work has also investigated combining scalable video, with multi-connections in [14] and in comparison with H.264/AVC in [5]. However, the data-dependencies between layers in H.264/SVC medium grained scalability are a concern, as, unlike in FGS, enhancement layer packets may successfully arrive but be unable to be reconstructed if key pictures also fail to arrive. Instead, data-partitioning in this paper can be viewed as a simplified form of SNR or quality layering [24] which can be protected against congestion drops through duplicate slices. Temporal scalability is avoided because of its impact on media synchronization, especially lip synchronization.

There remains the problem of channel errors, which in this paper are guarded against in two ways. Firstly, to reduce temporal error propagation in H.264/AVC it is possible to place intra-refresh macroblocks (MBs) amongst the normally inter-coded MBs of P-pictures. These MBs are placed with naturally encoded inter-coded MBs within partition-B slices. We have investigated the correct way to provision intra-refresh MBs. Because such intra-coded MBs are not as efficient to code as when predictive inter-coding is applied, the maximum level of intra-refresh MBs should be assessed to avoid excess overhead.

Secondly, a more important form of protection is the use of application-layer FEC, which in this paper is through rateless channel coding [23]. Raptor coding [29] in particular is a systematic variety of rateless code that does not share the high error floors [25] of prior rateless codes. It also has O(n) decoding computational complexity. A point to note is that rateless codes are a probabilistic channel code in the sense that reconstruction is not guaranteed. Because Raptor coding is indeed rateless, it is possible to adaptively vary the amount of redundant data according to an estimate of the channel conditions. It is

Table 1
NAL unit types

NAL unit type	Class	Content of NAL unit
0	–	Unspecified
1	VCL	Coded slice
2	VCL	Coded slice partition A
3	VCL	Coded slice partition B
4	VCL	Coded slice partition C
5	VCL	Coded slice of an IDR picture
6–12	Non-VCL	Suppl. info., Parameter sets, etc.
13–23	–	Reserved
24–31	–	Unspecified

also possible to piggyback additionally generated redundant data if the estimate is insufficient to allow reconstruction of the packet or the belief propagation decoding algorithm fails to run to completion. However, further details are reserved for Section 4.

We have investigated video streaming in the context of IPTV. IPTV services include live TV program-on-demand, which can be 'start-again' TV, as well as 'catch-up' TV [7], both of which require unicast streaming. Video-on-demand can also be provided as a part of an IPTV service. However, if this form of IPTV is to be extended to mobile devices then wireless access [36] presents a challenging environment for video streaming, especially if sport scenes are streamed, as these often contain rapid motion, leading to high temporal coding complexity. Error bursts can disrupt the fragile compressed bitstream (because of predictive coding dependencies), while error control should not increase end-to-end latency because of display and decode deadlines. However, as intelligent content management of IPTV moves popular material nearer to the end-user [34], it becomes feasible to combine application-layer channel coding with retransmission of additional redundant data without incurring prohibitive delays to real-time video display.

The remainder of this paper is organized as follows. Section 2 discusses the two forms of error resilience source coding employed in the proposed protection scheme. The following Section briefly introduces rateless channel coding. Section 4 discusses the protection schemes active features: channel coding adaptation, retransmission requests, and slice duplication. Section 5 describes the simulation configuration used to evaluate the scheme. Section 6 provides results according to duplicate packet and intra-refresh provision. Finally, Section 7 draws some conclusions.

2. Error resilience measures

This Section describes the two source coding error resilience measures used to protect the data before channel coding.

3. Data-partitioning

The H.264/AVC codec conceptually separates the Video Coding Layer (VCL) from the Network Abstraction Layer (NAL). The VCL specifies the core compression features, while the NAL supports delivery over various types of network. In a communication channel, the quality-of-service is affected by the two parameters of bandwidth and the probability of error. Therefore, as well as video compression efficiency, which is provided for through the VCL layer, adaptation to communication channels should be carefully considered. The concept of the NAL, together with the error-resilience features in H.264/AVC, allows communication over a variety of different channels. Table 1 is a summarized list of different NAL unit types. NAL units 1 to 5 contain different VCL data that will be described later. NAL units 6 to 12 are non-VCL units containing additional information such as parameter sets and supplemental information.

Each frame in H.264/AVC can be divided into several slices; each of which contains a flexible number of MBs. Variable Length Coding (VLC) that is entropy coding of the compressed data takes place as the final stage of the hybrid codec. In H.264/AVC, arithmetic coding replaced other forms of entropy coding in earlier codecs. In each slice, the arithmetic coder is aligned and its predictions are reset. Hence, every slice in the frame is independently decodable. Therefore, slices can be considered as points of decoder resynchronization that prevent error propagation to the entire picture. Each slice is placed within a separate NAL unit (see Table 1). The slices of an Instantaneous Decoder Refresh- (IDR-)[1] or I-picture (i.e. a picture with all intra slices) are located in type 5 NAL units, while those belonging to a non-IDR or I-picture (P- or B-pictures) are placed in NAL units of type 1, and in types 2 to 4 when data-partitioning mode is active, as now explained.

In type 1 and type 5 NALs, MB addresses, MVs and the transform coefficients of the blocks, are packed into the packet, in the order they are generated by the encoder. In Type 5, all parts of the compressed bitstream are equally important, while in type 1, the MB addresses and MVs are much more important than the (integer-valued) Discrete Cosine Transform (DCT) coefficients. In the event of errors in that type of packet, the fact that symbols appearing earlier in the bit-stream suffer less from errors than those which come later[2] means that bringing the more important parts of the video data (such as headers and MVs) ahead of the less important data or separating the more important data altogether for better protection against errors can significantly reduce channel errors. In the standard video codecs, this practice is known as data partitioning.

However, in H.264/AVC when data partitioning is enabled, every slice is divided into three separate partitions and each partition is located in either of type-2 to type-4 NAL units, as listed in Table 1. A NAL unit of type 2, also known as partition A, comprises the most important information of the compressed video bit stream of P- and B-pictures, including the MB addresses, MVs, and essential headers. If any MBs in these pictures are intra-coded, their DCT coefficients are packed into the type-3 NAL unit, also known as partition-B. Type 4 NAL, also known as partition-C, carries the DCT coefficients of the motion-compensated inter-picture coded MBs.

In the context of H.264/AVC data-partitioning, it should be pointed out that though partition-A is independent of partitions B and C, *constrained intra prediction* should be set [24] to make partition-B independent of partition-C. However, partition-C cannot be made independent of partition-B if the Context-Adaptive Variable Length Coding (CAVLC) option is set. As data-partitioning is currently only enabled in the H.264/AVC Extended profile, whereas the alternative Context Adaptive Binary Arithmetic Coding (CABAC) option is not available in the Extended profile, it seems that partition-C is inevitably dependent on partition-B.

3.1. Intra-refresh macroblocks

In the proposed robust scheme, an issue is how to protect against the temporal error propagation that

---

[1] An IDR picture is confusedly equivalent to an I-picture in previous standards. An I-picture in H.264/AVC allows predictive references beyond the boundary of a GOP.

[2] Because of the cumulative effect of VLC, symbols nearer the slice synchronization marker suffer less from errors than those that appear later in a bitstream.

can occur whenever predictively-coded P-frames are lost. A traditional way to do this is to insert periodic, intra-coded I-frames, usually every 12 or 15 frames that is every half-second according to frame rate. The spatially-encoded macroblocks (MBs) of the I-frame halt any temporal error propagation and act as anchor points for a future set of frames. Unfortunately, the insertion of I-frames leads to sudden data transmission increases due to the coding inefficiency of spatial referencing. It can also lead to buffering delays as the multiple packets associated with I-frames are formed and transmitted. Therefore, distributed insertion of intra-refresh MBs is an alternative. In the JM codec implementation of the H/264/AVC standard, two main methods of distributed insertion are available: either random placement of IR MBs within each frame; or placing a line of intra-refresh MBs within each P-frame on a cyclic basis.

In the latter forced intra-refresh method, the line size can be increased to a region or slice [32] in order to control the rate that the total picture area is refreshed. Against this suggestion must be balanced the overhead from including a complete line or region of MBs, as such MBs are costly to encode. In fact, in this paper it is suggested that despite the apparent advantages of the cyclic line method, at least in respect to data-partitioned video compression, random selection of intra-refresh MBs is preferable. Random intra refresh may on occasion select existing intra-coded MBs but it has the advantage that in the JM implementation of H.264/AVC the overhead from intra-coded MBs is readily controllable. This is because intra-refresh MBs are not the only form of intra-coded MBs, as encoders will insert such MBs when new areas of a picture are revealed, as may especially happen when there is rapid motion within a sequence. Thus, the forced intra-refresh method does not account for areas of the region that may already have been intra-coded. It also has another weakness in that future motion prediction may occur from regions of a prior picture yet to be refreshed. This defect can be remedied, possibly by restricting the range of the MVs within the refreshed region [17] or by observing the direction of motion within a sequence [28]. However, these alternatives [17,28] add coding complexity to the intra-refresh process and may be unnecessary when processing a low-motion video sequence. Another possibility is to adaptively alter the extent of MB provision [21] according to scene content and channel conditions. This is most suited to live encoding and is not a general method.

One advantage of forced intra-refresh is that it provides a natural channel-swapping (or zapping) point at the start of each refresh cycle, just as periodic I-frames provided. In catch-up TV systems, for example the BBC's iPlayer, such a facility is not required, as no channels are swapped. If for channel swapping, gradual decoder refresh is performed instead of periodic refresh then growing the refresh area from an isolated region is preferable to random MB refresh, as the subjective visual effect is better. This facility was proposed to the Joint Video team developing the H.264/AVC codec [21] but is not currently implemented, perhaps because a method of signaling a switching point is the subject of a patent.

A point to note is the different way that random intra-refresh MBs are specified in the H.264/AVC JM 14.2 implementation compared to that of cyclic IR line intra update. For random intra-refresh MBs, a maximum percentage of intra-refresh can be specified, which percentage includes existing naturally encoded intra-coded MB. If the given quota of intra-refresh MB is already largely occupied by naturally-encoded intra-coded MBs, then only a small amount of extra randomly inserted MBs will be added. In contrast, if a line of intra-refresh MBs is inserted then these MBs are added in addition to those intra-coded MBs that have already been included by the encoder.

Figure 1 is a comparison between the relative sizes of the partitions according to QP for the video clip which is employed in Section 6's evaluation. The test sequence was *Football*, which is a scene with rapid movement and consequently has a high temporal coding complexity. This sequence is of a content type that quality of experience subjective testing indicates currently provides a difficult viewing

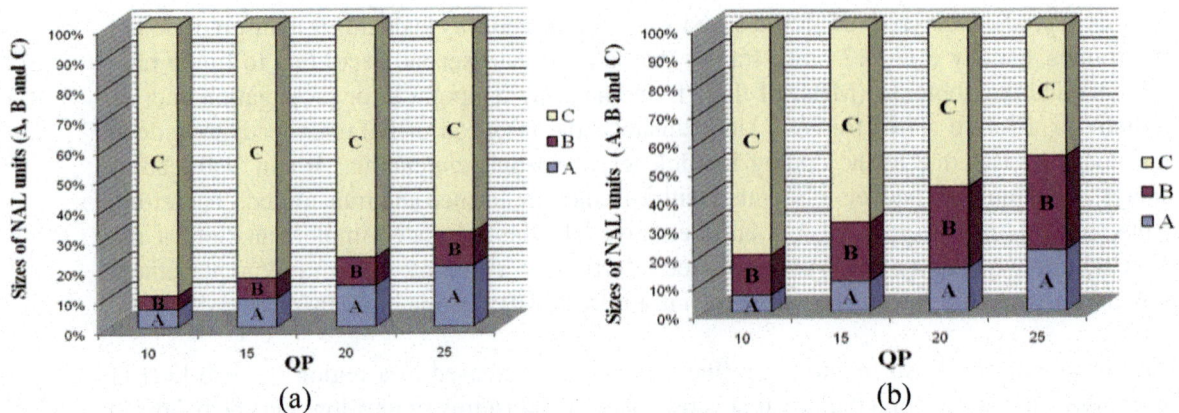

Fig. 1. Relative sizes of data partitions according to quantization parameter (QP) for the Football video sequence, with (a) 5% intra-coded refresh MBs, and (b) 25% intra-coded refresh MBs.

experience [1] on mobile devices. *Football* was Variable Bit Rate (VBR) encoded with 4:2:0 chroma sub-sampling at Common Intermediate Format (CIF) (352 × 288 pixels/picture), with a Group-of-Pictures (GOP) structure of IPPP... at 30 Hz, i.e. one initial I-picture followed by all predictive P-pictures. This arrangement removes the complexity of bi-predictive B-pictures at a cost in an increased bit-rate. Notice that the range of QP in H.264/AVC is 0–51 with higher values corresponding to higher compression ratios and lower quality video. From the Figure it is apparent that the partition-B's contribution increases as the percentage of intra-refresh MBs grows, making partition-B packets more vulnerable to congestion and channel error as a result. The size of partition-C bearing packets declines with increasing QP as a result of coarser quantization of transform coefficients.

Table 2 is a more complete comparison of the actual sizes of the partitions' contributions according to QP for the *Football* sequence. From the Table, again it is apparent that, as the percentage of intra-refresh MBs is increased, the size in bytes of partition-B increases for the same QP. Because more MBs are assigned to partition-B, the size of partition-C reduces. Because of the relatively large amount of naturally intra-encoded MBs, this effect is gradual until 25% of random intra-refresh MBs is added. A point to notice is that the total size of the stream declines significantly with coarser quantization due to a higher QP. Because intra-coding is less efficient, the total sizes increase as the percentage of intra-refresh MBs is increased, making the percentage of intra-refresh MBs a significant parameter to examine in any transmission scheme.

4. Rateless channel coding

We have used rateless channel coding, specifically Raptor coding, in the protection scheme. Rateless or Fountain coding [23], of which Raptor coding [29] is a subset, is ideally suited to a binary erasure channel in which either the lower-layer error-correcting code works or the channel decoder fails and reports that it has failed to the application layer. In erasure coding, flawed data symbols may be reconstructed from a set of successfully received symbols (if sufficient of these symbols are successfully received). The class of Fountain codes allows a continual stream of additional symbols to be generated in the event that the original symbols could not be decoded. It is the ability to easily generate new symbols that makes Fountain codes rateless. Decoding will succeed with small probability of decode failure if any of $k (1 +$

Table 2

Size of different data partitions' contribution in byte, as a mean over Football's frames at various QP for various levels of intra-refresh MBs

QP	2% Intra-refresh MB			
	A	B	C	Total
20	1842	2678	3889	8409
25	1687	1697	2533	5917
30	1459	1047	1496	4002
35	1117	572	688	2377

QP	5% Intra refresh MB			
	A	B	C	Total
20	1845	2767	3867	8479
25	1690	1763	2511	5964
30	1463	1082	1482	4027
35	1120	595	682	2397

QP	6% Intra-refresh MB			
	A	B	C	Total
20	1846	2810	3850	8506
25	1696	1793	2502	5991
30	1467	1098	1479	4044
35	1123	604	681	2408

QP	25% Intra-refresh MB			
	A	B	C	Total
20	1893	3450	3669	9012
25	1746	2216	2379	6341
30	1505	1346	1405	4256
35	1146	729	646	2521

QP	MB Line Intra Update			
	A	B	C	Total
20	1885	3385	3683	8953
25	3683	2160	2400	8243
30	1498	1312	1414	4224
35	1143	716	652	2511

ε) symbols are successfully received, where k is the number of source symbols and ε is a low percentage of coding overhead. For example, in [2] for rateless coding of Internet video streams that overhead was 5%. In its simplest form, the symbols are combined in an exclusive OR (XOR) operation according to the order specified by a randomized, low-density generator matrix and, in this case, the probability of decoder failure is $\partial = 2^{-k\varepsilon}$, which for large k approaches the Shannon limit. The random sequence must be known to the receiver but this is easily achieved through knowledge of the sequence seed.

In general, encoding of rateless codes is accomplished as follows: Choose d_i randomly from some distribution of degrees, where $\rho_{di} = Pr[\text{degree } d_i]$; Pr is the probability of a given event. Choose d_i random information symbols R_i among the k information symbols. These R_i symbols are then XORed together to produce a new composite symbol, which forms one symbol of the transmitted packet. Thus, if the symbols are bytes then all of the R_i byte's bits are XORed with all of the bits of the other randomly selected bytes in turn. It is not necessary to specify the random degree or the random symbols chosen if it is assumed that the (pseudo-)random number generators of sender and receiver are synchronized.

Symbols are processed at the decoder as follows. If a symbol arrives with degree greater than one it is buffered. If a clean symbol arrives with degree one then it is XORed with all symbols in which it was used in the encoding process. This reduces the degree of each of the symbols to which the degree-one symbol is applied. When a symbol is eventually reduced to degree one, it too can be used in the decoding process. Notice that a degree-one symbol is a symbol for which no XORing has taken place. Notice also that for packet erasure channels a clean degree-one symbol (a packet) is easily established as such. For byte-erasures the physical layer FEC can be reasonable expected to isolate clean symbols or blocks of clean symbols.

In the decoding process, the robust Soliton distribution [23] is employed as the degree distribution, as this produces degree-one symbols at a convenient rate for decoding. It also avoids isolated symbols that are not used elsewhere. Two tuneable parameters c and δ serve to form the expected number of useable degree one symbols. Set

$$S = c \ln \left(\frac{k}{\delta} \right) \sqrt{k} \tag{1}$$

where c is a constant close to 1 and δ is a bound on the probability that decoding fails to complete. Now define

$$
\begin{aligned}
\tau(d) &= \frac{S}{k} \frac{1}{d} \text{ for } d = 1, 2, \ldots (k/S) - 1 \\
&= \frac{S}{k} \ln \left(\frac{S}{\delta} \right) \text{ for } d = k/S \\
&= 0 \text{ for } d > k/S
\end{aligned}
\tag{2}
$$

as an auxiliary positive-valued function to give the robust Soliton distribution:

$$\mu(d) = \frac{\rho(d) + \tau(d)}{z} \tag{3}$$

where z normalizes the probability distribution to unity and is given by:

$$z = \sum_d (\rho(d) + \tau(d)). \tag{4}$$

Then, in order to model Raptor coding, we employed the following statistical model [22]:

$$
\begin{aligned}
P_f(m, k) &= 1 && \text{if } m < k, \\
&= 0.85 \times 0.567^{m-k} && \text{if } m \geqslant k,
\end{aligned}
\tag{5}
$$

where $P_f(m, k)$ is the failure probability of the code with k source symbols if m symbols have been received. Notice that the authors of [22] remark and show that for $k > 200$ the model almost perfectly models the performance of the code. In the protection scheme, the symbol size was set to bytes within a packet. Clearly, if instead 200 packets are accumulated before the rateless decoder can be applied (or at least Eq. (5) is relevant) there is a penalty in start-up delay for the video stream and a cost in providing sufficient buffering at the mobile stations.

5. Protection scheme

This Section establishes the protection scheme in the context of WiMAX wireless access.

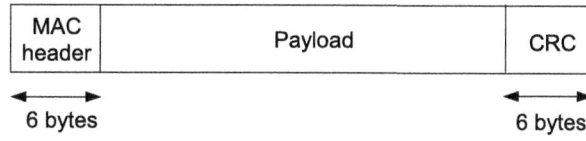

Fig. 2. General format of a MAC PDU with optional CRC.

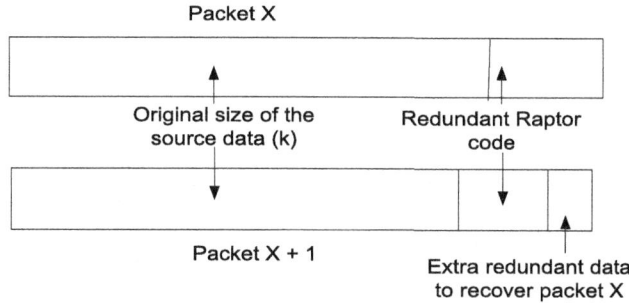

Fig. 3. Division of payload data in a packet (MPDU) between source data, original redundant data and piggybacked data for a previous erroneous packet.

5.1. Packetization

To try to ensure corrupted packets are reconstructed if the original estimated redundant data is insufficient for reconstruction, the IPTV scheme works by retransmitting piggybacked redundant data. To reduce latency, the number of such retransmissions, after an ARQ over the uplink, was limited to one. The application-layer receives only clean symbols (bytes) from any corrupted packets. This latter function is performed by PHYsical-layer FEC which passes up correctly received blocks of data (through a Cyclic Redundancy Check (CRC)) but suppresses erroneous data. Thus, in IEEE 802.16e [3], a binary, non-recursive convolutional encoder with a constraint length of 7 and a native coding rate of 1/2 operates at the physical layer. Upon receipt of the correctly received data, decoding of the information symbols is attempted, which will fail with a probability given by Eq. (5) for $k > 200$.

Figure 2 shows the general format of a WiMAX packet, with an optional CRC that could be used to check successful reconstruction or not of the packet after application of the rateless code (assuming cross-layer access to the MAC CRC). Figure 3 shows how ARQ triggered retransmissions work. In the Figure, the payload of packet X is corrupted to such an extent that it cannot be reconstructed. Therefore, in packet X+1 some extra redundant data is included up to the level that decode failure is no longer practically certain. If the extra redundant data is insufficient to reconstruct the original packet's payload, the packet is simply dropped. Otherwise, of course, it is passed to the H.264/AVC decoder.

5.2. Adaptation

The instantaneous probability of byte loss, BL (taken from a probability distribution with mean BL_{mean}) is used to calculate the amount of redundant data adaptively added to the payload. The details of how BL is found are given in Section 5.2. If the original packet length is L, then the redundant data is given simply by

$$R = L \times BL + (L \times BL^2) + (L \times BL^3)\ldots$$
$$= L/(1 - BL) - L,$$

(6)

which adds successively smaller additions of redundant data, based on taking the previous amount of redundant data multiplied by *BL*.

If it turns out that the packet cannot be reconstructed, despite the provision of redundant data, then additional redundant data are added to the next packet (*Extra redundant data* in Fig. 3).

It is implied from Eq. (5) that if less than k symbols (bytes) in the payload are successfully received then a further $k - m + e$ extra redundant bytes can be sent to reduce the risk of failure. In the evaluation tests, e was set to four, resulting in a risk of failure of 8.7% (from Eq. (5)) in reconstructing the original packet if the extra redundant data successfully arrives. This reduced risk arises because of the exponential decay of the risk that is evident from Eq. (5).

5.3. Slice duplication

In the protection scheme, one or more of the data-partitions is duplicated so that dropped packets can be replaced. During bursts it is possible that a packet and its duplicate replacement are both affected by channel noise. Thus, extra redundant data are transmitted in an attempt to reconstruct the packet for the least affected of the two packets (the original and copy).

Rather than duplicating packets, as mentioned in Section 1, it is possible to send a reduced-quality version of a slice, which in H.264/AVC is referred to as a redundant picture slice [37]. However, employing a coarser quantization than in the main stream, can lead to drift between the encoder and decoder, as the encoder never knows which version of the slice has been decoded. Besides, replacing one partition with a redundant slice with a different QP to the other partitions would not even permit reconstruction. It should, however, be noted that in some circumstances, employing redundant picture slices is acceptable. For example, in [26] for an ad hoc network with routing path diversity, redundant picture slices were sent across one of the routes, as a guard against loss of packets in another route.

A possibility [37] is to use correctly-received reference pictures for reconstruction of redundant pictures rather than the reference pictures used by primary pictures. The decoder is able to select from a set of potential replacement redundant pictures according to the possibility of correct reconstruction. Alternatively, in [8], MBs were selected for their relative impact on reconstruction and placed within FMO slices, at some increase in computational complexity.

6. Simulation model

This Section discusses the simulation settings for the evaluation of the protection scheme in the following Section.

6.1. WiMAX modeling

To establish the behavior of rateless coding under WiMAX the well-known ns-2 simulator was augmented with a module from the Chang Gung University, Taiwan [33] that has proved an effective way of modeling IEEE 802.16e's behavior.

To evaluate the scheme, transmission over WiMAX was carefully modeled. The PHY layer settings selected for WiMAX simulation are given in Table 3. The antenna heights are typical ones taken from the standard [10]. The antenna is modeled for comparison purposes as a half-wavelength dipole, whereas a sectored set of antenna on a mast might be used in practice to achieve directivity and, hence, better performance. The IEEE 802.16 TDD frame length was set to 5 ms, as only this value is supported in

Table 3
IEEE 802.16e parameter settings

Parameter	Value
PHY	OFDMA
Frequency band	5 GHz
Bandwidth capacity	10 MHz
Duplexing mode	TDD
Frame length	5 ms
Max. packet length	1024 B
Raw data rate	10.67 Mbps
IFFT size	1024
Modulation	16-QAM 1/2
Guard band ratio	1/8
MS transmit power	245 mW
BS transmit power	20 W
Approx. range to SS	1 km
Antenna type	Omni-directional
Antenna gains	0 dBD
SS antenna height	1.2 m
BS antenna height	30 m

OFDMA = Orthogonal Frequency Division Multiple Access, QAM = Quadrature Amplitude Modulation, TDD = Time Division Duplex.

the WiMAX forum simplification of the standard [30]. The data rate results from the use of one of the mandatory coding modes [10] for a TDD downlink/uplink sub-frame ratio of 3:1. The WiMAX base station (BS) was assigned more bandwidth capacity than the uplink to allow the BS to respond to multiple mobile subscriber stations (SSs). Thus, the parameter settings in Table 3, such as the modulation type and physical-layer coding rate, are required to achieve a data-rate of 10.67 Mbps over the downlink.

In order to introduce sources of traffic congestion, an always available FTP source was introduced with TCP transport to a second SS. Likewise, a CBR source with packet size of 1000 B and inter-packet gap of 0.03 s was also downloaded to a third SS. While the CBR and FTP traffic occupies WiMAX's non-rtPS (non-real-time polling service) queue, rather than the rtPS queue, they still contribute to packet drops in the rtPS queue for the video, if the packet rtPS buffer is already full or nearly full, while the nrtPS queue is being serviced. Sender buffer sizes were set to fifty packets (WiMAX MAC Protocol Data Units (PDUs) [3]), as is normal to reduce buffer waiting time and reduce energy consumption at the SSs. For clarity of interpretation of the results, this paper does not model horizontal handovers, though it should be noted these are a real possibility that can disrupt video streams, and in [6] there is a proposal to speed-up connection set-up in WiMAX handovers. Related work on efficient key exchange during mobile roaming can be found in [15].

6.2. Channel model

A two-state Gilbert-Elliott channel model [38] was used in the physical layer of the simulation to simulate the channel model for WiMAX. To model the effect of slow fading at the packet-level, the *PGG* (probability of being in a good state) was set to 0.95, *PBB* (probability of being in a bad state) = 0.96, *PG* (probability of packet loss in a good state) = 0.02 and *PB* (probability of packet loss in a bad state) = 0.01 for the Gilbert-Elliott parameters. Additionally, it is still possible for a packet not to be dropped in the channel but, nonetheless to be corrupted through the effect of fast fading (or other sources of noise and interference). This byte-level corruption was modeled by a second Gilbert-Elliott model, with the same parameters (applied at the byte level) as that of the packet-level model except that *PB* (probability of byte loss) was increased to 0.165. The Gilbert-Elliott scheme, though it is simple, has been widely adopted [13], as it is thought to realistically model the error bursts that are apparent to the application layer and, more significantly, can be particularly damaging to compressed video streams, because of the predictive nature of source coding. Therefore, the impact of error bursts should be assessed [20] in video streaming applications.

In the adaptive scheme, the probability of channel byte loss serves to predict the amount of redundant data to be added to the payload. If *PGB* and *PBG* are the probabilities of going from good to bad state and from going from bad to good state respectively, then

$$\pi_G = PBG/(PBG + PGB) \tag{7}$$

$$\pi_B = PGB/(PBG + PGB) \tag{8}$$

are the steady state probabilities of being in the good and bad states. Consequently, the mean probability of channel corruption is given by

$$BL_{mean} = PG.\pi_G + PB.\pi_B \tag{9}$$

that is the mean of a Uniform distribution in this case.

The WiMAX standard specifies that a station should provide channel measurements that can form a basis for channel quality estimates. These are either Received Signal Strength Indicators or may be Carrier-to-Noise-and-Interference Ratio measurements made over modulated carrier preambles. In a simulation, assuming perfect channel knowledge of the channel conditions when the original packet was transmitted establishes an upper bound beyond which the performance of the adaptive scheme cannot improve. However, we have included measurement noise to test the robustness of the scheme. Measurement noise was modelled as a zero-mean Gaussian (normal) distribution and added up to a given percentage to the packet loss probability estimate.

6.3. Video configuration

The *Football* video sequence was also employed for the WiMAX downlink tests. As a GOP structure of IPPP... was employed, it is necessary to protect against error propagation in the event of inter-coded P-picture slices being lost. By default, to ensure higher quality video, 5% intra-coded MBs (randomly placed) were included in each frame (apart for the first I-picture) to act as anchor points in the event of slice loss. The JM 14.2 version of the H.264/AVC codec software was employed, with the Evalvid framework [16] used to reconstruct sequences, according to reported packet loss patterns from the simulator, and to assess the objective video quality (PSNR) relative to the input YUV raw video. Lost partition-C slice packets were compensated for by motion-copy error concealment at the decoder using the MVs in partition-A. From Section 2, for high- to medium-quality video, the size of partition-A is relatively smaller in length than the other two partitions and, therefore, less likely to suffer from channel error. Similarly, partition-B is generally smaller than partition-C, provided that the intra-refresh MB contribution is kept to a low percentage (as herein).

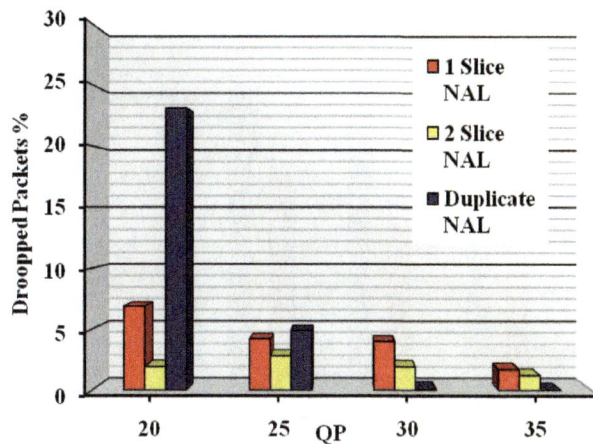

Fig. 4. Dropped packet rates when streaming *Football*, with different slicing configurations.

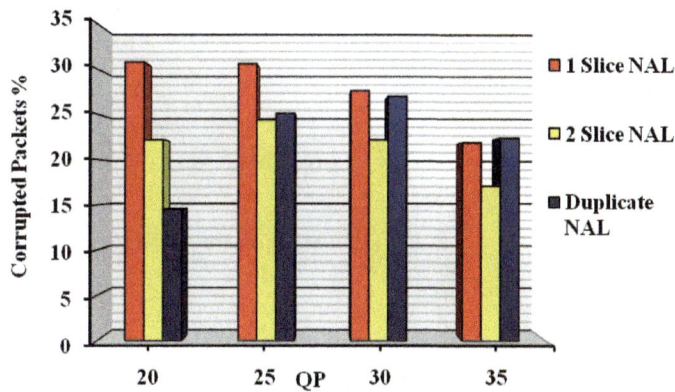

Fig. 5. Corrupted packet rates when streaming *Football*, with different slicing arrangements.

7. Evaluation

The protection scheme was evaluated for the impact of duplicate slices and the rate of intra-refresh MBs. For the former the full Gilbert-Elliott model was applied (refer to Section 5.2), while for the latter the secondary model of byte error only was applied. Three types of erroneous packets were considered: 1) packet drops at the BS sender buffer 2) packet drops through channel noise and interference, and 3) corrupted packets that were received but affected by Gilbert-Elliott channel noise to the extent that they could not be immediately reconstructed without an ARQ triggered retransmission of piggybacked redundant data. Notice that if the retransmission of additional redundant data still fails to allow the original packet to be reconstructed then the packet is simply dropped. The Raptor code Eq. (5) was applied to decide if a packet could be recovered, given the number of bytes that were declared to be in error.

7.1. Impact of duplicate packets

In Figs 4 to 8, three schemes are compared, all of which adaptively include redundant data according to the description of Section 4.2. As mentioned in Section 5.2, in the mean 5% additive Gaussian

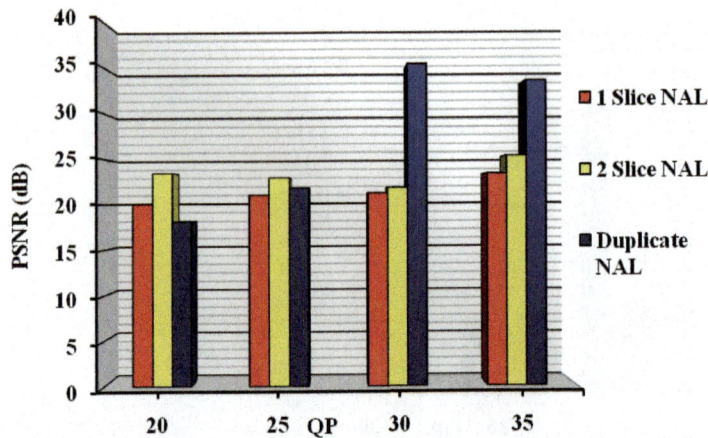

Fig. 6. Mean video quality when streaming *Football*, with different slicing arrangements.

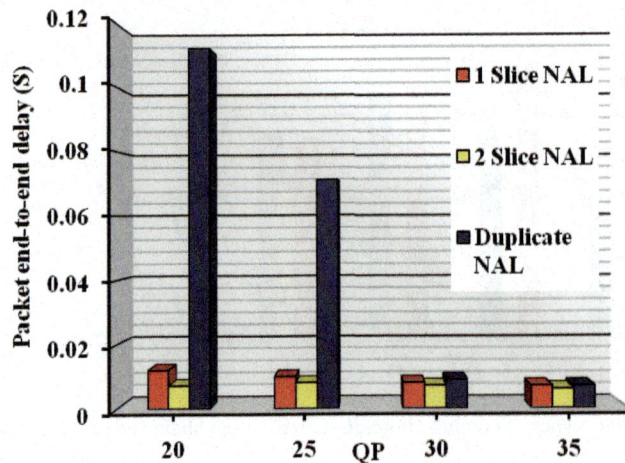

Fig. 7. Mean single transmission packet end-to-end delay when streaming *Football*, with different slicing arrangements.

measurement noise was applied to the estimate of *BL* in Eq. (5) to make the adaptive schemes more realistic. The scheme labeled *1 slice NAL* treats a picture as a single slice but creates the three data partitions (A, B, and C) described in Section 2. The scheme labeled *2 slice NAL* geometrically divides each picture into two horizontal slices. Each of the two slices is split into three data partitions and as before each forms a NAL unit. Each NAL unit occupies an IP/UDP/RTP packet, which, after packetization as a MAC Service Data Unit [3], occupies a single WiMAX MAC PDU. A number of header compression schemes are available to significantly reduce the impact of the IP/UDP/RTP overhead. The main effect of the two slice scheme is to reduce packet sizes. The scheme labeled *Duplicate NAL* also includes the duplicate slice packets, which in the following Figures means redundant partition-A, -B, and -C packets (all packets replicated). The *Redundant NAL* scheme is a single slice per picture scheme and, hence, no advantage is gained from smaller packet sizes. 5% intra-refresh data were added to each picture, increasing the size of partition-B packets (refer to Fig. 1). In these Figures, the Duplicate NAL scheme extends to all partitions. This does *not* amount to a change in bitrate because the packets are simply replicated. However, the end-to-end packet delay will obviously increase because of the interleaving of

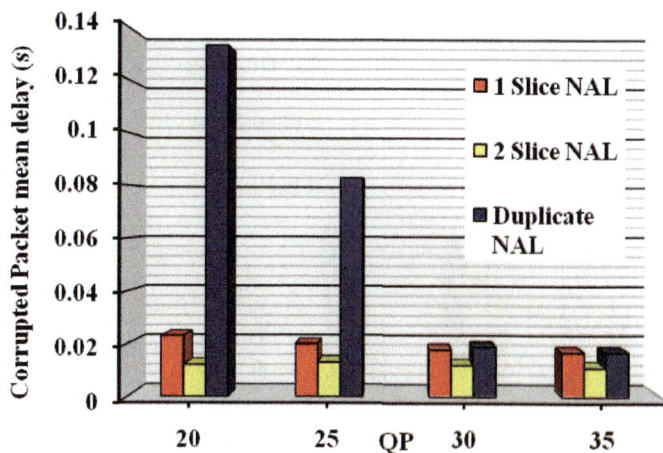

Fig. 8. Mean re-transmitted packet end-to-end delay when streaming *Football*, with different slicing arrangements.

the duplicate slice packets. Notice also that the number of packets sent for the two-slice scheme is the same as for the redundant slice packet scheme.

Though only a moderate percentage of one slice scheme packets (6.9%) are dropped outright, Fig. 4, from the combined effects of congestion and channel conditions, this results at QP = 20 to a PSNR of only 19.98 dB, Fig. 6, that is, it is at an unacceptable level. Unfortunately, though the percentage of packets dropped reduces (because of the smaller packet sizes), the sending quality is reduced at higher QPs, with the result that the objective quality at the receiver display remains below 25 dB, that is approximately equivalent to a poor rating in the ITU P.800's [12] mean opinion score scale. Packet size has a significant effect, as in the two-slice scheme video quality is increased. However, reducing the packet size is still insufficient (video quality below 25 dB) when faced with the combined effect of packet drops and channel conditions, Fig. 5.

In the duplicate NAL scheme, retransmission of extra redundant data was scheduled for all corrupted packets, even if two packets duplicated each other. This is because it is not possible to know in advance whether the extra redundant data will arrive for any one of the two packets. This provision has a significant effect in improving the video quality at higher QPs. The reason is that retransmitting extra redundant data by two alternative means increases the chance that a packet can be reconstructed. Overall video quality is approximately equivalent (above 31 dB) to the ITU P.800 'good' category at the higher QPs illustrated.

At lower QPs for the duplicate NAL scheme, higher packet drop rates occur (23.1% and 4.1%). The packet size at QP = 20 is large and the effect of sending such packets in duplication contributes to the high loss rate. Therefore, at low QP it appears that redundant slice provision is *not* an effective guard against packet drops Unfortunately also, for lower QPs total end-to-end delay for both normal, Fig. 7, and corrupted packets, Fig. 8, is high and there is a possibility of interruptions to the display, i.e. freeze frame effects. The throughput of the WiMAX channel is likely to remove the risk of display interruptions at higher QP delay times.

Figures 9 to 11 present an analysis of packet drops from packet loss alone, i.e. by application of the Gilbert-Elliot packet loss model of Section 4.2. From this analysis it is apparent that at lower QPs, packet drop numbers are distributed in inverse order of data priority. This should not be surprising as higher-priority packets (partition-A and -B packets) are smaller at lower QPs and, hence, in the mean, spend less time exposed to the channel. A surprising feature of these Figures is that in the duplicate

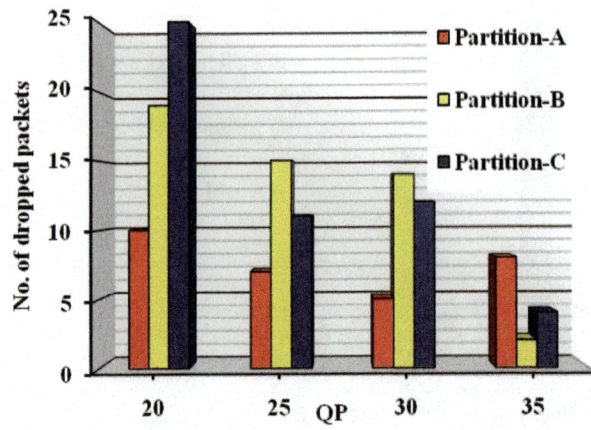

Fig. 9. Analysis of dropped packets according to partition type for one slice per picture.

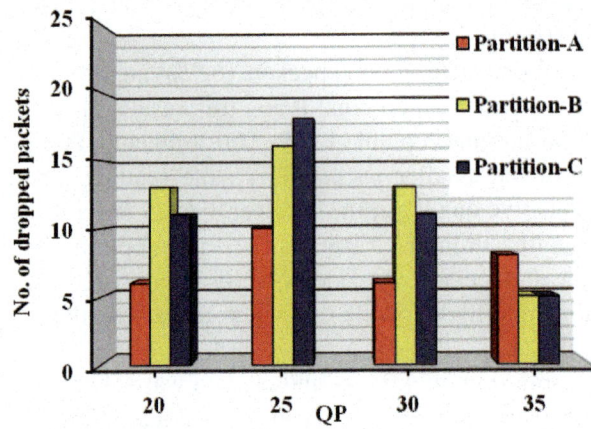

Fig. 10. Analysis of dropped packets according to partition type for two slices per picture.

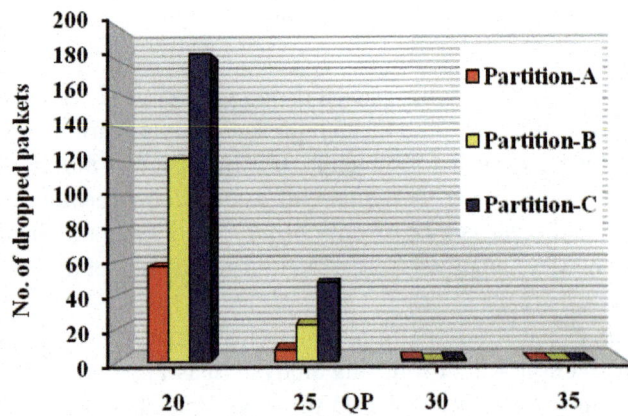

Fig. 11. Analysis of dropped packets according to partition type with duplicate slices.

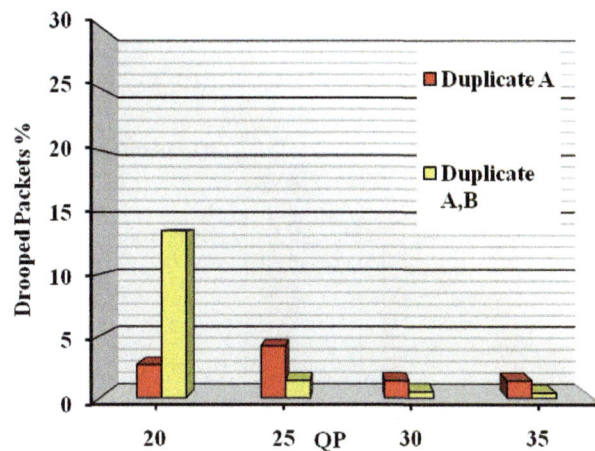

Fig. 12. Packet drop rate when streaming Football, with alternative duplicate NAL protection schemes.

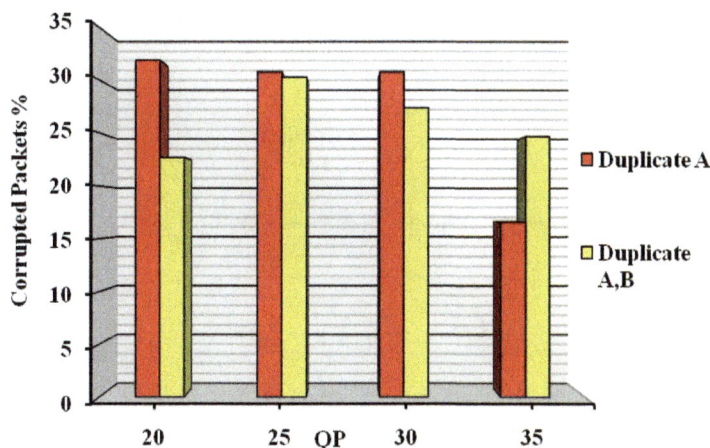

Fig. 13. Corrupted packet rate when streaming Football, with alternative duplicate NAL protection schemes.

NAL scheme at higher QPs, Fig. 11, few if any packets are dropped when traversing the channel, while at lower QPs large numbers of packets are lost through buffer overflow. Because the packet sizes are small at higher QPs, packet drops are generally less than 2% (except for QP = 30 in the one-slice NAL scheme).

Naturally it was of interest whether it was possible to employ only duplicate partition-A slice packets or only duplicate partition-A and partitition-B slice packets. These two options are examined in Figs 12 to 16. Figures 17 and 18, perform a partition-type analysis for the number of dropped packets in the selective duplication schemes and equally show that higher-priority data packets benefit from size-dependent packet losses. However, though other performance metrics are favorable, and though duplication of both partition-A and partition-B slice packets is preferable, results show that in this scenario for the *Football* video sequence, duplicate protection of partition-C is also necessary to reconstruct the stream to a satisfactory quality. We attribute this to the high motion present in the *Football* reference sequence, as this implies that succeeding frames may be very different. As a result, error concealment by motion vectors alone may be insufficient.

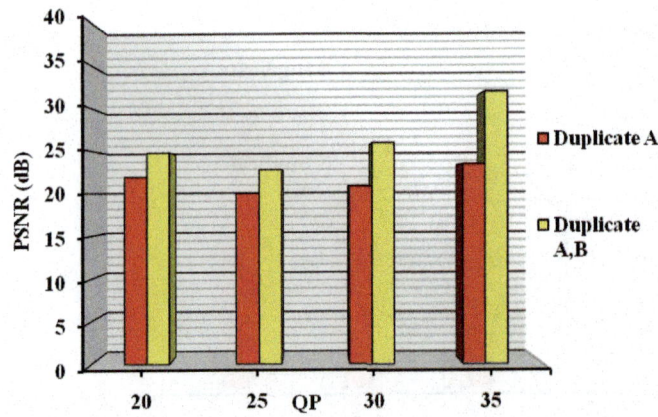

Fig. 14. Mean video quality when streaming Football, with alternative duplicate NAL protection schemes.

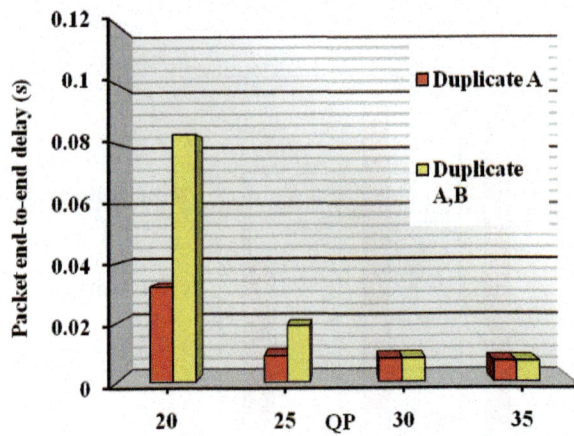

Fig. 15. Mean single transmission packet end-to-end delay when streaming Football, with alternative duplicate NAL protection schemes.

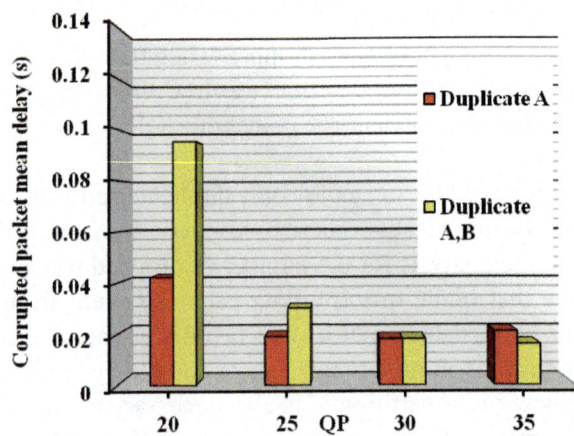

Fig. 16. Mean re-transmission packet end-to-end delay when streaming Football, with alternative duplicate NAL protection schemes.

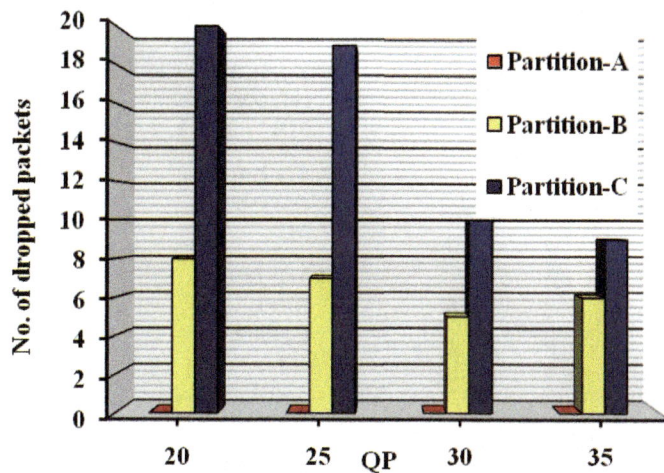

Fig. 17. Analysis of dropped packet numbers according to partition type when streaming Football, with duplicate partition-A packets.

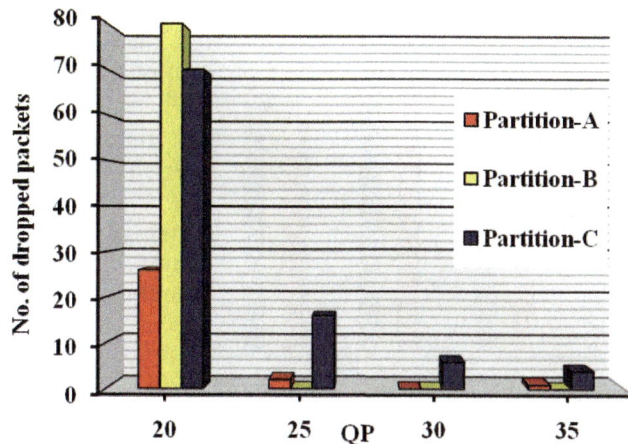

Fig. 18. Analysis of dropped packet numbers according to partition type when streaming Football, with duplicate partition-A and partition-B packets.

7.2. Impact of intra-refresh macroblock provision

Figures 19 to 21 show the result of retaining the 'duplicate NAL' scheme of Figs 4 to 6 but varying the provision of intra-refresh MBs. Increasing the provision of intra-refresh MB above 5% to 6% and higher in the case of MB line intra update, increases the throughput and, hence, the bandwidth requirements in respect to co-existing traffic. 6% rather than 5% MB refresh is chosen because without naturally encoded intra-refresh MBs, then one line of MBs corresponds to about 6% of a CIF picture. From the PSNR results it can be seen that reducing the intra-refresh MB percentage to 2% actually improves the PSNR at QPs 30 and 35. The main effect of reducing the percentage of intra-refresh MBs is that the size of partition-B-bearing packets is reduced. In turn, this makes these packets less likely to be affected by channel conditions, especially burst errors arising from the simulated Gilbert-Elliott model.

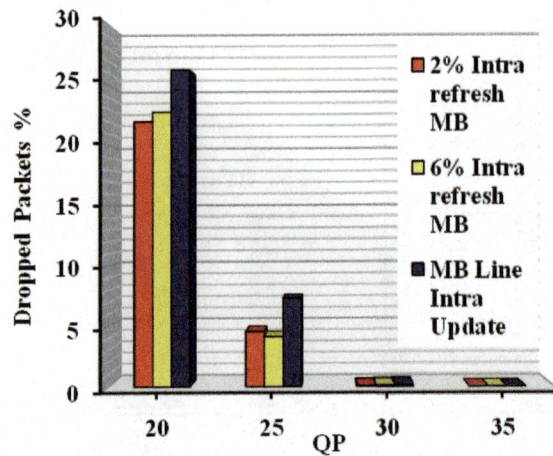

Fig. 19. Dropped packet rates when streaming *Football*, with duplicate NAL protection extending to all partitions and different levels of intra-refresh MB protection.

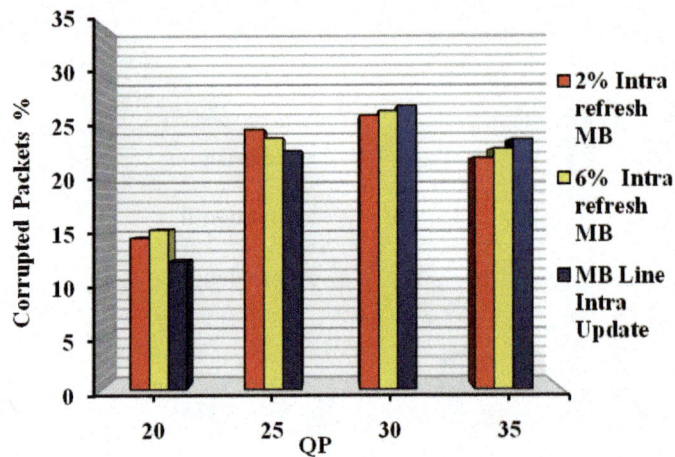

Fig. 20. Corrupted packet rates when streaming *Football*, with duplicate NAL protection extending to all partitions and different levels of intra-refresh MB protection.

8. Conclusion

Though a form of hybrid ARQ has been applied in this paper and though an advanced form of packetization (data-partitioning) was combined with adaptive rateless coding, it is still very possible that unacceptable video quality can occur when streaming over a broadband wireless channel. This is particularly so if there is high motion in the streamed video, which unfortunately includes many sports sequences. Therefore, duplication of the stream seems very necessary in harsh channel conditions, though if the channel estimation indicates that conditions are benign, then the duplicate stream can simply be turned off. The careful choice of quantization parameter for VBR video is necessary, as too low a choice will result in large packets that are prone to loss or corruption in one of several ways. A low level of intra-refresh provision is adequate and the low level results in reduced risk of packet loss. The adaptive scheme presented can be turned around in that instead of repairing lower priority data packets, these packets can be discarded if channel conditions do not merit retransmission.

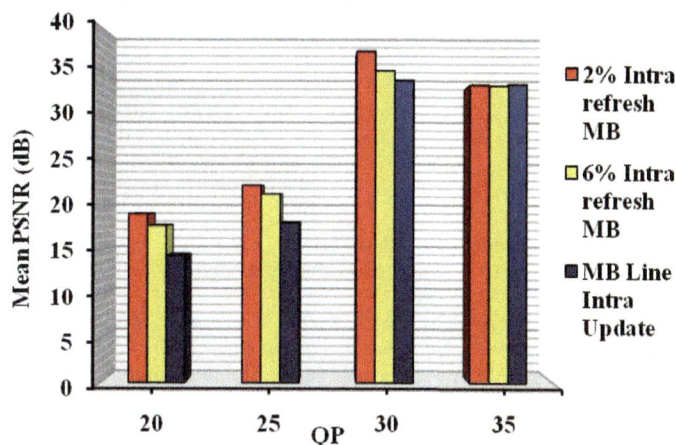

Fig. 21. Mean video quality when streaming *Football*, with duplicate NAL protection extending to all partitions and different levels of intra-refresh MB protection.

References

[1] F. Agboma and A. Liotta, Addressing user expectations in mobile content delivery, *Mobile Information Systems* **3**(3/4) (2007), 153–164.

[2] S. Ahmad, R. Hamzaoui and M. Al-Akaidi, Robust live unicast video streaming with rateless codes, Proc. Int'l PacketVideo Workshop, 2007, pp. 87–84.

[3] J.G. Andrews, A. Ghosh and R. Muhamed, Fundamentals of WiMAX: Understanding broadband wireless networking, Prentice Hall, Upper Saddle River, NJ, 2007.

[4] B. Barmada, M.M. Mahdi, E.V. Jones and M. Ghanbari, Prioritiised transmission of data-partitioned H.264 video with hierarchical QAM, *IEEE Sig Proc Letter* **12**(8) (2005), 577–580.

[5] J. Casampere, P. Sanshez, T. Villameriel and J. Del Ser, Performance evaluation of H.264/MPEG-4 Scalable Video Coding over IEEE 802.16e networks, Proc. IEEE Int'l Symp. on Broadband Multimedia Systems and Broadcasting, 2009, pp. 1–6.

[6] Y.-F. Ciou, F.-Y. Leu, Y.-L. Huang and K. Yim, A handover security mechanism employing the Diffie-Hellman key exchange approach for the IEEE 802.16e wireless networks, *Mobile Information Systems* **7**(3) (2011), 241–269.

[7] N. Degrande, K. Laevens and D. De Vleeschauwer, Increasing the user perceived quality for IPTV services, *IEEE Communs Mag* **46**(2) (2008), 94–100.

[8] P. Ferré, D. Agrafiotis and D. Bull, A video error resilience redundant slices algorithm and its performance relative to other fixed redundancy schemes, *Image Communication* **25**(3) (2010), 163–178.

[9] O. Hillested, A. Perkis, V. Genc, S. Murphy and J. Murphy, Adaptive H.264/MPEG-4 SVC video over IEEE 802.16 broadband wireless access networks, Proc. Packet Video, 2007, pp. 26–35.

[10] IEEE 802.16e-2005, *IEEE Standard for Local and Metropolitan Area Networks*, Part 16: Air Interface for Fixed and mobile Broadband Wireless Access Systems, 2005.

[11] O. Issa, W. Li and H. Liu, Performance evaluation of TV over broadband wireless access networks, *IEEE Trans Broadcasting* **56**(2) (2010), 201–210.

[12] ITU-T Rec. P.800, Methods for subjective testing of video quality, 1996.

[13] Ch. Jiao, L. Schwiebert and B. Xu, On modeling the packet error statistics in bursty channels, *IEEE Conf on Local Computer Networks*, 2002, pp. 534–541.

[14] H.-H. Juan, H.-C. Huang, C.Y. Huang and T. Chiang, Cross-layer mobile wireless MAC designs for the H.264/AVC scalable video coding, *Wireless Networks* **16**(1) (2008), 113–123.

[15] H.G. Kim and J.H. Lee, Diffie-Hellman key based authentication in proxy mobile IPv6, *Mobile Information Systems* **6**(1) (2010), 107–121.

[16] J. Klaue, B. Rathke and A. Wolisz, EvalVid – A framework for video transmission and quality evaluation, Proc. Int'l Conf. on Modeling Techniques and Tools for Computer Performance, 2003, pp. 255–272.

[17] E. Krause et al., Method and apparatus for refreshing motion compensated sequential video images, US 5,057,916, United States Patent Office, 1991.

[18] L.M. Lee, H.-J. Park, S.G. Choi and J.K. Choi, Adaptive hybrid transmission mechanism for on-demand mobile IPTV over WiMAX, *IEEE Trans Broadcasting* **55**(2) (2009), 468–477.

[19] A.H. Li, S. Kittitornkun, Y.H. Hu, D.S. Park and J.D. Villasenor, Data partitioning and reversible variable length codes for robust video communications, *Proc. of Data Compression Conf.*, 2002, pp. 460–489.

[20] Y.-J. Liang, J.G. Apostolopoulos and B. Girod, Analysis of packet loss for compressed video: Effect of burst losses and correlation between error frames, *IEEE Trans Circ Systems Video Technol* **18**(7) (2008), 861–874.

[21] Y.-J. Liang, K. El-Maleh and S. Manjunath, Upfront intra-refresh decision for low-complexity wireless video telephony, *IEEE Int Symposium on Circuits and Systems* (2006).

[22] M. Luby, T. Gasiba, T. Stockhammer and M. Watson, Reliable multimedia download delivery in cellular broadcast networks, *IEEE Trans Broadcasting* **53**(1) (2007), 235–246.

[23] D.J.C.MacKay, Fountain codes, *IEE Proc.: Communications* **152**(6) (2005), 1062–1068.

[24] S. Mys, P. Lambert and W. De Neve, SNR scalability in H.264/AVC using data partitioning, *Proc. Pacific Rim Conf. in Multimedia*, 2006, pp. 329–338.

[25] R. Palanki and J. Yedidai, Rateless codes on noisy channels., *Proc. Int'l Symp Inform Theory* 2004, p. 37.

[26] N. Qadri, M. Altaf, M. Fleury, and M. Ghanbari, *Mobile Information Systems* **6**(3) (2010), 259–280.

[27] I. Radulovic, Y.-K. Wang, S. Wenger, A. Hallapuro, M.H. Hannuksela and P. Frossard, Multiple description H.264 video coding with redundant pictures, Proc. Int'l Workshop on Mobile Video, 2007, pp. 37–42.

[28] R.M. Schreier and A. Rothermel, Motion adaptive intra refresh for the H.264 video coding standard, *IEEE Trans Consumer Electronics* **52**(1) (2006), 249–253.

[29] A. Shokorallahi, Raptor codes, *IEEE Trans Information Theory* **52**(6) (2006), 2551–2567.

[30] C. So-In, R. Jain and A.-K. Tamini, Capacity evaluation of IEEE 802.16e WiMAX, J. of Computer Systems, Networks, and Communications, [online], 2010, p. 12.

[31] A.M. Tourapis, K. Sühring and G. Sullivan, H.264/14496-10 AVC Reference Software Manual, 31st Meeting of the Joint Video Team, London, 2009.

[32] T.D. Tran, L.-K. Liu and P.H. Westering, Low-delay MPEG-2 video coding, *Proc SPIE – Int Soc Opt Eng (USA)* **3309** (1997), 510–516.

[33] F.C.D. Tsai et al., The design and implementation of WiMAX module for ns-2 simulator, Workshop on ns2, article no. 5, 2006.

[34] D. de Vleeschauwer and K. Laevens, Performance of caching systems for IPTV on-demand services, *IEEE Trans Broadcasting* **55**(2) (2009), 491–501.

[35] S. Wenger, H.264/AVC over IP, *IEEE Trans Circuits Syst Video Technol* **13**(7) (2003), 645–655.

[36] M. Wu, S. Makharia, H. Liu, D. Li and S. Mathur, IPTV multicast over wireless LAN using merged hybrid ARQ with staggered adaptive FEC, *IEEE Trans Broadcasting* **55**(2) (2009), 363–374.

[37] C. Zhu, Y.K. Wang, M. Hannuksela and H. Li, Error resilient video coding using redundant pictures, *Proc IEEE Int'l Conf on Image Processing*, 2006, pp. 801–804.

[38] M. Zorzi and R. Rao, On the statistics of block errors in bursty channels, *IEEE Trans Commun* **45**(6) (1997), 660–667.

Permissions

The contributors of this book come from diverse backgrounds, making this book a truly international effort. This book will bring forth new frontiers with its revolutionizing research information and detailed analysis of the nascent developments around the world.

We would like to thank all the contributing authors for lending their expertise to make the book truly unique. They have played a crucial role in the development of this book. Without their invaluable contributions this book wouldn't have been possible. They have made vital efforts to compile up to date information on the varied aspects of this subject to make this book a valuable addition to the collection of many professionals and students.

This book was conceptualized with the vision of imparting up-to-date information and advanced data in this field. To ensure the same, a matchless editorial board was set up. Every individual on the board went through rigorous rounds of assessment to prove their worth. After which they invested a large part of their time researching and compiling the most relevant data for our readers.

The editorial board has been involved in producing this book since its inception. They have spent rigorous hours researching and exploring the diverse topics which have resulted in the successful publishing of this book. They have passed on their knowledge of decades through this book. To expedite this challenging task, the publisher supported the team at every step. A small team of assistant editors was also appointed to further simplify the editing procedure and attain best results for the readers.

Apart from the editorial board, the designing team has also invested a significant amount of their time in understanding the subject and creating the most relevant covers. They scrutinized every image to scout for the most suitable representation of the subject and create an appropriate cover for the book.

The publishing team has been an ardent support to the editorial, designing and production team. Their endless efforts to recruit the best for this project, has resulted in the accomplishment of this book. They are a veteran in the field of academics and their pool of knowledge is as vast as their experience in printing. Their expertise and guidance has proved useful at every step. Their uncompromising quality standards have made this book an exceptional effort. Their encouragement from time to time has been an inspiration for everyone.

The publisher and the editorial board hope that this book will prove to be a valuable piece of knowledge for researchers, students, practitioners and scholars across the globe.

List of Contributors

Abdulbaset Gaddah and Thomas Kunz
Department of Systems and Computer Engineering, Carleton University, Ottawa, Canada

Carlos Gañán, Jose L. Muñoz, Oscar Esparza, Jorge Mata-Díaz and Juanjo Alins
Telematics Department, Universitat Politècnica de Catalunya, Barcelona, Spain

Jonathan Loo
Computer Communications Department, Middlesex University, London, UK

P. Nicopolitidisa, K. Christidis, G.I. Papadimitriou and A.S. Pomportsis
Department of Informatics, Aristotle University of Thessaloniki, Box 888, 54124, Thessaloniki, Greece

P. G. Sarigiannidis
Department of Engineering Informatics and Telecommunications, University of Western Macedonia, 50100, Kozani, Greece

Thierry Delot
University Lille North of France, Valenciennes, France
INRIA Lille Nord Europe, Villeneuve d'Ascq, France

Sergio Ilarri
IIS Department, University of Zaragoza, Zaragoza, Spain

Sylvain Lecomtea and Nicolas Cenerarioa
University Lille North of France, Valenciennes, France

Péter Fülöp, Sándor Imre, Sándor Szabó and Tamás Szálka
Budapest University of Technology and Economics, Department of Telecommunication, 2, Magyar tudósok körútja, Budapest 1117, Hungary

Antonio J. Jaraa and Antonio F. Skarmeta
Department of Information and Communications Engineering, Computer Science Faculty at the University of Murcia, Murcia, Spain

Socrates Varakliotis and Peter Kirstein
Department of Computer Science, University College London, London, UK

Rashid Bin Muhammad
Department of Computer Science, Kent State University, Kent, OH, USA

Sergio Ilarri, Eduardo Mena and Roberto Yus
Department of Computer Science and Systems Engineering, University of Zaragoza, Zaragoza, Spain

Arantza Illarramendi
Basque Country University, San Sebastián, Spain

Maider Laka and Gorka Marcos
Vicomtech Research Center, San Sebastián, Spain

Feilong Tang
Department of Computer Science and Engineering, Shanghai Jiao Tong University, Shanghai 200240, China
School of Computer Science and Engineering, The University of Aizu, Fukushima 965-8580, Japan

Ilsun You
School of Information Science, Korean Bible University, 16 Danghyun 2-gil, Nowon-gu, Seoul, South Kroea

Minyi Guo
Department of Computer Science and Engineering, Shanghai Jiao Tong University, Shanghai 200240, China

Song Guo and Long Zheng
School of Computer Science and Engineering, The University of Aizu, Fukushima 965-8580, Japan

Laith Al-Jobouri, Martin Fleury and Mohammed Ghanbari
Department of CSEE, University of Essex, Essex, UK

CPSIA information can be obtained
at www.ICGtesting.com
Printed in the USA
BVOW10*1101040716

454361BV00002B/140/P